VOLUME ONE HUNDRED AND TWENTY

ADVANCES IN CANCER RESEARCH

AEG-1/MTDH/LYRIC Implicated in Multiple Human Cancers

VOLUME ONE HUNDRED AND TWENTY

ADVANCES IN CANCER RESEARCH

AEG-1/MTDH/LYRIC Implicated in Multiple Human Cancers

Edited by

DEVANAND SARKAR

Department of Human & Molecular Genetics,
VCU Institute of Molecular Medicine,
VCU Massey Cancer Center,
Virginia Commonwealth University,
School of Medicine, Richmond, Virginia, USA

PAUL B. FISHER

Professor and Chair,
Department of Human & Molecular Genetics,
Director, VCU Institute of Molecular Medicine,
Thelma Newmeyer Corman Chair in Cancer Research,
VCU Massey Cancer Center,
Virginia Commonwealth University, School of Medicine,
Richmond, Virginia, USA

ELSEVIER

AMSTERDAM • BOSTON • HEIDELBERG • LONDON
NEW YORK • OXFORD • PARIS • SAN DIEGO
SAN FRANCISCO • SINGAPORE • SYDNEY • TOKYO
Academic Press is an imprint of Elsevier

Academic Press is an imprint of Elsevier
525 B Street, Suite 1800, San Diego, CA 92101-4495, USA
225 Wyman Street, Waltham, MA 02451, USA
32 Jamestown Road, London, NW1 7BY, UK
The Boulevard, Langford Lane, Kidlington, Oxford, OX5 1GB, UK
Radarweg 29, PO Box 211, 1000 AE Amsterdam, The Netherlands

First edition 2013

ISBN: 978-0-12-401676-7
ISSN: 0065-230X

For information on all Academic Press publications
visit our website at store.elsevier.com

Printed and bound in USA

13 14 15 16 12 11 10 9 8 7 6 5 4 3 2 1

CONTENTS

CONTRIBUTORS

Swadesh K. Das
Department of Human and Molecular Genetics, Virginia Commonwealth University, School of Medicine, Richmond, Virginia, USA

Santanu Dasgupta
Department of Human and Molecular Genetics, and VCU Massey Cancer Center, Virginia Commonwealth University, School of Medicine, Richmond, Virginia, USA

Rob DeSalle
Division of Invertebrate Zoology, American Museum of Natural History, and Department of Biology, New York University, New York, New York, USA

Luni Emdad
Department of Human and Molecular Genetics, and VCU Massey Cancer Center, Virginia Commonwealth University, School of Medicine, Richmond, Virginia, USA

Paul B. Fisher
Department of Human and Molecular Genetics; VCU Institute of Molecular Medicine, and VCU Massey Cancer Center, Virginia Commonwealth University, School of Medicine, Richmond, Virginia, USA

Bin Hu
Department of Human and Molecular Genetics, Virginia Commonwealth University, School of Medicine, Richmond, Virginia, USA

Dong-Chul Kang
Ilsong Institute of Life Science, Hallym University, Anyang, Kyonggi-do, Republic of Korea

Yibin Kang
Department of Molecular Biology, Princeton University, Princeton, and Cancer Institute of New Jersey, New Brunswick, New Jersey, USA

Kamel Khalili
Department of Neurology and Neuroscience, Weill-Cornell Medical Center-New York Presbyterian Hospital, New York, NY USA

Seok-Geun Lee
Cancer Preventive Material Development Research Center, Institute of Korean Medicine, College of Korean Medicine, Kyung Hee University, Seoul, Republic of Korea

Kimberly K. Leslie
Department of Obstetrics and Gynecology, and Holden Comprehensive Cancer Center, The University of Iowa, Iowa City, Iowa, USA

Xiangbing Meng
Department of Obstetrics and Gynecology, and Holden Comprehensive Cancer Center, The University of Iowa, Iowa City, Iowa, USA

Evan K. Noch
Department of Neurology and Neuroscience, Weill-Cornell Medical Center-New York Presbyterian Hospital, New York, NY USA

Devanand Sarkar
Department of Human and Molecular Genetics; VCU Institute of Molecular Medicine, and VCU Massey Cancer Center, Virginia Commonwealth University, School of Medicine, Richmond, Virginia, USA

Kristina W. Thiel
Department of Obstetrics and Gynecology, The University of Iowa, Iowa City, Iowa, USA

Liling Wan
Department of Molecular Biology, Princeton University, Princeton, New Jersey, USA

PREFACE

Cancer is a heterogeneous and constantly evolving disease involving genetic and epigenetic changes in the cancer cells themselves as well as structural and architectural alterations in their microenvironment. Concerted efforts over the past two decades have resulted in significant strides in our understanding of the molecular pathogenesis of cancer leading to the development of diverse modalities of treatment including small molecule inhibitors of specific target genes implicated in the carcinogenic process. However, as an erudite chameleon cancer, initially responsive to a specific agent develops secondary mutations and/or epigenetic changes culminating in cancer recurrence with renewed vigor. If diagnosed early, the primary tumor might be managed by surgical removal followed by a combination of aggressive chemo- and/or radiotherapy. However, as the disease becomes locally invasive or distally metastatic, the disease becomes more and more difficult to manage and it is the metastatic cancer that is responsible for mortality in more than 90% cases. Comprehending the mechanism of metastasis is thus mandatory to develop effective strategies to manage this diverse disease and provide significant disease-free survival benefits to patients. Cancer represents a collective term encompassing a conglomeration of diseases that present with specific hallmarks. However, each cancer is unique in its genetic makeup resulting in differences in behavior and progression, which limit effective application of a standard single modality of treatment in all cancer indications. These complexities represent major challenges and impediments to developing effective cancer therapies. Theoretically, identification of molecule(s) that might play a prominent regulatory role in all or most cancers would be a significant advancement facilitating development of therapeutic approaches providing a panacea for all cancers. Astrocyte-elevated gene-1 (AEG-1), also known as metadherin (MTDH) and LYRIC, fulfills many of the aforementioned criteria to serve as a *bona fide* molecule that might be targeted to establish a lasting therapeutic response.

Since its initial cloning in 2002, a plethora of studies have been performed in a wide variety of cancers clearly establishing AEG-1/MTDH/LYRIC overexpression. These cancers include tumors of every organ system where AEG-1/MTDH/LYRIC is overexpressed in 40–50% of patients, and in some cases, such as in hepatocellular carcinoma (HCC), in ~90% cases. Such an overexpression of an oncogene in all types of cancers has not been documented for any other gene. Although AEG-1/MTDH/LYRIC is ubiquitously expressed, the expression level is very low in normal tissues while its expression gradually increases as the cancer becomes more and more progressed with the highest expression detected in invasive and metastatic tumors. A clear negative correlation between AEG-1 expression levels and poor prognosis of patients has been established with AEG-1-overexpressing patients showing significantly decreased overall survival and increased recurrence rates. These clinicopathological correlation studies have paved the way for inclusion of AEG-1/MTDH/LYRIC in MammaPrint, the first and only FDA-approved individualized metastasis risk assessment assay for breast cancer that includes a unique 70-gene signature. Thus, AEG-1/MTDH/LYRIC has made its mark as a diagnostic/prognostic biomarker for cancer.

In initial cloning endeavors, AEG-1/MTDH/LYRIC was identified as a gene promoting metastasis of breast cancer. However, with more and more phenotypic studies in diverse cancers, it is evident that AEG-1/MTDH/LYRIC affects all the hallmarks of cancer. AEG-1/MTDH/LYRIC itself could not immortalize normal liver cells, as was evident from a hepatocyte-specific transgenic mouse model that did not develop spontaneous cancer. However, it induced transformation of normal immortal cells, either alone or in combination with Ha-ras, and markedly augmented proliferation, resistance to apoptosis and anoikis, invasion, angiogenesis, metastasis, and chemoresistance of cancer cells. The contribution of AEG-1/MTDH/LYRIC to different aspects of aggressive cancers demonstrates potential organ specificity. In breast cancer, AEG-1/MTDH/LYRIC knockdown markedly abrogates metastatic ability of the cells while marginally affecting growth of the primary tumor. However, in other cancers, such as HCC, prostate cancer, or malignant glioma, AEG-1/MTDH/LYRIC knockdown equally affects both primary tumor growth and metastasis. Thus, a treatment approach targeting AEG-1/MTDH/LYRIC needs to take the type of cancer into consideration, and a combinatorial approach targeting both primary tumor growth and metastatic events might be mandatory to ensure significant objective response in patients.

The mechanisms by which AEG-1/MTDH/LYRIC is overexpressed in cancer have been well delineated demonstrating multiple layers of regulation including genomic amplification, and transcriptional, posttranscriptional, and posttranslational controls. However, the mechanism by which AEG-1/MTDH/LYRIC itself induces its oncogenic effects requires further clarification. The absence of any known domains or motifs in

AEG-1/MTDH/LYRIC complicates the process of predicting AEG-1/MTDH/LYRIC function. In this context, analyzing downstream gene expression, interacting partners and intracellular signaling pathways have identified potential mechanisms of action. Activation of NF-κB by direct interaction with the p65 subunit was identified as one major mechanism of proinvasive function of AEG-1/MTDH/LYRIC. Activation of PI3K/Akt pathway has been implicated to confer antiapoptotic and proangiogenic functions of AEG-1/MTDH/LYRIC. However, the mechanism by which AEG-1/MTDH/LYRIC activates the PI3K/Akt pathway still requires elucidation. Multiple studies have identified Staphylococcal nuclease domain containing-1 (SND1), a nuclease in the RNA-induced silencing complex (RISC), as a major interacting partner of AEG-1/MTDH/LYRIC. The combined contribution of SND1 and AEG-1 increases RISC activity, thereby facilitating oncogenic miRNA-mediated degradation of tumor suppressor mRNAs and eventually carcinogenesis. AEG-1/MTDH/LYRIC induces protective autophagy that contributes to chemoresistance. Consequently, AEG-1/MTDH/LYRIC might modulate a variety of intracellular events to confer an aggressive carcinogenic process.

A major challenge in studying AEG-1/MTDH/LYRIC is its subcellular localization. AEG-1/MTDH/LYRIC protein contains a transmembrane domain as well as multiple nuclear localization signals. Indeed, separate laboratories initially cloned AEG-1/MTDH/LYRIC either as a tight-junction protein or an endoplasmic reticulum/nucleus/nucleolus-resident protein. In cancer cells, AEG-1/MTDH/LYRIC is predominantly detected in the cytoplasm and in the cell membrane. The membrane-anchored AEG-1/MTDH/LYRIC contains a lung-homing domain through which it facilitates adhesion of breast cancer cells to lung endothelial cells promoting metastasis. However, AEG-1/MTDH/LYRIC-interacting molecules on endothelial cells remain to be identified. The cytoplasmic AEG-1/MTDH/LYRIC seems to facilitate organization of protein complexes, potentially serving as a scaffold protein, thereby promoting RNAi-mediated gene regulation and protein translation. The nuclear AEG-1/MTDH/LYRIC regulates transcription, such as by NF-κB or YY1. The function of AEG-1/MTDH/LYRIC in the ER or in the tight junction remains to be determined. Thus, more in-depth studies need to be performed to better understand AEG-1/MTDH/LYRIC function in different subcellular compartments.

Even though a large number of studies have been performed in cancer cells, the primary function(s) AEG-1/MTDH/LYRIC serves in normal cells remains to be determined. In normal cells, its predominant localization is in the nucleus and nucleolus suggesting that the main function of

AEG-1/MTDH/LYRIC might be transcriptional regulation. Studies in primary astrocytes suggest a potential role of AEG-1/MTDH/LYRIC in wound healing. Analysis of an AEG-1/MTDH/LYRIC knockout mouse is essential to obtain in-depth insights into the mode of action of this intriguing molecule *in vivo*. AEG-1/MTDH/LYRIC is only present in vertebrates where it is highly conserved across species. Evolutionary studies need to be performed to determine the origin of this gene and whether there is similarity with any ancient gene of lower organism to understand the reason for preserving this gene through evolution. The conservation of the sequence suggests that AEG-1/MTDH/LYRIC might serve an important function in normal cells, and some studies in cancer cells have suggested a potential role of AEG-1/MTDH/LYRIC in overcoming stress, which might be applicable to normal cells as well.

While mechanistic studies using cell lines and mouse models will continue to unravel the functional aspects of AEG-1/MTDH/LYRIC, more effort is needed to develop strategies for AEG-1/MTDH/LYRIC inhibition for translational applications. Initial immunological studies using AEG-1/MTDH/LYRIC as a vaccine look promising and need to be stringently confirmed. Crystal structure of AEG-1/MTDH/LYRIC needs to be resolved to design small molecules or peptides that might perturb the interaction of AEG-1/MTDH/LYRIC with its partners thereby blocking its function. Routine screening for AEG-1/MTDH/LYRIC needs to be implemented in the clinics since AEG-1/MTDH/LYRIC expression level will not only provide prognostic information but also determine responsiveness to chemotherapeutic agents.

This volume contains seven chapters providing a synopsis of the current literature on AEG-1/MTDH/LYRIC and future directions of AEG-1/MTDH/LYRIC-related studies. Chapter 1 describes the initial cloning and characterization of AEG-1/MTDH/LYRIC. Chapter 2 focuses on clinicopathological studies on AEG-1/MTDH/LYRIC. Chapters 3–5 describe the regulatory role of AEG-1/MTDH/LYRIC on angiogenesis, metastasis, and chemoresistance, respectively. Chapter 6 describes AEG-1/MTDH/LYRIC effect on central nervous system that includes brain cancers as well as neurodegenerative diseases, HIV infection, and migraine. Chapter 7 focuses on the role of AEG-1/MTDH/LYRIC in HCC. We hope that this book will evoke interest on AEG-1/MTDH/LYRIC and stimulate more research to study this multifunctional intriguing molecule.

Devanand Sarkar
Paul B. Fisher

CHAPTER ONE

AEG-1/MTDH/LYRIC, the Beginning: Initial Cloning, Structure, Expression Profile, and Regulation of Expression

Seok-Geun Lee[*,1,2], **Dong-Chul Kang**[†,1,2], **Rob DeSalle**[‡,§], **Devanand Sarkar**[¶], **Paul B. Fisher**[¶]

[*]Cancer Preventive Material Development Research Center, Institute of Korean Medicine, College of Korean Medicine, Kyung Hee University, Seoul, Republic of Korea
[†]Ilsong Institute of Life Science, Hallym University, Anyang, Kyonggi-do, Republic of Korea
[‡]Division of Invertebrate Zoology, American Museum of Natural History, New York, New York, USA
[§]Department of Biology, New York University, New York, New York, USA
[¶]Department of Human and Molecular Genetics, VCU Institute of Molecular Medicine, VCU Massey Cancer Center, Virginia Commonwealth University, School of Medicine, Richmond, Virginia, USA
[1]Corresponding authors: e-mail address: seokgeun@khu.ac.kr; dckang@hallym.ac.kr
[2]These authors equally contributed

Contents

Advances in Cancer Research, Volume 120
ISSN 0065-230X
http://dx.doi.org/10.1016/B978-0-12-401676-7.00001-2

Abstract

Since its initial identification as a HIV-1-inducible gene in 2002, astrocyte elevated gene-1 (AEG-1), subsequently cloned as metadherin (MTDH) and lysine-rich CEACAM1 coisolated (LYRIC), has emerged over the past 10 years as an important oncogene providing a valuable prognostic marker in patients with various cancers. Recent studies demonstrate that AEG-1/MTDH/LYRIC is a pleiotropic protein that can localize in the cell membrane, cytoplasm, endoplasmic reticulum (ER), nucleus, and nucleolus, and contributes to diverse signaling pathways such as PI3K–AKT, NF-κB, MAPK, and Wnt. In addition to tumorigenesis, this multifunctional protein is implicated in various physiological and pathological processes including development, neurodegeneration, and inflammation. The present review focuses on the discovery of AEG-1/MTDH/LYRIC and conceptualizes areas of future direction for this intriguing gene. We begin by describing how AEG-1, MTDH, and LYRIC were initially identified by different research groups and then discuss AEG-1 structure, functions, localization, and evolution. We conclude with a discussion of the expression profile of AEG-1/MTDH/LYRIC in the context of cancer, neurological disorders, inflammation, and embryogenesis, and discuss how AEG-1/MTDH/LYRIC is regulated. This introductory discussion of AEG-1/MTDH/LYRIC will serve as the basis for the detailed discussions in other chapters of the unique properties of this intriguing molecule.

1. INTRODUCTION

Astrocyte elevated gene-1 (*AEG-1*), also called metadherin (*MTDH*) and lysine-rich CEACAM1 coisolated (*LYRIC*), was first identified and cloned by subtraction hybridization of genes expressed at elevated levels in primary human fetal astrocytes (PHFAs) infected by human immunodeficiency virus-1 (HIV-1) (Su et al., 2002). Subsequently, three independent groups, in addition to ours, reported cloning and initial characterization of the full-length cDNA of *AEG-1* between 2002 and 2005 (Britt et al., 2004; Brown & Ruoslahti, 2004; Kang et al., 2005; Su et al., 2002; Sutherland, Lam, Briers, Lamond, & Bickmore, 2004). AEG-1 is also referred to as MTDH, LYRIC, or 3D3/LYRIC in the literature depending on the context of its discovery and postulated function.

Association of AEG-1/MTDH/LYRIC with the cancer phenotype was evident from its initial characterization in which AEG-1/MTDH/LYRIC was shown to be overexpressed in various types of cancerous cell lines and a mediator of metastasis of murine breast cancer cells to the lungs (Brown & Ruoslahti, 2004; Kang et al., 2005). AEG-1/MTDH/LYRIC expression is elevated in almost all types of cancers and has been shown to enhance proliferation, survival, and metastatic capability of cancer cells through multiple mechanisms (Emdad et al., 2007; Sarkar et al., 2009; Ying, Li, & Li, 2011; Yoo, Emdad, et al., 2011). AEG-1/MTDH/LYRIC is also involved in glioma-associated neurodegeneration potentially through its ability to enhance glutamate excitotoxicity as seen in HIV-1-associated neuropathy (Lee et al., 2011). In addition to its tumor-promoting activity, AEG-1/MTDH/LYRIC is also implicated in migraine, inflammation, and development (Anttila et al., 2010; Jeon et al., 2010; Sarkar et al., 2008; Yoo, Emdad, et al., 2011).

AEG-1/MTDH/LYRIC participates in diverse signaling pathways culminating in multiple cellular responses. AEG-1/MTDH/LYRIC regulates signaling pathways including PI3K/AKT, NF-κB, MEK/ERK, and WNT/β-catenin and is regulated by c-Myc and vice versa, forming a positive feedback loop (Emdad et al., 2006; Lee, Su, Emdad, Sarkar, & Fisher, 2006; Yoo et al., 2009). It is hypothesized that the tumor-promoting activity of AEG-1/MTDH/LYRIC is a consequence of activation of defined signaling pathways and the positive feedback between AEG-1/MTDH/LYRIC and c-Myc. Moreover, AEG-1/MTDH/LYRIC appears to function in microRNA (miRNA) processing through interaction with SND1 (Staphylococcal nuclease and tudor domain containing 1), a component of RISC (RNA-induced silencing complex) (Yoo, Santhekadur, et al., 2011). What is very apparent is that AEG-1/MTDH/LYRIC associates with multiple biological phenomena and further research will undoubtedly identify a broad spectrum of processes correlated with and potentially regulated by AEG-1/MTDH/LYRIC.

Despite the canonical role of AEG-1/MTDH/LYRIC in tumor progression and potentially in development of other diseases, the molecular basis of the numerous functions of AEG-1/MTDH/LYRIC remains to be defined. Thus, this thematic volume is very timely to overview the current status of AEG-1/MTDH/LYRIC research in detail and to provide conceptual basis for new insights into AEG-1/MTDH/LYRIC functions. As a part of this undertaking, we are providing a brief chronological history and summary of the contexts of the initial clonings and characterization of AEG-1/MDTH/LYRIC. In addition, we describe the expression patterns of AEG-1/MTDH/LYRIC and its regulation, which are crucial for understanding the precise roles of AEG-1/MTDH/LYRIC in regulating normal and abnormal physiology.

2. INITIAL CLONING

Over the past two decades, transcriptome analysis has provided a robust method for expanding gene discovery. The advent of new methodologies, including microarrays and various forms of differential hybridization (including subtraction hybridization), has permitted the efficient identification of differentially expressed sequence tags (ESTs) in both normal and pathogenic conditions, as well as generating tissue-specific profiles. In conjunction with progress in various genome projects and full-length cDNA cloning projects, transcriptome analysis has expanded its scope from searches for ESTs to characterization and functional analysis of these ESTs. *AEG-1/MTDH/LYRIC* was reported as an outcome of successful application of the then novel function-based gene discovery tools including RaSH (rapid subtraction hybridization) (Jiang, Kang, Alexandre, & Fisher, 2000; Kang et al., 2005; Simm et al., 2001; Su et al., 2002), *in vivo* phage display screening (Brown & Ruoslahti, 2004), and gene-trap screening of localization-specific genes (Sutherland et al., 2004). Based on its differential expression in multiple contexts and its unique expression pattern, AEG-1/MTDH/LYRIC was chosen for further study. This first section focuses on the settings and methodologies of cloning *AEG-1/MTDH/LYRIC* and the initial characterization efforts in the context of our current knowledge on AEG-1/MTDH/LYRIC.

2.1. Astrocyte elevated gene-1

HIV-associated dementia (HAD) develops in about 20% of HIV-1-infected individuals (Gonzalez-Scarano & Martin-Garcia, 2005). HIV primarily infects microglia and rarely infects astrocytes but does not infect neurons in the central nervous system (Brack-Werner, 1999). Although HIV failed to efficiently infect astrocytes, infected astrocytes were still believed to participate in neuropathogenesis associated with HIV infection (Borjabad, Brooks, & Volsky, 2010; Wang et al., 2004). In an attempt to elucidate etiology of HIV-induced neurodegeneration, alteration of gene expression in HIV-infected astrocytes was analyzed by RaSH (Su, Chen, et al., 2003; Su et al., 2002). *AEG-1* was among the 15 AEGs whose expression was upregulated in PHFA upon infection of HIV-1 or treatment with gp120 (Su et al., 2002).

Cloning AEGs and also astrocyte suppressed genes (ASGs) was accomplished by RaSH that was invented in 2000 for discovery of genes distinctly expressed between HIV-1-infected and uninfected PHFA (Jiang et al., 2000; Su, Chen, et al., 2003; Su et al., 2002; Su, Kang, et al., 2003). RaSH employs directional subtraction of PCR-amplified cDNA libraries and vector selection of uniquely expressing unsubtracted target fragments (Boukerche, Su, Kang, & Fisher, 2004; Kang, Jiang, Su, Volsky, & Fisher, 2002). RaSH is a very simple and efficient method and does not require large amounts of mRNA as does conventional subtractive cDNA library techniques. Since its first demonstration of proof of principle and successful application to cloning AEGs and ASGs, RaSH was also found to have utility in various experimental settings including gene discovery associated with HIV-resistance of T lymphocytes (Simm et al., 2001) and genes associated with metastasis (Boukerche et al., 2004).

AEG-1 is a late onset gene that is expressed 3 ~ 7 days after HIV-1 infection (Su, Chen, et al., 2003; Su et al., 2002). *AEG-1* expression in PHFA is also increased by treatment with HIV envelope protein, gp120, and TNF-α and did not require productive infection of astrocytes (Su, Chen, et al., 2003; Su et al., 2002). The signaling pathways of *AEG-1* induction in astrocytes by HIV-1 infection and treatment of gp120 or TNF-α requires clarification. Studies aimed to elucidate the signaling pathways may consider cross talk of CD4/CXCR4 and TNFR signaling as a starting point, since treatment of both gp120 and TNF-α increases AEG-1 expression in the astrocytes (Rehman & Wang, 2009; Xia et al., 2008).

Characterization of AEG-1 function was possible after *AEG-1* full-length cDNA was cloned using the complete open reading frame (C-ORF) technique (Kang et al., 2005). The C-ORF technique was successfully applied to clone full-length cDNAs of unknown ESTs including *mda*-5 (melanoma differentiation associated gene-5) and *hPNPase*$^{old\text{-}35}$ (human polynucleotide phosphorylase) (Kang et al., 2002; Leszczyniecka et al., 2002). C-ORF utilizes a degenerate stem-loop annealing primer (dSLAP) that consists of a stem-loop structure and 3′ 12 random nucleotides (Kang & Fisher, 2005, 2007). The design of dSLAP is to promote annealing the primer to the 3′ end of reverse transcribed cDNAs and to provide primer site for second strand cDNA synthesis. Application of C-ORF yielded C-ORF of *AEG-1*, which could then be used for subsequent detailed functional characterization.

AEG-1 mRNA consists of 3611 nucleotides, excluding the poly A tail, and encodes a 582-amino acid protein with calculated molecular weight 64 kDa and pI 9.33 (Kang et al., 2005). Northern blotting indicated at least three transcripts for *AEG-1* (9, 4, and 1.5 kb) that might be generated by alternative splicing and/or different transcription start sites. Antibody directed against whole AEG-1 protein detected two closely migrating bands of relative mobility ~86,000 in cell lysates from either untransfected or expression vector AEG-1-HA transfected cells. However, only the higher migrating band was detected in immunoblotting with anti-HA antibody. Since the HA-tag is located at the C-terminal end of the molecule, the lower migrating band is most probably generated by C-terminal truncation of AEG-1.

Membrane topology of AEG-1 is predicted as either type Ib or II and remains controversial. Although the topology of AEG-1 needs resolution, immunofluorescent microscopy with antibody against the whole AEG-1 protein clearly demonstrated colocalization of AEG-1 with the ER marker calreticulin and speckled distribution in the nucleus (Emdad et al., 2006; Kang et al., 2005). Localization on the plasma and the nuclear membranes was not obvious in these experiments.

The astrocytic glutamate transporter EAAT2 mediates uptake of the excitatory neurotransmitter glutamate that is responsible for neuronal transmission and is released in the synapses (Borjabad et al., 2010; Wang et al., 2004). Glutamate has the capacity to damage neurons by a process called glutamate excitotoxicity, if the glutamate remains at high levels in the synaptic cleft (Choi, 1988). HIV infection decreases EAAT2 expression in astrocytes followed by reduced uptake of glutamate (Wang et al., 2004). To investigate how glutamate transport is regulated by HIV-1 in astrocytes, we cloned the *EAAT2* promoter (Su, Leszczyniecka, et al., 2003). We demonstrated that glutamate transport was regulated by HIV-1 at the *EAAT2* promoter level and found that regulators of glutamate transport functioned by either enhancing or decreasing *EAAT2* promoter activity (Lee et al., 2008; Rothstein et al., 2005; Su, Leszczyniecka, et al., 2003; Wang et al., 2004). To investigate the mechanism by which AEG-1 expression might affect *EAAT2* promoter function, we used reporter assays. AEG-1 significantly decreased *EAAT2* promoter activity, but not *EAAT1* promoter activity, suggesting that AEG-1 association with HAD occurred by downregulation of EAAT2 expression followed by reduction in glutamate transport (Kang et al., 2005; Lee et al., 2011).

AEG-1 expression was upregulated in glioma cells, which motivated us to examine its expression in additional types of cancer cells. AEG-1 was expressed higher in cancer cells in comparison with immortal normal counterparts (Kang et al., 2005). Subsequently, tumor-promoting activity of AEG-1 was examined by soft agar colony formation assay in immortalized melanocytes transfected with AEG-1 and/or T24 Ha-ras expression vectors (Kang et al., 2005). Expression of AEG-1 by itself and coexpression with T24 Ha-ras significantly increased soft agar colony formation of immortalized melanocytes, suggesting tumor-promoting activity of AEG-1 by itself and in collaboration with Ha-ras. Later, AEG-1 was found to be a downstream of oncogenic Ha-ras signaling pathway (Lee et al., 2006).

Expression of AEG-1 in astrocytes infected with HIV or treated with TNF-α or gp120 suggested a potential association of AEG-1 upregulation with HAD (Kang et al., 2005; Su, Chen, et al., 2003). Three key features of AEG-1, that is, localization, relationship with different cancers and HAD-association were determined in the initial efforts to characterize the molecule (Kang et al., 2005). Although the relationship between AEG-1 and neurodegeneration requires further investigation, its tumor-promoting activity has been well established by numerous critical studies (Yoo, Emdad, et al., 2011).

2.2. Metadherin

Although AEG-1 was found to be associated with cancer serendipitously after screening for genes overexpressed in HIV-infected astrocytes, *MTDH*, the murine ortholog of *AEG-1*, was cloned in an attempt to identify cell surface molecules mediating metastasis of cancer cells (Brown & Ruoslahti, 2004). A domain of MDTH was involved in lung metastasis of mouse 4T1 mammary tumors and designated as a lung-homing domain (LHD) (a.a. 378–440). AEG-1/MDTH was subsequently found to be overexpressed in human breast cancer (Hu et al., 2009; Kornegoor et al., 2012; Li et al., 2008; Tokunaga et al., 2012) and other cancers (reviewed in Yoo, Emdad, et al., 2011).

MDTH, metastasis adhesion protein, was identified using *in vivo* phage screening (Brown & Ruoslahti, 2004). Breast cancer predominantly metastasizes to lung, bone, brain, and liver. In order to find proteins implicated in metastasis of breast cancer cells to the lung, a phage display library was prepared with cDNAs enriched for secreted and transmembrane proteins of 4T1. Phage that specifically localized to the 4T1 lung metastases was selected and cloned for further analysis (Brown & Ruoslahti, 2004). MTDH was predicted to be a type II membrane protein the C-terminal of which is located in the extracellular space. The authors defined the location and function of the LHD of MTDH (Brown & Ruoslahti, 2004).

Since MTDH was cloned as a mediator of metastasis, its LHD was predicted to be located in the extracellular region as anticipated by motif analysis. Location of MTDH in cancer cells was investigated with various methods including FACS and immunofluorescent microscopy (Brown & Ruoslahti, 2004). FACS demonstrated extracellular localization of the LHD tagged with the Myc epitope. In addition, MTDH was found in the edge of impermeabilized cells, whereas it was dispersed in the cytoplasm of permeabilized cells, suggesting that localization of MTDH LHD is extracellular.

Expression in cancer cells and biological functions of MTDH LHD were verified (Brown & Ruoslahti, 2004). Expressing MTDH in HEK293T cells enhanced lung localization of the cells. In addition, knockdown of MTDH or anti-MTDH antibody inhibited lung metastasis of 4T1 cells. These experiments clearly demonstrated that MTDH can mediate lung metastasis of 4T1 breast cancer cells.

Based on the functional data providing compelling evidence for *MTDH* as a lung metastasis-promoting gene, MTDH became the official nomenclature for *AEG-1* and *LYRIC*. A separate study also demonstrated that MTDH overexpression enhanced lung metastasis of human breast cancer cells (Hu et al., 2009). However, there remains a controversy about the topology of MTDH. Metastasis association of MTDH mandates type II topology of the protein. LHD of MTDH was shown to be extracellular by FACS and immunofluorescent microscopy, which strongly supports type II topology (Brown & Ruoslahti, 2004). However, type Ib topology of MTDH was also predicted and intracellular localization of its orthologs was confirmed by immunofluorescent microscopy (Britt et al., 2004; Kang et al., 2005; Sutherland et al., 2004). Thus, controversies on the membrane topology, localization, and dual functionality of AEG-1/MTDH remain to be resolved.

2.3. Lysine-rich CEACAM1 coisolated

Two additional groups of researchers reported an originally novel protein which they named LYRIC that stands for "lysine-rich CEACAM1 coisolated" (Britt et al., 2004) and 3D3/LYRIC (Sutherland et al., 2004). Unlike function-based cloning in the previous two attempts (Brown & Ruoslahti, 2004; Kang et al., 2005), LYRIC and 3D3/LYRIC were discovered through identification of proteins localized at tight junction and the nuclear subcompartment, respectively.

Rat LYRIC was identified as an interacting protein with CEACAM1, and the cDNA was cloned by expression library screening (Britt et al., 2004). The existence of a transmembrane domain (TMD), a nuclear localization signal (NLS), and a potential ATP/GTP-binding site was predicted from the deduced protein sequence. Organ-specific expression of the transcript was examined by Northern blotting analysis, which revealed two transcripts (3.5 and 2.7 kb) with organ-specific distribution. Three splice variants, which lack either of the two or both exons, were also cloned.

Although interaction of LYRIC with CEACAM1 was not pursued further, its colocalization with tight junction was analyzed intensively in various cells and under different growth conditions with a monoclonal antibody MAb 52.15 (Britt et al., 2004). Endogenous LYRIC displayed membrane localization around the cell perimeter, while FLAG-tagged LYRIC displayed diffuse cytoplasmic and perinuclear distribution. LYRIC showed a honeycomb staining pattern in epithelial cells. LYRIC was found colocalized with tight junction component ZO-1 in cultured epithelial cells, tissues, and tumor grafts that formed tight junctions. The LYRIC staining pattern was lost by disruption of the tight junction upon calcium removal and restored by reformation of the junction, which is quite compelling evidence for colocalization of LYRIC with ZO-1 at tight junctions.

Mouse 3D3/LYRIC was cloned independently from AEG-1 and MTDH in an attempt to screen proteins located at distinct subcompartments of the nucleus using a gene trapping approach (Sutherland et al., 2004). A reporter gene with an upstream splice acceptor site was transfected into target cells and allowed to fuse with trapped genes through splicing. Then, localization of the trapped gene was screened by staining for the reporter gene. The N-terminal of LYRIC was trapped and found to localize at the nuclear membrane as discrete patches. Northern blot analysis also revealed existence of multiple transcripts, and TMD of type Ib topology and an NLS were also predicted from the amino acid sequence.

By contrast, mouse 3D3/LYRIC showed a different localization from the rat one and even from that of mouse MTDH (Sutherland et al., 2004). The mouse 3D3/LYRIC was found localized at the ER and nucleus by immunofluorescent microscopy. Endogenous 3D3/LYRIC localized in the nucleolus and in unidentified intranuclear suborganelles in a spotty pattern. Further analysis revealed existence of LYRIC in dense fibrillar components of the nucleolus. Western blotting of protein samples from fractionated organelles confirmed nucleolar localization of 3D3/LYRIC. Furthermore, the smaller of the two 3D3/LYRIC bands detected by Western blotting of endogenous protein predominantly existed in the nucleolus. Moreover, a 3D3/LYRIC mutant lacking the TMD was also localized in the nucleolus. 3D3/LYRIC was suggested to be modified based on these results, which might be necessary for locating a transmembrane protein such as 3D3/LYRIC to the nuclear suborganelles including the nucleolus.

2.4. Confirmation of initial observations relative to AEG-1/MTDH/LYRIC

We briefly overviewed the four articles reporting the initial characterization results on AEG-1/MTDH/LYRIC (Britt et al., 2004; Brown & Ruoslahti, 2004; Kang et al., 2005; Sutherland et al., 2004). All four laboratories reported similar physicochemical properties of AEG-1/MTDH/LYRIC, including amino acid composition, calculated molecular weight, and pI and TMD. Existence of predicted NLS was described in two of the papers (Britt et al., 2004; Sutherland et al., 2004). Although the number and sizes of the transcripts and their tissue distribution were different, a ~3.5 kb mRNA band was reported in the various studies. Mobility of AEG-1/MTDH/LYRIC in SDS-PAGE was slower than expected probably due to the high pI value of the protein. Doublets around ~85 kDa that showed distinct organelle distribution were demonstrated in two reports (Kang et al., 2005; Sutherland et al., 2004).

Localization and membrane topology of AEG-1/MTDH/LYRIC are critical factors for proper functioning of the molecule. However, with respect to the localization and membrane topology of AEG-1/MTDH/LYRIC, there are remaining unresolved issues. Both type Ib and II membrane topology of AEG-1/MTDH/LYRIC are predicted (Britt et al., 2004; Brown & Ruoslahti, 2004; Kang et al., 2005; Sutherland et al., 2004). Type Ib topology with ER/nucleus localization supports tumor-promoting activity of AEG-1/MTDH/LYRIC by stimulating AEG-1/MTDH/LYRIC-associated signaling processes (Kang et al., 2005; Sutherland et al., 2004), whereas type II is preferred for the proposed metastasis-mediating role of this molecule (Brown & Ruoslahti, 2004). The mechanism underlying nuclear localization of AEG-1/MTDH/LYRIC requires further clarification to explain the nuclear functions of AEG-1/MTDH/LYRIC. Further experimentation is required to understand the dual topology and multifunctionality of AEG-1/MTDH/LYRIC. Existence of transcript variants and their functions is another issue requiring clarification.

An association between AEG-1/MTDH/LYRIC and cancer was suggested during the early studies of AEG-1/MTDH (Brown &

Ruoslahti, 2004; Kang et al., 2005) and this assumption has now been firmly established through numerous in-depth subsequent studies (Emdad et al., 2007; Sarkar et al., 2009; Ying et al., 2011; Yoo, Emdad, et al., 2011). AEG-1/MTDH/LYRIC has been documented to function as an oncogene (Emdad et al., 2009). Understanding how this unique gene functions as an oncogene, including defining and characterizing its many potential interactive partners, will undoubtedly provide further insights into the mechanism of action of AEG-1/MTDH/LYRIC. This unique oncogenic property of AEG-1/MTDH/LYRIC and those studies confirming that suppressing expression of this gene inhibits various transformation-associated properties supports the possibility of AEG-1/MTDH/LYRIC as a significant drug target and diagnostic/prognostic marker, which was not anticipated during its initial cloning (Su et al., 2002).

3. STRUCTURE OF AEG-1/MTDH/LYRIC

AEG-1/MTDH/LYRIC cDNAs encode 582, 581, and 579 a.a. proteins of molecular weights ~64 kDa and pI ~9.3 in human, rat, and mouse, respectively (Britt et al., 2004; Brown & Ruoslahti, 2004; Kang et al., 2005; Sutherland et al., 2004). Sequence homology at nucleotide level of human *AEG-1/MTDH/LYRIC* cDNA to mouse and rat are 90% and 89%, respectively. Reported length of rat lyric cDNA (2234 bps) is much shorter than those of mouse (3575 bps) and human (3611 bps) at the 3′-untranslated region (UTR). Sequence identity and homology at protein level of human AEG-1/MTDH/LYRIC to mouse are 91% and 94% and to rat are 90% and 94%, respectively (Fig. 1.1). AEG-1/MTDH/LYRIC is lysine-rich protein from which the name LYRIC was derived (Britt et al., 2004; Sutherland et al., 2004). The highest sequence mismatches are found at amino acids of 93–105, 135–155, and 515–531, and two of these areas of high sequence mismatches are located outside any known domains, although core sequences in the protein are conserved.

Proteins with defined domains and motifs are readily amenable to functional analysis. When originally identified and cloned, AEG-1/MTDH/LYRIC was reported not to have any known functional domains and motifs except an N-terminal TMD, three NLSs, and a potential ATP/GTP-binding site (Britt et al., 2004; Brown & Ruoslahti, 2004; Kang et al., 2005; Sutherland et al., 2004). However, intensive functional studies on AEG-1/MTDH/LYRIC so far have defined at least five additional domains

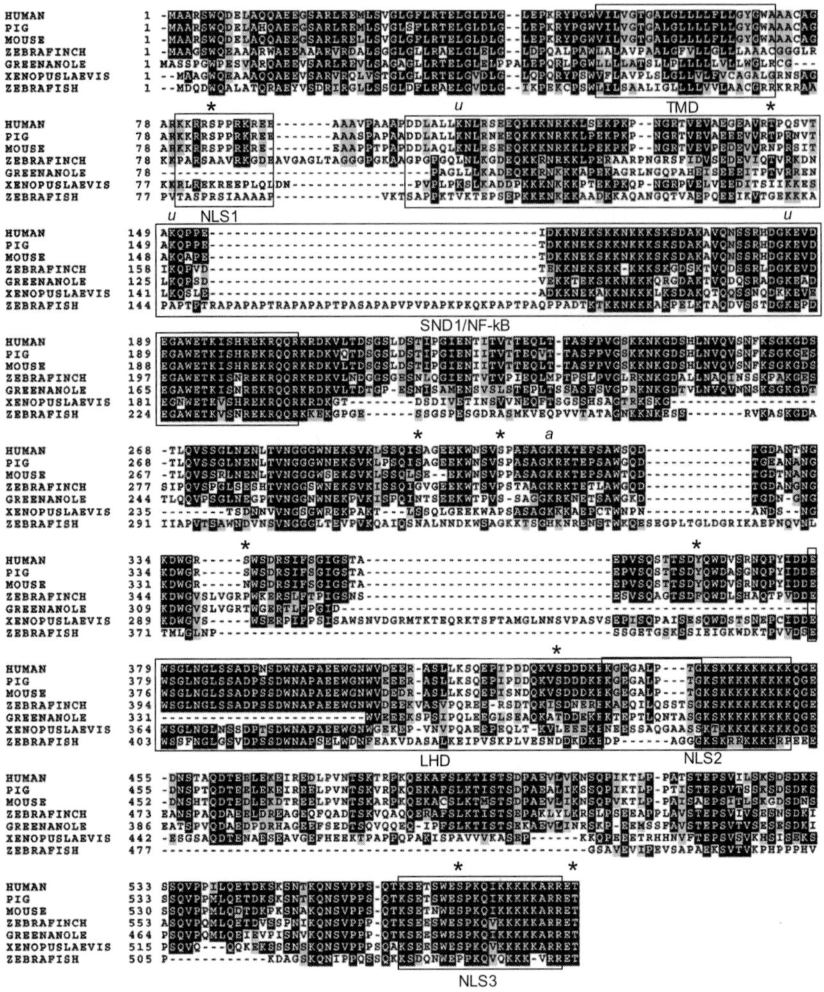

```
                                                                                 *                    TMD       *
HUMAN          1 -MAARSWQDELAQQAEEGSARLREMLSVGLGFLRTELGLDLG--LEPKRYPGWVILVGTGALGLLLLFLLGYGWAAACAG
PIG            1 -MAARSWQDELAHQAEEGSARLREMLSVGLSFLRTELGLDLG--LEPKRYPGWVILVGTGALGLLLLFLLGYGWAAACAG
MOUSE          1 -MAARSWQDELAQQAEEGSARLRELLSVGLGFLRTELGLDLG--LEPKRYPGWVILVGTGALGLLLLFLLGYGWAAACAG
ZEBRAFINCH     1 -MAAGSWQEAAARWAEEAAARVRDALSGGLGLLRAELGLELG--LDPQALPAWLALAVPAALGFVLLGLLLAAACGGGLR
GREENANOLE     1 MASSPGWPESVARQAEEVSARLREVLSAGAGLLRTELGLELPPALEPQRLPGWLLLLATSLLPLLLLVLLWGLRCG---
XENOPUSLAEVIS  1 --MAAGWQEAAAQQAEEVSARVRQLVSTGLGLLRTELGVDLG--LQPQRYPSWVFLAVPLSLGLLVLFVCAGALGRNSAG
ZEBRAFISH      1 --MDQDWQALATQRAEYVSDRIRGLLSSGLDFLRAELGVDLG--IKPEKCPSWLILSAALIGLLLVVLAACGRRKRRAA

                     *                                u
HUMAN         78 ARKKRRSPPRKREE-------AAAVPAAAPDDLALLKNLRSEEQKKKNRKKLSEKPKP--NGRTVEVAEGEAVRTPQSVT
PIG           78 ARKKRRSPPRKREE-------AAASPAPAADDLALLKNLRNEEQKKKNRKKLPEKPKP--NGRTVEVAEEEVVRTPRNVT
MOUSE         78 ARKKRRSPPRKRE--------EAAPPTPAPDDLAQLKNLRSEEQKKKNRKKLPEKPKP--NGRTVEVPEDEVVRNPRSIT
ZEBRAFINCH    78 KKPARSAAVRKGDEAVGAGLTAGGGPGKAAGPGPGQLNLKGDEQKKRNRKKLPERAARPNGRSFIDVSEDEVIQTVRKDN
GREENANOLE    78 -----------------------------PAGLLLKADEQKKRNKKKAPEKAGRLNGQPAHEISEEEITPTVRREN
XENOPUSLAEVIS 77 KKRLREKREEPLQLDN---------------PVPLPKSLKADDPKKKNKKKPTEKPKQP-NGRPVELVEEDITSIIKKES
ZEBRAFISH     77 PVTASPRSIAAAAP------------VKTSAPPKTVKTEPSEPKKKNKKKAADKKAQANGQTVAEPQEEIKVTGEKKKA
                 u   NLS1                                                                          u
HUMAN        149 AKQPPE----------------------------------------IDKKNEKSKKNKKKSKSDAKAVQNSSRHDGKEVD
PIG          149 AKQPPE----------------------------------------TDKKNEKSKKNKKKSKSDAKAVQNSSRHDGKEVD
MOUSE        148 AKQAPE----------------------------------------TDKKNEKSKKNKKKSKSDAKAVQNSSRHDGKEVD
ZEBRAFINCH   158 IKQPVD----------------------------------------TEKKNEKSKK-KKKSKGDSKTVQDSSRLDGKEVD
GREENANOLE   125 LKQPSD----------------------------------------VEKKTEKSKKNKKKQRGDAKTVQDQSRADGKEAD
XENOPUSLAEVIS 141 LKQSLE----------------------------------------ADKKNEKAKKNKKKLKSDAKQTQQSSNQDKKEVE
ZEBRAFISH    144 PAPTPTRAPAPAPTRAPAPAPTPASAPAPVPVPAPKPKQKPAPTPAQPPADTKTKKNKKKAKPELKTAQDVSSTDGKEPD
                                               SND1/NF-kB
HUMAN        189 EGAWETKISHREKRQQRKRDKVLTDSGSLDSTIPGIENTITVTTEQLT-TASFPVGSKKNKGDSHLNVQVSNFKSGKGDS
PIG          189 EGAWETKISHREKRQQRKRDKVQTDSGSLDSTIPGIENIITVTTEQVT-TASFPVGSKKNKGDSHLNVQVSNFKSGKGES
MOUSE        188 EGAWETKISHREKRQQRKRDKVLTDSGSLDSTIPGIENIITVTTEQLT-TASFPVGSKKNKGDSHLNVQVSNFKSGKGDS
ZEBRAFINCH   197 EGAWETKISNREKRQQRKRDKVLNDGGSGESNLQGIENTVTVPIEQLMPTPSLPVGLRKNKGDALLNAQINSSKPAKGES
GREENANOLE   165 EGAWETKISNREKRQQRKRDKVLTDTGP-ESNISAMENSVSLSTEPLTSSASFSVGPRKNKGDTVLNVQVNNSKSGKGDT
XENOPUSLAEVIS 181 EGNWETKVSHREKRQQRKRDKGT--------DSDIVETINSVVNEQFTSGSSHSAGTRKSKG----------------
ZEBRAFISH    224 EGAWETKVSNREKRQQRKKEKGPGE------SSGSPESGDRASMKVEQPVVTATAGNKKNKESS-------RVKASKGDA
                                            *         *     a
HUMAN        268 -TLQVSSGLNENLTVNGGGWNEKSVKLSSQISAGEEKWNSVSPASAGKRKTEPSAWSQD-------------TGDANTNG
PIG          268 -TLQVSSGLNENLTVNGGGWNEKSVKLPSQISAGEEKWNSVSPASAGKRKTEPSAWGQD-------------TGEANANG
MOUSE        267 -TLQVSSRLNENLTVNGGGWSEKSVKLSSQLSE--EKWNSVPPASAGKRKTEPSAWTQD-------------TGDTNANG
ZEBRAFINCH   277 SIPQVSPGLSESHTVNGGSWNEKSVKISSQIGVGEEKWTSVPSTAAGKRKTETLAWGQD-------------TGDANGNG
GREENANOLE   244 TLQQVPSGLNEGPTVNGGNWNEKPVKISPQINTSEEKWTPVS-SAGGKRKNETSAWGKD-------------TGDN-GNG
XENOPUSLAEVIS 235 --------TSDNNVVNGSGWREKPAKT---LSSQLGEEKWAPSASAGKKKAEPCTWNPN---------------ANDS--NG
ZEBRAFISH    291 IIAPVTSAWNDVNSVNGGGLTEVPVKQAIQSNALNNDKWSAGKKTSGHKNRENSTWKQESEGPLTGLDGRIKAEPNQVNL
                          *                                                         *
HUMAN        334 KDWGR-----SWSDRSIFSGIGSTA---------------------------------EPVSQSTTSDYQWDVSRNQPYIDDE
PIG          334 KDWGR-----SWSDRSIFSGIGSTA---------------------------------EPVSQSTTSDYQWDASGNQPYIDDE
MOUSE        331 KDWGR-----NWSDRSIFSGIGSTA---------------------------------EPVSQSTTSDYQWDVSRNQPYIDDE
ZEBRAFINCH   344 KDWGVSLVGRPWKERSLFTPIGSNS---------------------------------ESVSQAGTSDFQWDLSHAQTPVDDE
GREENANOLE   309 KDWGVSLVGRTWGERTLFPGID------------------------------------------------------------
XENOPUSLAEVIS 289 KDWGVS-----WSERPIPPSISAWSNVDGRMTKTEQRKTSFTAMGLNNSVPASVSEPISQPAISESQWDSTSNEPCIDDE
ZEBRAFISH    371 TMLGLNP---------------------------------------------------SSGETGSKSSIEIGKWDKTPVVDSE
                                                         *
HUMAN        379 WSGLNGLSSADPNSDWNAPAEEWGNWVDEER-ASLLKSQEPIPDDQKVSDDDKEKGEGALP---TGKSKKKKKKKKKQGE
PIG          379 WSGLNGLSSADPSSDWNAPAEEWGNWVEEER-ASLLKSQEPIPDDQKVSDDDKEKGEGALP---TGKSKKKKKKKKKQGE
MOUSE        376 WSGLNGLSSADPSSDWNAPAEEWGNWVDEDR-ASLLKSQEPISNDQKVSDDDKEKGEGALP---TGKSKKKKKKKKKQGE
ZEBRAFINCH   394 WSGLNGLSSADPSSDWNAPAEEWGNWVDEEKVASVPQREE-RSDTQKISDNEREKAEQILQSSTSGKSKKKKKKKKKQGE
GREENANOLE   331 ---------------------------WVEEEKSPSIPQLEEGLSEAQKATDDEKEKTEPTLQNTASGKSKKKKKKKKKQGE
XENOPUSLAEVIS 364 WSGLNGLNSSDPTSDWNAPAEEWGNWGEKEP-VNVPQAEEPEQLT-KVLEEEKENEESSAQGAASSKTKKKKKKKKKQGE
ZEBRAFISH    403 WSSFNGLGSVDPSSDWNAPSELWDNFEAKVDASALKEIPVSKPLVESNDDKDKEDP------AGGGKSKRRKKKKRPEEE
                                            LHD                              NLS2
HUMAN        455 -DNSTAQDTEELEKEIREDLPVNTSKTRPKQEKAFSLKTISTSDPAEVLVKNSQPIKTLP-PATSTEPSVILSKSDSDKS
PIG          455 -DNSPTQDTEELEKEIREELPVNTSKVRPKQEKAFSLKTISTSDPAEALIKSSQPIKTLP-PTISTEPSVTSSKSDSDKS
MOUSE        452 -DNSHTQDTEDLEKDTREELPVNTSKARPKQEKACSLKTMSTSDPAEVLIKNSQPVKTLP-PAISAEPSITLSKGDSDNS
ZEBRAFINCH   473 EANSPAQDAEELDREAGEQFQADTSKVQAQQERAFSLKTISTSEPAKLYLKRSLPSEEAPPLAVSTEPSVIVSESNSDKI
GREENANOLE   386 EATSPVQDAEDPDRHAGEEFSEDTSQVQQEQ-IPFSLKTISTSEKAEVLINRSKP-EEMSSFAVSTEPSVTVSESESDKI
XENOPUSLAEVIS 442 -ESGSAQDTENAESEAVGEFHEEKTPAPPQPAKISPAVVVKASEP------KKQPEEETRHHNVFTEPSVSVKHSISEKS
ZEBRAFISH    477 ---------------------------------------------GSAVEVIPEVSAPAEKSVTVKPHPPPHV
                                              *              *
HUMAN        533 SSQVPPILQETDKSKSNTKQNSVPPS-QTKSETSWESPKQIKKKKKARRET
PIG          533 SSQVPPMLQETDKSKSNTKQNSVPPS-QTKSETSWESPKQIKKKKKARRET
MOUSE        530 SSQVPPMLQDTDKPKSNAKQNSVPPS-QTKSETNWESPKQIKKKKKARRET
ZEBRAFINCH   553 ASQVPQMLQETDVSSPNIKQNSVPPP-QTKSEESWESPKQVKKKKKARRET
GREENANOLE   464 PSQVPQMLQEIEVPISNVKQNSVPPP-QTKSEESWESPKQIKKKKKARRET
XENOPUSLAEVIS 515 PSQVQ---QQKEKSSSNSKQNSVPPPSQAKSEESWESPKQVKKKKKARRET
ZEBRAFISH    505 P-----------KDAGSKQNIPPQSSQKKSDQNWEPPKQVQKKK-VRRET
                                                      NLS3
```

Figure 1.1 Sequence alignment and posttranslational modification of AEG-1/MTDH/LYRIC. Amino acid sequences of AEG-1/MTDH/LYRIC of seven representative vertebrates were aligned by ClustralW2 (http://www.ebi.ac.uk/Tools/services/web/toolform.ebi?tool=clustalw2) and curated with BOXSHADE 3.21 (http://www.ch.embnet.org/software/BOX_form.html). Identities in the alignment are shown in inverted characters and similarities in shaded ones. Running PhosphoSitePlus (http://www.phosphosite.org/homeAction.do) identified 13 posttranslational modification sites (*, phosphorylation; *u*, ubiquitination; *a*, acetylation). Transmembrane domain (TMD), nuclear localization signals (NLS-1, 2, and 3), SND1/NF-κB-interacting domain (SND1/NF-κB), and lung-homing domain (LHD) are demarked by boxes. Unmarked sites here for legibility are PLZF-interacting domains (a.a. 1–285 and 487–582) and a BCCIP-interacting domain (a.a. 72–169) of human AEG-1/MTDH/LYRIC.

that participate in protein–protein interactions (Fig. 1.1) (Yoo, Emdad, et al., 2011).

AEG-1/MTDH/LYRIC was predicted to have a TMD at a.a. 51–72 in human and its membrane topology was considered as a type Ib or II (Britt et al., 2004; Brown & Ruoslahti, 2004; Kang et al., 2005; Sutherland et al., 2004). FACS analysis of MTDH tagged with Myc epitope at the LHD suggested extracellular distribution of the C-terminal, which supports type II topology (Brown & Ruoslahti, 2004). However, type Ib topology is supported by functional analysis of the molecule (Britt et al., 2004; Kang et al., 2005; Sutherland et al., 2004). Therefore, as indicated, the topology issue remains to be resolved. GFP-fusion protein of ΔTMD-LYRIC localized at the nucleolus and a protein smaller than full-length LYRIC was enriched in the nucleolus fraction, suggesting truncation of the molecule at around TMD for nuclear distribution (Sutherland et al., 2004). However, N-terminal including TMD (a.a. 1–71) was required for AEG-1/MTDH/LYRIC-induced NF-κB activation, although p65 subunit of NF-κB interacted at a.a. 101–205 of AEG-1/MTDH/LYRIC (Sarkar et al., 2008). In addition, a protease potentially targeting and releasing AEG-1/MTDH/LYRIC from the membrane has yet to be identified. Support for the nuclear function of AEG-1/MTDH/LYRIC, either in nucleolus or in nucleoplasm, requires additional experiments.

Three NLSs and a potential ATP/GTP-binding site were predicted for AEG-1/MTDH/LYRIC (Britt et al., 2004; Sutherland et al., 2004). Although NLSs in AEG-1/MTDH/LYRIC were well characterized in subsequent studies, binding of ATP/GTP to the predicted sites remains to be confirmed. AEG-1/MTDH/LYRIC distributes both in the cytoplasm, specifically in the ER, and nucleus, and the three putative NLSs in AEG-1/MTDH/LYRIC (NLS-1, 2, and 3 at a.a. 79–91, 432–451, and 561–580, respectively) appear to function differently in the nuclear localization of AEG-1/MTDH/LYRIC. NLS-3 (a.a. 546–582) is a primary determinant of AEG-1/MTDH/LYRIC nuclear localization, while an extended NLS-1 (a.a. 78–130) regulates its nucleolar localization (Thirkettle, Girling, et al., 2009). AEG-1/MTDH/LYRIC is ubiquitinated at NLS-2 extension (a.a. 415–486), and the NLS-2 ubiquitination is postulated to direct its cytoplasmic distribution that occurs in cancer cells (Thirkettle, Girling, et al., 2009).

AEG-1/MTDH/LYRIC multifunctionality might be ascribed to the existence of multiple binding domains with diverse proteins (Yoo, Emdad, et al., 2011). LHD defined in AEG-1/MTDH/LYRIC (a.a. 378–440) that confers metastatic capabilities of breast cancer cells to the lung might promote interaction of the cells with lung microvasculature, although its interacting partner has not been defined (Brown & Ruoslahti, 2004). AEG-1/MTDH/LYRIC translocates to the nucleus and activates NF-κB by its interaction with p65 subunit of NF-κB through a.a. 101–205 region in TNF-α treated cells. By interacting with cyclic AMP-response element-binding protein (CREB)-binding protein (CBP), AEG-1/MTDH/LYRIC facilitates NF-κB-CBP complex on the IL-8 promoter in inflammatory responses (Emdad et al., 2006; Sarkar et al., 2008). AEG-1/MTDH/LYRIC also interacts with sumoylated PLZF (promyelocytic leukemia zinc finger protein) through two regions (a.a. 1–285 and 487–582) (Thirkettle, Mills, Whitaker, & Neal, 2009). PLZF is an inhibitor of c-Myc transcription and interaction of AEG-1/MTDH/LYRIC with PLZF prevents its recruitment to the c-Myc promoter, which could result in upregulation of c-Myc transcription. BRCA2- and CDKN1A-interacting protein alpha (BCCIPα) that interacts with $p21^{mda\text{-}6/cip1}$-CDK complex and enhances $p21^{mda\text{-}6/cip1}$ activity is another binding partner of AEG-1/MTDH/LYRIC (Ash, Yang, & Britt, 2008). Interaction of BCCIPα with AEG-1/MTDH/LYRIC facilitates degradation of BCCIPα. AEG-1/MTDH/LYRIC was also found to bind SND1 and participate in RISC in a yeast two-hybrid screening (Yoo, Santhekadur, et al., 2011). Interaction with SND1 increased RISC activity in degradation of tumor suppressor mRNAs (Wang et al., 2012; Yoo, Santhekadur, et al., 2011). These interactions of AEG-1/MTDH/LYRIC with PLZF, BCCIPα, and SND1 could promote cell proliferation, which could be translated into tumor-promoting activity of AEG-1/MTDH/LYRIC in addition to a metastasis-mediating role.

In addition to the experimentally verified domains, numerous posttranslational modification sites are predicted and reported via high-throughput proteomic analysis (Hornbeck et al., 2012; Xue et al., 2008). Among the posttranslational modification sites curated in PhosphoSitePlus (http://www.phosphosite.org/homeAction.do), 13 residues are reported to be phosphorylated (nine sites), ubiquitinated (three sites), or acetylated (one site) more than five times in high-throughput mass spectrometry as indicated in Fig. 1.1 (Hornbeck et al., 2012). Two phosphorylation sites (S568 and T582) are well conserved in vertebrates, and evolutionary conservation is also verified for two ubiquitination sites (K150 and K185) and one acetylation site (K314). However, biological significance of the posttranslational modification remains to be investigated.

4. EVOLUTION OF AEG-1/MTDH/LYRIC

A phylogenetic analysis of DNA sequences for the *AEG-1/MTDH/LYRIC* gene from a wide range of vertebrates is shown in Fig. 1.2. This tree was generated by Bayesian inference using a GTR-Invariants-Gamma model with 1 million generations and 20% burn-in (Ronquist et al., 2012). There are three immediately obvious and important observations that can be made from this tree and a consideration of the phylogenetic occurrence of the *AEG-1/MTDH/LYRIC* gene in animals (Fig. 1.2). First, these are the only major animal taxa in the current whole genome sequence database where *AEG-1/MTDH/LYRIC* exists. This result indicates that *AEG-1/MTDH/LYRIC* arose in the common ancestor of all jawed vertebrates over 500 million years ago. The second striking result from this tree is the extremely long branches in fish (light blue group), amphibians and archosaurs (birds and reptiles—purple group), and in specific species scattered throughout the mammals. The only groups of mammals that do not show extremely long branches are the primates (black group) and the carnivores (red group). Usually, when this pattern emerges in phylogenetic analysis, it suggests a strong degree of purifying natural selection in the short branched clades in the tree. By quantifying the base to tip lengths of various clades (table inset in Fig 1.2), we suggest that the primates and carnivores have fourfold slower rates of change in their AEG-1/MTDH/LYRIC genes. Compared to archosaurs and bony fish, primates and carnivores have 10- to 30-fold differences in group depth. Some of these differences in rates of change are probably due to the age of the groups under consideration.

5. EXPRESSION PROFILE

AEG-1/MTDH/LYRIC mRNA expression was ubiquitously detected in all human normal tissues with relatively higher expression in the heart, skeletal muscle, liver, and endocrine glands such as adrenal gland and thyroid (Kang et al., 2005). However, 10 years after its initial cloning, AEG-1/MTDH/LYRIC is now appreciated as an important oncogene overexpressed in various types of human cancers analyzed so far such as brain tumor, breast cancer, hepatocellular carcinoma (HCC), colorectal cancer, neuroblastoma, non-small cell lung cancer (NSCLC), etc., and its overexpression in tumor cells enhances characteristics of malignant aggressiveness including increased tumor growth, invasion and metastasis, angiogenesis, and

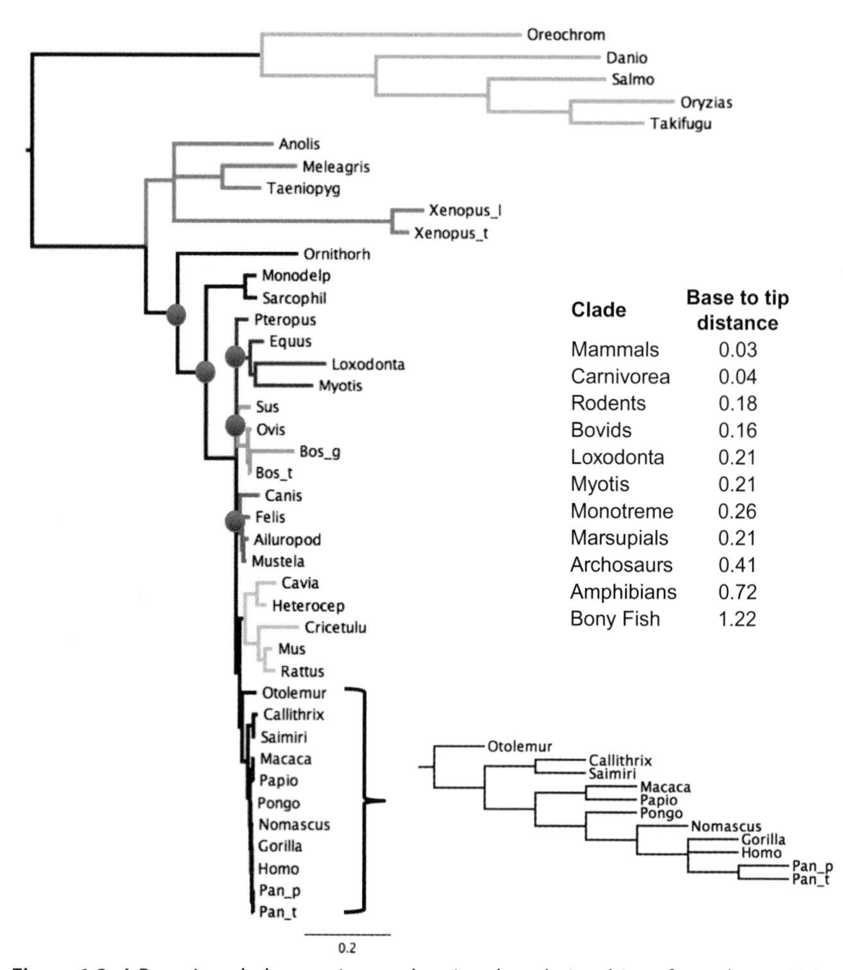

Clade	Base to tip distance
Mammals	0.03
Carnivorea	0.04
Rodents	0.18
Bovids	0.16
Loxodonta	0.21
Myotis	0.21
Monotreme	0.26
Marsupials	0.21
Archosaurs	0.41
Amphibians	0.72
Bony Fish	1.22

Figure 1.2 A Bayesian phylogenetic tree showing the relationships of vertebrate *AEG-1/MTDH/LYRIC* genes. The model used in the analysis was a GTR with invariants gamma. One million generations were run, upon which convergence of chains was assessed to be adequate. The tree is the result of removing a burn-in of 20%. All nodes in the tree except for those marked by blue circles have posterior probabilities of 0.99 or better. The nodes marked by blue circles had posterior probabilities between 0.5 and 0.8, suggesting that these nodes are not particularly robust. The different colored branches in the tree indicate well established taxonomic groups except for the dark blue which contains elephants and bats and horse. The more conventionally accepted groups are as follows: light blue, bony fish; purple, archosaurs and amphibians; light green, bovids and close relatives; red, carnivores; orange, rodents; black, primates. The inset table shows the base to tip distance of the indicated groups in the tree. (See Page 1 in Color Section at the back of the book.)

chemoresistance (Yoo, Emdad, et al., 2011). In addition, recent studies have shown differential expression patterns of AEG-1/MTDH/LYRIC in other physiological and pathological processes (Yoo, Emdad, et al., 2011). In this section, we review the expression patterns of AEG-1/MTDH/LYRIC in various physiological and pathological circumstances.

5.1. Cancer

5.1.1 Brain tumors and neuroblastoma

Recent studies compared the expression of AEG-1/MTDH/LYRIC in various types of brain tumors (Emdad et al., 2010; Lee et al., 2011; Liu et al., 2010; Xia et al., 2010). Western blotting analysis in 98 patient samples of various brain tumors (25 glioblastoma multiforme (GBM), 18 astrocytoma, 18 meningioma, 19 oligodendroglioma, and 18 other types such as ependymoma, ganglioglioma, peripheral neural sheath tumor, etc.) compared to nine normal brain tissues indicated a 3- to 10-fold higher expression of AEG-1/MTDH/LYRIC in >90% cases (Emdad et al., 2010). Immunofluorescence staining in 33 GBM patient tissues, 2 grade III astrocytoma tissues, and 5 normal brain samples in a tissue microarray revealed higher expression of AEG-1/MTDH/LYRIC in the patient tissues (Lee et al., 2011). In addition, expression of AEG-1/MTDH/LYRIC in various glioma cell lines was elevated compared to that in primary normal astrocytes and oligodendroglial cells (Emdad et al., 2010; Liu et al., 2010; Xia et al., 2010). Immunohistochemical analysis of AEG-1/MTDH/LYRIC in 296 archived glioma patient tissues including 39 grade I cases, 121 grade II cases, 88 grade III cases, and 48 grade IV cases compared with that in normal tissues indicated that AEG-1/MTDH/LYRIC was upregulated in 89.5% of patients with a significant correlation with clinicopathologic grades of glioma (Liu et al., 2010). In oligodendroglioma, both mRNA and protein levels of AEG-1/MTDH/LYRIC were elevated in patient samples and oligodendroglioma cell lines compared to adjacent noncancerous brain tissues and primary oligodendroglial cells, and immunohistochemical analysis of 75 patient samples showed increased AEG-1/MTDH/LYRIC expression in 68% of cases with a significant correlation with histological grades of oligodendroglioma and with shorter survival time (Xia et al., 2010). Furthermore, the increased expression of AEG-1/MTDH/LYRIC played a critical role in invasion of glioma cells through activation of matrix metalloproteinases (MMP-2 and MMP-9) (Emdad et al., 2010; Liu et al., 2010). In neuroblastoma, the most common sympathetic nervous tumor, the increased expression of AEG-1/MTDH/LYRIC was observed in 6 of 10 patient samples and neuroblastoma cell lines compared to normal peripheral nerve tissues and normal cell counterparts (Lee et al., 2009). Analysis of AEG-1/MTDH/LYRIC in more archived neuroblastoma patient tissues (32 patients) indicated that AEG-1/MTDH/LYRIC expression was present in all samples with higher expression in 75% cases and was strongly correlated with various clinicopathologic stages and reduced survival of patients (Liu, Liu, Han, Zhang, & Sun, 2011). These studies have documented that AEG-1/MTDH/LYRIC expression is elevated in various neoplasms of the nervous system indicating its critical role in tumor invasion and suggest that it would be worthwhile systematically investigating its roles in other tumor types of the nervous system.

5.1.2 Head and neck tumors

Three independent recent studies revealed increased AEG-1/MTDH/LYRIC expression as a novel prognostic marker for salivary gland carcinoma and tongue carcinoma progression and patient survival (Deng & Feng, 2011; Ke et al., 2012; Liao et al., 2011). Immunohistochemical analysis of AEG-1/MTDH/LYRIC in tissues of 141 salivary gland carcinoma patients (Liao et al., 2011), 45 tongue squamous cell cancer patients (Deng & Feng, 2011), and 93 tongue squamous cell carcinoma patients (Ke et al., 2012) showed its higher expression than that in each peritumoral normal tissues, and the increased AEG-1/MTDH/LYRIC expression was strongly correlated with the clinicopathologic classifications and poor survival of the patients. In addition, the increased expression of *AEG-1/MTDH/LYRIC* was detected at the level of mRNA in the cancer patients (Ke et al., 2012; Liao et al., 2011).

5.1.3 Breast cancer

One of the first tumor types in which increased AEG-1/MTDH/LYRIC expression was appreciated was breast cancer (Brown & Ruoslahti, 2004). The group who cloned AEG-1/MTDH/LYRIC as MTDH demonstrated that AEG-1/MTDH/LYRIC is overexpressed in breast cancer cell lines, breast cancer patient tissues, and breast tumor xenografts, and that its overexpression is related with lung metastasis of breast cancer (Brown & Ruoslahti, 2004). Many subsequent studies with over 1000 breast cancer patient tissues from six independent groups including both female and male noticed that AEG-1/MTDH/LYRIC overexpression was detected in more than 40% of patient tissues compared to normal tissues and was strongly correlated with the clinical staging and various pathological classifications of breast cancer indicating its roles in aggressive tumor growth, chemoresistance, invasion and metastasis, and angiogenesis, as well as with poor survival of the patients (Hu et al., 2009; Kornegoor et al., 2012; Li et al., 2009, 2008; Li, Li, et al., 2011; Su, Zhang, & Yang, 2010; Tokunaga et al., 2012). Increased AEG-1/MTDH/LYRIC expression in clinical tissues was detected at the level of both mRNA and protein, and genomic amplification has been reported as one of the mechanisms by which AEG-1/MTDH/LYRIC is overexpressed in breast cancer (Hu et al., 2009; Kornegoor et al., 2012; Li et al., 2008; Tokunaga et al., 2012). Full-length sequencing of *AEG-1/MTDH/LYRIC* using 108 breast cancer patient tissues and 100 normal tissues discovered nine novel variants of *AEG-1/MTDH/LYRIC*, and two of them were associated with the susceptibility of breast cancer, even though further studies with the variants are required (Liu, Zhang, et al., 2011). These results clearly suggest that AEG-1/MTDH/LYRIC overexpression is a valuable marker of breast cancer progression associated with poor survival of the patients.

5.1.4 Non-small cell lung cancer

The clinical importance of AEG-1/MTDH/LYRIC was also determined in NSCLC. Overexpression of AEG-1/MTDH/LYRIC was detected in NSCLC cell lines and patient tissues at the levels of both mRNA and protein, and immunohistochemical staining of AEG-1/MTDH/LYRIC in 267 NSCLC patient tissues revealed the expression of AEG-1/MTDH/LYRIC mostly in the cytoplasm of the tissues, and a strong correlation of AEG-1/MTDH/LYRIC overexpression was evident with clinicopathologic characteristics of NSCLC and poor clinical outcome of the patients (Song et al., 2009; Sun et al., 2012). In addition, analyzing mRNA expression of 11 genes involved in the EGFR and NF-κB pathways in 60 metastatic NSCLC patients and lung cancer cell lines revealed that *AEG-1/MTDH/LYRIC* expression was correlated with *BRCA1* expression and the overexpression of both genes was associated with poor survival of the patients, suggesting that the combination of *AEG-1/MTDH/LYRIC* and *BRCA1* expression could serve as a potential prognostic marker for the disease (Santarpia et al., 2011).

5.1.5 Liver and gallbladder cancers

Expression of AEG-1/MTDH/LYRIC was significantly higher in HCC patient tissues including hepatitis B virus-related HCC patients and HCC cell lines than in normal hepatocytes at the levels of both mRNA and protein, where the protein was mostly localized at the perinuclear region of the HCC cells. Moreover, its overexpression was correlated with clinicopathologic characters including cell proliferation, invasion, metastasis, chemoresistance, and angiogenesis, as well as with poor clinical outcome of the patients (Gong et al., 2012; Yoo et al., 2009; Zhou, Deng, et al., 2012; Zhu et al., 2011). AEG-1/MTDH/LYRIC overexpression in HCC was associated with elevated copy number and was correlated with regulation of epithelial–mesenchymal transition (EMT) markers, suggesting a mechanism of AEG-1/MTDH/LYRIC overexpression in HCC and its role in HCC metastasis by inducing EMT (Yoo et al., 2009; Zhu et al., 2011). Two independent groups also investigated the expression of AEG-1/MTDH/LYRIC in gallbladder carcinoma (GBC) (Liu & Yang, 2011; Sun et al., 2011). They found that AEG-1/MTDH/LYRIC expression was higher in GBC patient tissues and GBC cell lines than in normal counterpart tissues and cells at the level of both mRNA and protein, and its overexpression in patients was strongly correlated with cancer progression including proliferation, differentiation, and metastasis, and with poor prognosis of the patients (Liu & Yang, 2011; Sun et al., 2011).

5.1.6 Renal cancer

Expression of AEG-1/MTDH/LYRIC in renal cell carcinoma (RCC) patient tissues and RCC cell lines was markedly higher than that in the paired normal tissues and normal kidney cells at both mRNA and protein levels, and immunohistochemical analysis in 106 cases of RCC patient tissues showed overexpression of AEG-1/MTDH/LYRIC in 96 cases (94.1%) (Chen, Ke, Shi, Yang, & Wang, 2010). Statistical analysis of the results indicated a significant correlation of AEG-1/MTDH/LYRIC expression with tumor grade, clinical staging, and tumor progression including metastasis as well as with poor clinical outcome of the patients.

5.1.7 Bladder and prostate cancers

Prostate cancer is another first tumor type in which overexpression of AEG-1/MTDH/LYRIC was examined (Kikuno et al., 2007). Immunohistochemical staining of AEG-1/MTDH/LYRIC revealed its overexpression in 16 of 20 prostate cancer patient samples compared to benign prostatic hyperplasia cases (Kikuno et al., 2007). Another independent study using a large cohort of 206 patients confirmed higher expression of AEG-1 in prostate cancer compared with benign tissue and found that AEG-1 predominantly localized in the nucleus of benign prostatic hyperplasia tissues as well as in thyroid and lung while its cytoplasmic expression was predominantly detected in prostate cancer tissues (Thirkettle, Girling, et al., 2009). In addition, when compared with normal bones, 9 of 11 prostate bone metastases showed increased AEG-1/MTDH/LYRIC expression, which was exclusively distributed in the cytoplasm and membrane (Thirkettle, Girling, et al., 2009). These results indicated that increased expression and significant changes in the distribution of AEG-1/MTDH/LYRIC can predict Gleason grade and patient survival of prostate cancer. A recent study also found that significant levels of AEG-1/MTDH/LYRIC expression was detected in 65% of bladder cancer patients, but not in normal bladder tissues, and that increased expression of AEG-1/MTDH/LYRIC in bladder cancer tissues was correlated with tumor grade, clinical staging, and tumor progression as well as with poor clinical outcome of the patients (Zhou, Li, Wang, Yin, & Zhang, 2012).

5.1.8 Ovary and endometrium cancers

Recent studies have also investigated the expression of AEG-1/MTDH/LYRIC in ovarian cancer, the most lethal gynecological cancer (Borley, Wilhelm-Benartzi, Brown, & Ghaem-Maghami, 2012; Li et al., 2012; Li, Liu, et al., 2011; Meng, Luo, Ma, Hu, & Lou, 2011). Immunohistochemical staining of AEG-1/MTDH/LYRIC in 81 ovarian cancer patient tissues revealed that high expression of AEG-1/MTDH/LYRIC was detected in 66.7% of ovarian cancer patients in association with clinicopathologic features of the cancer and poor overall survival of the patients when compared with patients with lower expression of AEG-1/MTDH/LYRIC (Meng et al., 2011). Another study using 157 epithelial ovarian cancer patient tissues found that AEG-1/MTDH/LYRIC overexpression was detected in 87 patients and 95.4% and 47.1% of the patients with AEG-1/MTDH/LYRIC overexpression presented, respectively, peritoneal dissemination and lymph node metastasis, indicating AEG-1/MTDH/LYRIC as a valuable clinical marker for predicting ovarian cancer metastasis (Li, Liu, et al., 2011). In addition, immunohistochemical study with 131 stage III–IV ovarian serous carcinoma patient tissues indicated that AEG-1/MTDH/LYRIC overexpression was associated with poor prognosis and cisplatin resistance in advanced serous ovarian cancer (Li et al., 2012). A systemic review using 30 studies from the Cancer Genome Atlas ovarian cancer data sets also noticed that expression of only AEG-1/MTDH/LYRIC and insulin-like growth factor-1 receptor was significantly correlated with surgical debulking of the ovarian cancer (Borley et al., 2012).

Investigation of AEG-1 expression in 35 normal endometrium, 40 atypical endometrial hyperplasia tissues, and 274 endometrial cancer patient tissues showed that AEG-1/MTDH/LYRIC expression was gradually elevated in normal, atypical hyperplasia, and cancer tissues, and statistical analysis of the results indicated that overexpression of AEG-1/MTDH/LYRIC in the patients was significantly correlated with clinicopathological characteristics of endometrial cancer including stages, aggressive growth, invasion, and metastasis as well as with poor overall survival and disease-free survival compared with patients with lower AEG-1/MTDH/LYRIC expression (Song, Li, Lu, Zhang, & Geng, 2010). Together these results suggest that AEG-1/MTDH/LYRIC overexpression is a valuable prognostic marker for two types of cancers in the female genital tract.

5.1.9 Tumors in gastrointestinal tract

Overexpression of AEG-1/MTDH/LYRIC was associated with tumor progression and prognosis in cancers of the esophagus and stomach (Jian-bo et al., 2011; Yu et al., 2009). Expression of AEG-1/MTDH/LYRIC was markedly increased in esophageal cancer cell lines and surgical esophageal squamous cell carcinoma (ESCC) patient tissues compared with each normal counterpart at both transcriptional and translational levels (Yu et al., 2009). Immunohistochemical analysis and statistical analysis of AEG-1/MTDH/LYRIC expression in ESCC patient samples further revealed that overexpression of AEG-1/MTDH/LYRIC was detected in 47.6% cases in correlation with clinical stages, pathological characters, and poor survival of the patients (Yu et al., 2009). Overexpression of AEG-1/MTDH/LYRIC was also found in 66 of 106 gastric cancer patients in association with TNM stage and Ki-67 proliferation index, and AEG-1/MTDH/LYRIC was mainly localized in the cytoplasm of primary gastric cancer cells (Jian-bo et al., 2011). Furthermore, overexpression of AEG-1/MTDH/LYRIC was significantly associated with poor survival of the gastric cancer patients as an independent prognostic factor for the disease (Jian-bo et al., 2011). Colon cancer is another tumor type in the gastrointestinal tract associated with overexpression of AEG-1/MTDH/LYRIC (Yoo, Emdad, et al., 2011). Investigation of AEG-1/MTDH/LYRIC expression at both mRNA and protein levels using over 1000 colon cancer patient tissues including adenomas and carcinomas and colon cancer cell lines compared with distant normal mucosa and adjacent nontumor tissues revealed that increased expression of AEG-1/MTDH/LYRIC in colon cancer was significantly associated with the advanced stages of various clinicopathological staging systems such as

UICC stage, TNM classification, Duke's stage, etc., as well as with poor survival of the patients (Gnosa et al., 2012; Jiang, Zhu, Zhu, & Piao, 2012; Song, Li, Li, & Geng, 2010; Wang et al., 2012; Zhang et al., 2012). AEG-1/MTDH/LYRIC was mainly detected in the cytoplasm of colon cancer cells, but nuclear staining of AEG-1/MTDH/LYRIC was more common in lesions with more advanced disease stages (Song, Li, Li, et al., 2010). In addition, a positive correlation in the context of increased expression or direct interaction of AEG-1/MTDH/LYRIC with β-catenin or SND1 in colon cancer patients indicated the mechanisms by which AEG-1/MTDH/LYRIC is involved in colon cancer progression (Wang et al., 2012; Zhang et al., 2012).

5.1.10 Bone tumors

Expression of AEG-1/MTDH/LYRIC in osteosarcoma, the most common histological form of primary bone cancer, was examined using 62 osteosarcoma patient tissues and 20 normal bones (Wang et al., 2011). Statistical analysis of the data with immunohistochemical staining of AEG-1/MTDH/LYRIC in the tissues found that AEG-1 was overexpressed in osteosarcoma tissues in strong correlation with gender, clinical stages, classification, metastasis, differentiation, and poor survival of the patients (Wang et al., 2011).

5.1.11 Lymphoid neoplasms

Elevated expression of AEG-1/MTDH/LYRIC at both mRNA and protein levels was also evident in 76.7% cases of diffuse large B-cell lymphoma (DLBCL) and 80.6% cases of T-cell non-Hodgkin's lymphoma (T-NHL) patient tissues (Ge et al., 2012; Yan, Zhang, Chen, & Zhang, 2012). Furthermore, overexpression of AEG-1/MTDH/LYRIC in DLBCL cells increased expression and nuclear translocation of β-catenin as seen in colon cancer, and AEG-1/MTDH/LYRIC in T-NHL was mostly detected in the cytoplasm (Ge et al., 2012; Yan, Zhang, et al., 2012; Zhang et al., 2012).

5.1.12 Serum anti-AEG-1 autoantibody

As described above, overexpression of AEG-1/MTDH/LYRIC has been observed in various types of human cancer in association with poor clinical outcome. Another intriguing study recently reported the existence of anti-AEG-1 autoantibodies in the serum samples of 49% (248 of 483) cancer patients including patients with breast cancer, HCC, rectal cancer, lung cancer, or gastric cancer while no detection of AEG-1/MTDH/LYRIC antibody in the serum of 230 normal individuals (Chen et al., 2012). These results suggest that AEG-1/MTDH/LYRIC autoantibody in the serum could be a potential diagnostic biomarker for various cancer patients even though further studies are required.

5.2. Brain with neurological disorders

5.2.1 HIV-associated dementia and glioma-induced neurodegeneration

As described in the cloning section, *AEG-1/MTDH/LYRIC* was first identified as an elevated gene in PHFA infected with HIV-1, or treated with HIV-1 gp120 or TNF-α (Su, Chen, et al., 2003; Su et al., 2002). However, expression of AEG-1/MTDH/LYRIC in the HAD patients and the role of AEG-1/MTDH/LYRIC in HIV-related pathogenesis in the brain remain to be determined. Impaired glutamate uptake by aberrant EAAT2 expression in astrocytes induces glutamate excitotoxicity, which causes neuronal damage and various neurodegenerative diseases including HAD and glioma-induced neurodegeneration (Brack-Werner, 1999; Choi, 1988; Gonzalez-Scarano & Martin-Garcia, 2005; Kim et al., 2011; Su, Leszczyniecka, et al., 2003). Hyperactivation of glutamate receptors, specifically *N*-methyl-D-aspartate receptors (NMDARs), increases Ca^{2+} influx into the postsynaptic cell to activate apoptotic pathways (Choi, 1988; Duchen, 2012). Our groups reported that AEG-1/MTDH/LYRIC repressed the expression of EAAT2 which is an astroglial glutamate transporter responsible for glutamate clearance in the neuronal synapses, suggesting a potential role of AEG-1/MTDH/LYRIC in the disease by regulating glutamate excitotoxicity in the brain (Kang et al., 2005; Lee et al., 2011). In addition, expression of AEG-1/MTDH/LYRIC was increased in glioma patients, and the increased AEG-1/MTDH/LYRIC reduced EAAT2 expression and also caused a reduction of glutamate uptake by glial cells, resulting in neuronal cell death in the patients with glioma (Lee et al., 2011). Other groups also reported that glioma cells release excitotoxic levels of glutamate, which promotes neuronal cell death, peritumoral seizures and tumor-associated epilepsy, as well as malignant glioma progression (Watkins & Sontheimer, 2012). Together, these results suggest that overexpression of AEG-1/MTDH/LYRIC could be involved in HIV-associated dementia as well as in glioma-induced neurodegeneration by regulating EAAT2 expression as summarized in Fig. 1.3.

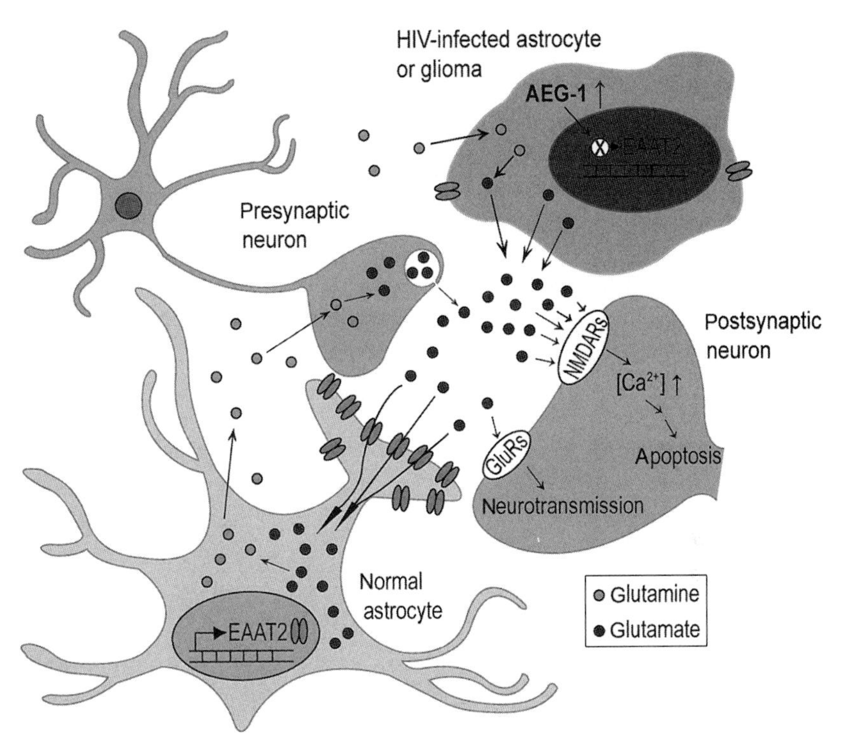

Figure 1.3 Molecular mechanism of AEG-1 function in HAD- and glioma-induced neurodegeneration. The EAAT2 in normal astrocytes is responsible for clearing extracellular glutamate to minimize excitotoxicity, and astrocytes convert glutamate to glutamine via the glutamine synthetase. Increased AEG-1 in HIV-infected astrocytes and glioma reduces EAAT2 expression and boosts synaptic glutamate concentrations. Increased synaptic glutamate overactivates glutamate receptors (GluRs), especially NMDA receptors (NMDARs) in postsynaptic neurons, resulting in induction of high levels of Ca^{2+} influx and apoptotic cell death. (See Page 2 in Color Section at the back of the book.)

5.2.2 Huntington disease

Although role of AEG-1/MTDH/LYRIC was not determined in the pathogenesis of Huntington disease (HD), a recent study observed a significant increase of AEG-1/MTDH/LYRIC in the brain tissues of HD patients compared with normal brain tissues, its localization in the ER and nucleolus, and induction of AEG-1/MTDH/LYRIC by ER stress in neurons, suggesting a potential role of AEG-1/MTDH/LYRIC in HD-associated ER stress (Carnemolla et al., 2009). Further studies are required to confirm these observations and to define the mechanism involved in upregulation in HD.

5.2.3 Migraine

A recent genome-wide association study (GWAS) of migraine with 2731 cases from three European headache clinics and 10,747 population-matched controls established AEG-1/MTDH/LYRIC as the first genetic marker for the disease (Anttila et al., 2010). Another following GWAS with 2446 cases and 8534 controls in six population-based European cohorts revealed a modest gene-based significant association between migraine and AEG-1/MTDH/LYRIC (Ligthart et al., 2011). Although additional investigations are required to confirm, these results suggest that AEG-1/MTDH/LYRIC-mediated downregulation of EAAT2 may represent a mechanism by which AEG-1/MTDH/LYRIC contributes to the pathogenesis of migraine.

5.2.4 Reactive astrogliosis

A brain injury mouse model showed induction of AEG-1/MTDH/LYRIC in astrocytes at the wound sites, and the injury in human astrocytes induced specific compartmentalization into the nucleolus of AEG-1/MTDH/LYRIC (Vartak-Sharma & Ghorpade, 2012). In addition, the increased expression of AEG-1/MTDH/LYRIC was involved in the process of wound healing by inducing proliferation and migration of astrocytes. These results indicate a role of AEG-1/MTDH/LYRIC in mediating reactive astrogliosis as well as in regulating astrocyte responses to brain injury.

5.3. Inflammatory status

AEG-1/MTDH/LYRIC has been known to activate NF-κB, a key regulator of pro-inflammatory cytokines, suggesting a potential role of AEG-1/MTDH/LYRIC in inflammation, which is a crucial process in the pathogenesis of cancer progression and neurodegeneration (Sarkar et al., 2008; Yoo, Emdad, et al., 2011). A recent study found that AEG-1/MTDH/LYRIC was induced by lipopolysaccharide (LPS) via the NF-κB pathway in human promonocytic cells and also mediated the LPS-induced NF-κB activation and upregulation of TNFα and prostaglandin E2, suggesting a potential role of AEG-1/MTDH/LYRIC in LPS-induced inflammatory responses as a target gene of LPS (Khuda et al., 2009). In addition, AEG-1/MTDH/LYRIC was induced in the process of wound healing in response to brain injury, which pro-inflammatory cytokines and chemokines mediate during injury and the related disease progression (Vartak-Sharma & Ghorpade, 2012). Together these results suggest that

AEG-1/MTDH/LYRIC could be a key factor inducing inflammatory responses for progression of various diseases.

5.4. Development

We recently analyzed the expression patterns of AEG-1/MTDH/LYRIC during mouse development (Jeon et al., 2010). AEG-1/MTDH/LYRIC was detected in mid-to-hindbrain, frontonasal processes, limbs, and pharyngeal arches in the early developmental period E9.5–10.5, and its expression was increased in the brain, olfactory and skeletal systems, skin, hair follicles, and the liver at specific stages during E12.5–18.5, indicating a prominent role of AEG-1/MTDH/LYRIC during normal development of the organs. Colocalization of AEG-1/MTDH/LYRIC with Ki-67 suggested its role in cell proliferation during embryogenesis.

6. REGULATION OF GENE EXPRESSION

6.1. Induction of AEG-1/MTDH/LYRIC

AEG-1/MTDH/LYRIC was originally identified as one of several elevated genes in PHFA infected with HIV-1 or treated with HIV-1 gp120 or TNF-α, suggesting that AEG-1/MTDH/LYRIC could be involved in HIV-mediated pathogenesis in the brain and that HIV-1, gp120, and TNF-α are inducers of AEG-1/MTDH/LYRIC (Su, Chen, et al., 2003; Su et al., 2002). In addition, TNF-α induced nuclear translocation of AEG-1/MTDH/LYRIC and its interaction with p65 resulted in NF-κB activation (Emdad et al., 2006). However, the mechanisms by which these inducers increase the expression of AEG-1/MTDH/LYRIC in astrocytes and its nuclear translocation have not been clarified yet. Another recent study using an *in vivo* mouse model of reactive astrogliosis using needle injection revealed increased AEG-1/MTDH/LYRIC expression in astrocytes comprising a glial scar tissue around the needle tract, indicating that brain injury/trauma induces AEG-1/MTDH/LYRIC expression in astrocytes during astrogliosis (Vartak-Sharma & Ghorpade, 2012). In an HD mouse model, tunicamycin, an ER stress inducer, increased AEG-1/MTDH/LYRIC expression in neurons, suggesting its role in the pathogenesis of HD (Carnemolla et al., 2009). Although interesting inducers of AEG-1/MTDH/LYRIC in astrocytes and neurons of the brain have been uncovered, further studies are required to clarify the mechanisms by which these inducers contribute to AEG-1/MTDH/LYRIC expression.

AEG-1/MTDH/LYRIC has been recognized as an emerging oncoprotein upregulated in various types of human cancers as described in Section 5.1. AEG-1/MTDH/LYRIC is upregulated in cancer by a variety of mechanisms. AEG-1/MTDH/LYRIC was uncovered as a downstream target of oncogenic H-ras, which is one of the most crucial proteins in cellular transformation, tumor progression, and metastasis (Lee et al., 2006). H-ras markedly induced AEG-1/MTDH/LYRIC through the PI3K–AKT signaling pathway augmenting binding of c-Myc to E-box elements in the *AEG-1/MTDH/LYRIC* promoter, resulting in regulation of *AEG-1/MTDH/LYRIC* transcription (Lee et al., 2006). These results suggested that AEG-1/MTDH/LYRIC is one of the crucial mediators of H-ras- and c-Myc-induced oncogenesis. Since H-ras, PI3K/AKT, and c-Myc function as oncogenes in a large variety of cancers, this regulatory mechanism might explain why AEG-1/MTDH/LYRIC is overexpressed in all types of cancers with gradual increase in its expression as the disease progresses. Another study found that hypoxia and glucose deprivation induced AEG-1/MTDH/LYRIC expression in glioma cells for cell survival (Noch, Bookland, & Khalili, 2011). The PI3K–AKT pathway mediated the hypoxic induction of AEG-1/MTDH/LYRIC through stabilization of HIF-1α. AEG-1/MTDH/LYRIC induction by glucose deprivation was dependent on reactive oxygen species (ROS), and the increased AEG-1/MTDH/LYRIC prevented ROS production. These results suggest that AEG-1/MTDH/LYRIC induction is necessary for cancer cell survival under stress conditions such as hypoxia and glucose deprivation (Noch et al., 2011).

As described in Section 5.3, LPS is another inducer of AEG-1/MTDH/LYRIC in human promonocytic cells (Khuda et al., 2009). LPS induced AEG-1/MTDH/LYRIC via the NF-κB pathway, and AEG-1/MTDH/LYRIC was also required for LPS-induced NF-κB activation, indicating a positive feedback loop between AEG-1/MTDH/LYRIC and NF-κB (Khuda et al., 2009). LPS-mediated AEG-1 induction and the positive feedback regulation between AEG-1/MTDH/LYRIC and NF-κB were also examined in breast cancer cells (Zhao et al., 2011). In addition, AEG-1/MTDH/LYRIC mediated the LPS-induced IL-8 and MMP-9 production, which are critical molecules in LPS-induced invasion and metastasis (Zhao et al., 2011). Since inflammation has been implicated in migration and metastasis of breast cancer cells, these results suggest that AEG-1/MTDH/LYRIC may be involved in inflammation-induced tumor progression (Zhao et al., 2011).

6.2. Repression of AEG-1/MTDH/LYRIC

Repression of AEG-1/MTDH/LYRIC has not been evident in any pathophysiological condition. Instead, increased AEG-1/MTDH/LYRIC in pathological conditions such as cancer indicates that AEG-1/MTDH/LYRIC could be a potential therapeutic target for these diseases. Two recent studies found that the potential anticancer agents ursolic acid and cryptotanshinone repressed expression of AEG-1/MTDH/LYRIC in ovarian cancer and prostate cancer, respectively (Lee et al., 2012; Song et al., 2012). The mechanism by which ursolic acid repressed AEG-1/MTDH/LYRIC expression was not clarified, but HIF-1α was a mediator in cryptotanshinone-induced AEG-1/MTDH/LYRIC repression (Lee et al., 2012; Song et al., 2012). In addition, cadmium chloride known to induce cell death in breast cancer cells reduced AEG-1/MTDH/LYRIC expression and NF-κB activity in breast cancer cells, suggesting AEG-1/MTDH/LYRIC downregulation as a mediator for cadmium chloride-induced breast cancer cell death (Luparello, Longo, & Vetrano, 2012). Although three repressors of AEG-1/MTDH/LYRIC were uncovered, more studies are required to find more repressors as potential anticancer drugs, and furthermore investigations to understand more detailed action mechanisms of these agents would be valuable for establishing a better screening system to find anticancer drugs targeting AEG-1/MTDH/LYRIC.

6.3. AEG-1/MTDH/LYRIC as a target of miRNAs

miRNAs are short noncoding RNAs of 20–24 nucleotides that play important roles in normal development, differentiation, growth control, and human diseases, and especially in the context of cancer pathogenesis, where miRNAs are generally working as tumor suppressors and are downregulated in cancer cells (Jansson & Lund, 2012). Five miRNAs have been recognized as repressors of *AEG-1/MTDH/LYRIC* expression. The first miRNA identified as a regulator of *AEG-1/MTDH/LYRIC* expression was miR-26a in breast cancer (Zhang et al., 2011). miR-26a repressed AEG-1/MTDH/LYRIC expression by directly targeting 3′ UTR of *AEG-1/MTDH/LYRIC* mRNA, and moreover, a similar inverse correlation between the expression levels of miR-26a and AEG-1/MTDH/LYRIC was shown in breast cancer patient samples and breast cancer cell lines (Zhang et al., 2011). Second, two miRNAs, miR-203 and miR-22, suppressed expression of AEG-1/MTDH/LYRIC in p53-mutated colon cancer cells through the PI3K–AKT pathway via negative regulation of AKT2 and activation of PTEN, respectively, resulting in alleviation of chemoresistance of cancer cells, but not in the p53 wild-type cells (Li, Chen, Zhao, Kong, & Zhang, 2011; Li, Zhang, Zhao, Kong, & Chen, 2011). Another study found that one of the most well-known tumor suppressors miR-375 directly downregulated expression of AEG-1/MTDH/LYRIC by its binding to the 3′-UTR of *AEG-1/MTDH/LYRIC* mRNA, and similar expression patterns of decreased miR-375 and increased AEG-1/MTDH/LYRIC were observed in patient samples of several tumor types including HNSCC, liver cancer, breast cancer, and ESCC (He et al., 2012; Hui et al., 2011; Isozaki et al., 2012; Nohata et al., 2011; Ward et al., 2012). In addition, *AEG-1/MTDH/LYRIC* was identified as a target of miR-136 in glioma cells (Yang, Wu, et al., 2012). Furthermore, increased expression of AEG-1/MTDH/LYRIC by the suppressed miRNAs in cancer cells was significantly involved in tumorigenesis. These results clearly indicate that overexpression of miRNAs targeting *AEG-1/MTDH/LYRIC* might be a potential therapeutic strategy for inhibiting cancer cell growth.

7. CONCLUDING REMARKS

Between 2002 and 2004, four independent groups identified AEG-1/MTDH/LYRIC with different names. We first identified *AEG-1* as a HIV-1- and HIV-1 gp120-inducible gene from PHFA in 2002, and as a TNF-α-inducible gene. In 2004, three other groups identified 3D3/LYRIC as a transmembrane protein of the ER, nuclear envelope, and nucleolus in mouse cells using a gene-trap screening system, MTDH as a cell surface protein involved in breast cancer metastasis to the lung using a phage display strategy in a mouse model of breast cancer, and LYRIC as a tight junction-associated protein in polarized epithelial cells, respectively. These results suggested that AEG-1/MTDH/LYRIC could be important in both pathological and normal physiological processes. A recent study indicating that AEG-1/MTDH/LYRIC can induce protective autophagy supports a potential role in promoting cell survival (Bhutia et al., 2010). However, until now most researchers have focused on its oncogenic functions in cancer, even though some other biological functions and biochemical characteristics of AEG-1/MTDH/LYRIC including its amino acid sequences, localization, and posttranslational modification such as ubiquitination have been clarified. Especially, AEG-1/MTDH/LYRIC has been recognized as a potential prognostic marker for patients with various types of human cancer such as breast cancer, HCC, prostate cancer, glioma, etc. In addition, increased insights into the detailed mechanism on oncogenic functions of AEG-1/MTDH/LYRIC involved in cell proliferation, chemoresistance, angiogenesis, invasion, and metastasis clearly indicate that AEG-1/MTDH/LYRIC is a potentially good target for cancer treatment. Some recent results have already suggested that AEG-1/MTDH/LYRIC could be a target of therapeutic reagents and miRNAs with therapeutic potential for directly suppressing the cancer phenotype. Recent studies also reveal the importance of AEG-1/MTDH/LYRIC in other pathological processes such as neurodegeneration including migraine and HD, and inflammation. These results have aroused enhanced interests in AEG-1/MTDH/LYRIC, and a greater understanding of AEG-1/MTDH/LYRIC's normal and abnormal biological functions and basic biochemical characteristics including its structure will extend its roles in various other pathological and normal physiological processes. Additional reviews in this thematic issue focus on various aspects of AEG-1/MTDH/LYRIC and future research directions. There has been a decade of research on AEG-1/MTDH/LYRIC since its original identification as an AEG. We have made significant progress and learned a lot in a short time about this interesting molecule. Expanded research is justified and should usher in a greater understanding of the normal and abnormal functions of this protein which may serve as a viable target for targeting various pathogenic diseases, including cancer and neurodegeneration.

ACKNOWLEDGMENTS

We are indebted to the National Institutes of Health, including the National Cancer Institute Grant 1 R01 CA134721, National Institute of Neurological Disease and Stroke, 5 P01 NS31492, and the Thelma Newmeyer Corman Endowment to P. B. F., by National Cancer Institute Grant R01 CA138540 to D. S., by HRF-G-2012-6 to D. K. from Hallym University, and by a National Research Foundation of Korea grant (No. 2012-0005755) to S.-G. L. D. S. is a Harrison Scholar in the VCU Massey Cancer Center. P. B. F. holds the Thelma Newmeyer Corman Endowed Chair in Cancer Research at the VCU Massey Cancer Center.

REFERENCES

Anttila, V., Stefansson, H., Kallela, M., Todt, U., Terwindt, G. M., Calafato, M. S., et al. (2010). Genome-wide association study of migraine implicates a common susceptibility variant on 8q22.1. *Nature Genetics*, *42*(10), 869–873.

Ash, S. C., Yang, D. Q., & Britt, D. E. (2008). LYRIC/AEG-1 overexpression modulates BCCIPalpha protein levels in prostate tumor cells. *Biochemical and Biophysical Research Communications*, *371*(2), 333–338.

Bhutia, S. K., Kegelman, T. P., Das, S. K., Azab, B., Su, Z. Z., Lee, S. G., et al. (2010). Astrocyte elevated gene-1 induces protective autophagy. *Proceedings of the National Academy of Sciences of the United States of America, 107*(51), 22243–22248.

Borjabad, A., Brooks, A. I., & Volsky, D. J. (2010). Gene expression profiles of HIV-1-infected glia and brain: Toward better understanding of the role of astrocytes in HIV-1-associated neurocognitive disorders. *Journal of Neuroimmune Pharmacology, 5*(1), 44–62.

Borley, J., Wilhelm-Benartzi, C., Brown, R., & Ghaem-Maghami, S. (2012). Does tumour biology determine surgical success in the treatment of epithelial ovarian cancer? A systematic literature review. *British Journal of Cancer, 107*(7), 1069–1074.

Boukerche, H., Su, Z. Z., Kang, D. C., & Fisher, P. B. (2004). Identification and cloning of genes displaying elevated expression as a consequence of metastatic progression in human melanoma cells by rapid subtraction hybridization. *Gene, 343*(1), 191–201.

Brack-Werner, R. (1999). Astrocytes: HIV cellular reservoirs and important participants in neuropathogenesis. *AIDS, 13*(1), 1–22.

Britt, D. E., Yang, D. F., Yang, D. Q., Flanagan, D., Callanan, H., Lim, Y. P., et al. (2004). Identification of a novel protein, LYRIC, localized to tight junctions of polarized epithelial cells. *Experimental Cell Research, 300*(1), 134–148.

Brown, D. M., & Ruoslahti, E. (2004). Metadherin, a cell surface protein in breast tumors that mediates lung metastasis. *Cancer Cell, 5*(4), 365–374.

Carnemolla, A., Fossale, E., Agostoni, E., Michelazzi, S., Calligaris, R., De Maso, L., et al. (2009). Rrs1 is involved in endoplasmic reticulum stress response in Huntington disease. *The Journal of Biological Chemistry, 284*(27), 18167–18173.

Chen, X., Dong, K., Long, M., Lin, F., Wang, X., Wei, J., et al. (2012). Serum anti-AEG-1 autoantibody is a potential novel biomarker for malignant tumors. *Oncology Letters, 4*(2), 319–323.

Chen, W., Ke, Z., Shi, H., Yang, S., & Wang, L. (2010). Overexpression of AEG-1 in renal cell carcinoma and its correlation with tumor nuclear grade and progression. *Neoplasma, 57*(6), 522–529.

Choi, D. W. (1988). Glutamate neurotoxicity and diseases of the nervous system. *Neuron, 1*(8), 623–634.

Deng, N., & Feng, Y. (2011). Expression of EphA7 and MTDH and clinicopathological significance in the squamous cell cancer of the tongue. *Zhong Nan Da Xue Xue Bao Yi Xue Ban, 36*(12), 1195–1198.

Duchen, M. R. (2012). Mitochondria, calcium-dependent neuronal death and neurodegenerative disease. *Pflügers Archiv, 464*(1), 111–121.

Emdad, L., Lee, S. G., Su, Z. Z., Jeon, H. Y., Boukerche, H., Sarkar, D., et al. (2009). Astrocyte elevated gene-1 (AEG-1) functions as an oncogene and regulates angiogenesis. *Proceedings of the National Academy of Sciences of the United States of America, 106*(50), 21300–21305.

Emdad, L., Sarkar, D., Lee, S. G., Su, Z. Z., Yoo, B. K., Dash, R., et al. (2010). Astrocyte elevated gene-1: A novel target for human glioma therapy. *Molecular Cancer Therapeutics, 9*(1), 79–88.

Emdad, L., Sarkar, D., Su, Z. Z., Lee, S. G., Kang, D. C., Bruce, J. N., et al. (2007). Astroeyte elevated gene-1: Recent insights into a novel gene involved in tumor progression, metastasis and neurodegeneration. *Pharmacology and Therapeutics, 114*(2), 155–170.

Emdad, L., Sarkar, D., Su, Z. Z., Randolph, A., Boukerche, H., Valerie, K., et al. (2006). Activation of the nuclear factor kappaB pathway by astrocyte elevated gene-1: Implications for tumor progression and metastasis. *Cancer Research, 66*(3), 1509–1516.

Ge, X., Lv, X., Feng, L., Liu, X., Gao, J., Chen, N., et al. (2012). Metadherin contributes to the pathogenesis of diffuse large B-cell lymphoma. *PLoS One, 7*(6), e39449.

Gnosa, S., Shen, Y. M., Wang, C. J., Zhang, H., Stratmann, J., Arbman, G., et al. (2012). Expression of AEG-1 mRNA and protein in colorectal cancer patients and colon cancer cell lines. *Journal of Translational Medicine, 10*, 109.

Gong, Z., Liu, W., You, N., Wang, T., Wang, X., Lu, P., et al. (2012). Prognostic significance of metadherin overexpression in hepatitis B virus-related hepatocellular carcinoma. *Oncology Reports*, *27*(6), 2073–2079.

Gonzalez-Scarano, F., & Martin-Garcia, J. (2005). The neuropathogenesis of AIDS. *Nature Reviews Immunology*, *5*(1), 69–81.

He, X. X., Chang, Y., Meng, F. Y., Wang, M. Y., Xie, Q. H., Tang, F., et al. (2012). MicroRNA-375 targets AEG-1 in hepatocellular carcinoma and suppresses liver cancer cell growth in vitro and in vivo. *Oncogene*, *31*(28), 3357–3369.

Hornbeck, P. V., Kornhauser, J. M., Tkachev, S., Zhang, B., Skrzypek, E., Murray, B., et al. (2012). PhosphoSitePlus: A comprehensive resource for investigating the structure and function of experimentally determined post-translational modifications in man and mouse. *Nucleic Acids Research*, *40*(Database issue), D261–D270.

Hu, G., Chong, R. A., Yang, Q., Wei, Y., Blanco, M. A., Li, F., et al. (2009). MTDH activation by 8q22 genomic gain promotes chemoresistance and metastasis of poor-prognosis breast cancer. *Cancer Cell*, *15*(1), 9–20.

Hui, A. B., Bruce, J. P., Alajez, N. M., Shi, W., Yue, S., Perez-Ordonez, B., et al. (2011). Significance of dysregulated metadherin and microRNA-375 in head and neck cancer. *Clinical Cancer Research*, *17*(24), 7539–7550.

Isozaki, Y., Hoshino, I., Nohata, N., Kinoshita, T., Akutsu, Y., Hanari, N., et al. (2012). Identification of novel molecular targets regulated by tumor suppressive miR-375 induced by histone acetylation in esophageal squamous cell carcinoma. *International Journal of Oncology*, *41*(3), 985–994.

Jansson, M. D., & Lund, A. H. (2012). MicroRNA and cancer. *Molecular Oncology*, *6*(6), 590–610.

Jeon, H. Y., Choi, M., Howlett, E. L., Vozhilla, N., Yoo, B. K., Lloyd, J. A., et al. (2010). Expression patterns of astrocyte elevated gene-1 (AEG-1) during development of the mouse embryo. *Gene Expression Patterns*, *10*(7–8), 361–367.

Jian-bo, X., Hui, W., Yu-long, H., Chang-hua, Z., Long-juan, Z., Shi-rong, C., et al. (2011). Astrocyte-elevated gene-1 overexpression is associated with poor prognosis in gastric cancer. *Medical Oncology*, *28*(2), 455–462.

Jiang, H., Kang, D. C., Alexandre, D., & Fisher, P. B. (2000). RaSH, a rapid subtraction hybridization approach for identifying and cloning differentially expressed genes. *Proceedings of the National Academy of Sciences of the United States of America*, *97*(23), 12684–12689.

Jiang, T., Zhu, A., Zhu, Y., & Piao, D. (2012). Clinical implications of AEG-1 in liver metastasis of colorectal cancer. *Medical Oncology*, *29*(4), 2858–2863.

Kang, D. C., & Fisher, P. B. (2005). Complete open reading frame (C-ORF) technology: Simple and efficient technique for cloning full-length protein-coding sequences. *Gene*, *353*(1), 1–7.

Kang, D. C., & Fisher, P. B. (2007). Complete open reading frame (C-ORF) technique: Rapid and efficient method for obtaining complete protein coding sequences. *Methods in Molecular Biology*, *383*, 123–133.

Kang, D. C., Gopalkrishnan, R. V., Wu, Q., Jankowsky, E., Pyle, A. M., & Fisher, P. B. (2002). mda-5: An interferon-inducible putative RNA helicase with double-stranded RNA-dependent ATPase activity and melanoma growth-suppressive properties. *Proceedings of the National Academy of Sciences of the United States of America*, *99*(2), 637–642.

Kang, D. C., Jiang, H., Su, Z. Z., Volsky, D. J., & Fisher, P. B. (2002). RaSH-Rapid subtraction hybridization. In S. Lorkowski & P. Cullen (Eds.), *Analysig gene expression* (pp. 206–214). Germany: Wiley-VCH.

Kang, D. C., Su, Z. Z., Sarkar, D., Emdad, L., Volsky, D. J., & Fisher, P. B. (2005). Cloning and characterization of HIV-1-inducible astrocyte elevated gene-1, AEG-1. *Gene*, *353*(1), 8–15.

Ke, Z. F., He, S., Li, S., Luo, D., Feng, C., & Zhou, W. (2012). Expression characteristics of astrocyte elevated gene-1 (AEG-1) in tongue carcinoma and its correlation with poor prognosis. *Cancer Epidemiology*, *37*(2), 179–185.

Khuda, I. I., Koide, N., Noman, A. S., Dagvadorj, J., Tumurkhuu, G., Naiki, Y., et al. (2009). Astrocyte elevated gene-1 (AEG-1) is induced by lipopolysaccharide as toll-like receptor 4 (TLR4) ligand and regulates TLR4 signalling. *Immunology, 128*(1 Suppl.), e700–e706.

Kikuno, N., Shiina, H., Urakami, S., Kawamoto, K., Hirata, H., Tanaka, Y., et al. (2007). Knockdown of astrocyte-elevated gene-1 inhibits prostate cancer progression through upregulation of FOXO3a activity. *Oncogene, 26*(55), 7647–7655.

Kim, K., Lee, S. G., Kegelman, T. P., Su, Z. Z., Das, S. K., Dash, R., et al. (2011). Role of excitatory amino acid transporter-2 (EAAT2) and glutamate in neurodegeneration: Opportunities for developing novel therapeutics. *Journal of Cellular Physiology, 226*(10), 2484–2493.

Kornegoor, R., Moelans, C. B., Verschuur-Maes, A. H., Hogenes, M. C., de Bruin, P. C., Oudejans, J. J., et al. (2012). Oncogene amplification in male breast cancer: Analysis by multiplex ligation-dependent probe amplification. *Breast Cancer Research and Treatment, 135*(1), 49–58.

Lee, S. G., Jeon, H. Y., Su, Z. Z., Richards, J. E., Vozhilla, N., Sarkar, D., et al. (2009). Astrocyte elevated gene-1 contributes to the pathogenesis of neuroblastoma. *Oncogene, 28*(26), 2476–2484.

Lee, H. J., Jung, D. B., Sohn, E. J., Kim, H. H., Park, M. N., Lew, J. H., et al. (2012). Inhibition of hypoxia inducible factor alpha and astrocyte-elevated gene-1 mediates cryptotanshinone exerted antitumor activity in hypoxic PC-3 cells. *Evidence-Based Complementary and Alternative Medicine: eCAM, 2012*, 13 (Research Article).

Lee, S. G., Kim, K., Kegelman, T. P., Dash, R., Das, S. K., Choi, J. K., et al. (2011). Oncogene AEG-1 promotes glioma-induced neurodegeneration by increasing glutamate excitotoxicity. *Cancer Research, 71*(20), 6514–6523.

Lee, S. G., Su, Z. Z., Emdad, L., Gupta, P., Sarkar, D., Borjabad, A., et al. (2008). Mechanism of ceftriaxone induction of excitatory amino acid transporter-2 expression and glutamate uptake in primary human astrocytes. *The Journal of Biological Chemistry, 283*(19), 13116–13123.

Lee, S. G., Su, Z. Z., Emdad, L., Sarkar, D., & Fisher, P. B. (2006). Astrocyte elevated gene-1 (AEG-1) is a target gene of oncogenic Ha-ras requiring phosphatidylinositol 3-kinase and c-Myc. *Proceedings of the National Academy of Sciences of the United States of America, 103*(46), 17390–17395.

Leszczyniecka, M., Kang, D. C., Sarkar, D., Su, Z. Z., Holmes, M., Valerie, K., et al. (2002). Identification and cloning of human polynucleotide phosphorylase, hPNPase old-35, in the context of terminal differentiation and cellular senescence. *Proceedings of the National Academy of Sciences of the United States of America, 99*(26), 16636–16641.

Li, J., Chen, Y., Zhao, J., Kong, F., & Zhang, Y. (2011). miR-203 reverses chemoresistance in p53-mutated colon cancer cells through downregulation of Akt2 expression. *Cancer Letters, 304*(1), 52–59.

Li, C., Li, R., Song, H., Wang, D., Feng, T., Yu, X., et al. (2011). Significance of AEG-1 expression in correlation with VEGF, microvessel density and clinicopathological characteristics in triple-negative breast cancer. *Journal of Surgical Oncology, 103*(2), 184–192.

Li, C., Li, Y., Wang, X., Wang, Z., Cai, J., Wang, L., et al. (2012). Elevated expression of astrocyte elevated gene-1 (AEG-1) is correlated with cisplatin-based chemoresistance and shortened outcome in patients with stages III-IV serous ovarian carcinoma. *Histopathology, 60*(6), 953–963.

Li, C., Liu, J., Lu, R., Yu, G., Wang, X., Zhao, Y., et al. (2011). AEG -1 overexpression: A novel indicator for peritoneal dissemination and lymph node metastasis in epithelial ovarian cancers. *International Journal of Gynecological Cancer, 21*(4), 602–608.

Li, J., Yang, L., Song, L., Xiong, H., Wang, L., Yan, X., et al. (2009). Astrocyte elevated gene-1 is a proliferation promoter in breast cancer via suppressing transcriptional factor FOXO1. *Oncogene, 28*(36), 3188–3196.

Li, J., Zhang, N., Song, L. B., Liao, W. T., Jiang, L. L., Gong, L. Y., et al. (2008). Astrocyte elevated gene-1 is a novel prognostic marker for breast cancer progression and overall patient survival. *Clinical Cancer Research, 14*(11), 3319–3326.

Li, J., Zhang, Y., Zhao, J., Kong, F., & Chen, Y. (2011). Overexpression of miR-22 reverses paclitaxel-induced chemoresistance through activation of PTEN signaling in p53-mutated colon cancer cells. *Molecular and Cellular Biochemistry, 357*(1–2), 31–38.

Liao, W. T., Guo, L., Zhong, Y., Wu, Y. H., Li, J., & Song, L. B. (2011). Astrocyte elevated gene-1 (AEG-1) is a marker for aggressive salivary gland carcinoma. *Journal of Translational Medicine, 9*, 205.

Ligthart, L., de Vries, B., Smith, A. V., Ikram, M. A., Amin, N., Hottenga, J. J., et al. (2011). Meta-analysis of genome-wide association for migraine in six population-based European cohorts. *European Journal of Human Genetics, 19*(8), 901–907.

Liu, H. Y., Liu, C. X., Han, B., Zhang, X. Y., & Sun, R. P. (2011). AEG-1 is associated with clinical outcome in neuroblastoma patients. *Cancer Biomarkers, 11*(2), 115–121.

Liu, L., Wu, J., Ying, Z., Chen, B., Han, A., Liang, Y., et al. (2010). Astrocyte elevated gene-1 upregulates matrix metalloproteinase-9 and induces human glioma invasion. *Cancer Research, 70*(9), 3750–3759.

Liu, D. C., & Yang, Z. L. (2011). MTDH and EphA7 are markers for metastasis and poor prognosis of gallbladder adenocarcinoma. *Diagnostic Cytopathology, 41*(3), 199–205.

Liu, X., Zhang, N., Li, X., Moran, M. S., Yuan, C., Yan, S., et al. (2011). Identification of novel variants of metadherin in breast cancer. *PLoS One, 6*(3), e17582.

Luparello, C., Longo, A., & Vetrano, M. (2012). Exposure to cadmium chloride influences astrocyte-elevated gene-1 (AEG-1) expression in MDA-MB231 human breast cancer cells. *Biochimie, 94*(1), 207–213.

Meng, F., Luo, C., Ma, L., Hu, Y., & Lou, G. (2011). Clinical significance of astrocyte elevated gene-1 expression in human epithelial ovarian carcinoma. *International Journal of Gynecological Pathology, 30*(2), 145–150.

Noch, E., Bookland, M., & Khalili, K. (2011). Astrocyte-elevated gene-1 (AEG-1) induction by hypoxia and glucose deprivation in glioblastoma. *Cancer Biology & Therapy, 11*(1), 32–39.

Nohata, N., Hanazawa, T., Kikkawa, N., Mutallip, M., Sakurai, D., Fujimura, L., et al. (2011). Tumor suppressive microRNA-375 regulates oncogene AEG-1/MTDH in head and neck squamous cell carcinoma (HNSCC). *Journal of Human Genetics, 56*(8), 595–601.

Rehman, A. O., & Wang, C. Y. (2009). CXCL12/SDF-1 alpha activates NF-kappaB and promotes oral cancer invasion through the Carma3/Bcl10/Malt1 complex. *International Journal of Oral Science, 1*(3), 105–118.

Ronquist, F., Teslenko, M., van der Mark, P., Ayres, D. L., Darling, A., Hohna, S., et al. (2012). MrBayes 3.2: Efficient Bayesian phylogenetic inference and model choice across a large model space. *Systems Biology, 61*(3), 539–542.

Rothstein, J. D., Patel, S., Regan, M. R., Haenggeli, C., Huang, Y. H., Bergles, D. E., et al. (2005). Beta-lactam antibiotics offer neuroprotection by increasing glutamate transporter expression. *Nature, 433*(7021), 73–77.

Santarpia, M., Magri, I., Sanchez-Ronco, M., Costa, C., Molina-Vila, M. A., Gimenez-Capitan, A., et al. (2011). mRNA expression levels and genetic status of genes involved in the EGFR and NF-kappaB pathways in metastatic non-small-cell lung cancer patients. *Journal of Translational Medicine, 9*, 163.

Sarkar, D., Emdad, L., Lee, S. G., Yoo, B. K., Su, Z. Z., & Fisher, P. B. (2009). Astrocyte elevated gene-1: Far more than just a gene regulated in astrocytes. *Cancer Research, 69*(22), 8529–8535.

Sarkar, D., Park, E. S., Emdad, L., Lee, S. G., Su, Z. Z., & Fisher, P. B. (2008). Molecular basis of nuclear factor-kappa B activation by astrocyte elevated gene-1. *Cancer Research, 68*(5), 1478–1484.

Simm, M., Su, Z., Huang, E. Y., Chen, Y., Jiang, H., Volsky, D. J., et al. (2001). Cloning of differentially expressed genes in an HIV-1 resistant T cell clone by rapid subtraction hybridization, RaSH. *Gene, 269*(1–2), 93–101.

Song, Y. H., Jeong, S. J., Kwon, H. Y., Kim, B., Kim, S. H., & Yoo, D. Y. (2012). Ursolic acid from Oldenlandia diffusa induces apoptosis via activation of caspases and phosphorylation of glycogen synthase kinase 3 beta in SK-OV-3 ovarian cancer cells. *Biological and Pharmaceutical Bulletin, 35*(7), 1022–1028.

Song, H., Li, C., Li, R., & Geng, J. (2010). Prognostic significance of AEG-1 expression in colorectal carcinoma. *International Journal of Colorectal Disease, 25*(10), 1201–1209.

Song, H., Li, C., Lu, R., Zhang, Y., & Geng, J. (2010). Expression of astrocyte elevated gene-1: A novel marker of the pathogenesis, progression, and poor prognosis for endometrial cancer. *International Journal of Gynecological Cancer, 20*(7), 1188–1196.

Song, L., Li, W., Zhang, H., Liao, W., Dai, T., Yu, C., et al. (2009). Over-expression of AEG-1 significantly associates with tumour aggressiveness and poor prognosis in human non-small cell lung cancer. *The Journal of Pathology, 219*(3), 317–326.

Su, Z. Z., Chen, Y., Kang, D. C., Chao, W., Simm, M., Volsky, D. J., et al. (2003). Customized rapid subtraction hybridization (RaSH) gene microarrays identify overlapping expression changes in human fetal astrocytes resulting from human immunodeficiency virus-1 infection or tumor necrosis factor-alpha treatment. *Gene, 306*, 67–78.

Su, Z. Z., Kang, D. C., Chen, Y., Pekarskaya, O., Chao, W., Volsky, D. J., et al. (2002). Identification and cloning of human astrocyte genes displaying elevated expression after infection with HIV-1 or exposure to HIV-1 envelope glycoprotein by rapid subtraction hybridization, RaSH. *Oncogene, 21*(22), 3592–3602.

Su, Z. Z., Kang, D. C., Chen, Y., Pekarskaya, O., Chao, W., Volsky, D. J., et al. (2003). Identification of gene products suppressed by human immunodeficiency virus type 1 infection or gp120 exposure of primary human astrocytes by rapid subtraction hybridization. *Journal of Neurovirology, 9*(3), 372–389.

Su, Z. Z., Leszczyniecka, M., Kang, D. C., Sarkar, D., Chao, W., Volsky, D. J., et al. (2003). Insights into glutamate transport regulation in human astrocytes: Cloning of the promoter for excitatory amino acid transporter 2 (EAAT2). *Proceedings of the National Academy of Sciences of the United States of America, 100*(4), 1955–1960.

Su, P., Zhang, Q., & Yang, Q. (2010). Immunohistochemical analysis of Metadherin in proliferative and cancerous breast tissue. *Diagnostic Pathology, 5*, 38.

Sun, W., Fan, Y. Z., Xi, H., Lu, X. S., Ye, C., & Zhang, J. T. (2011). Astrocyte elevated gene-1 overexpression in human primary gallbladder carcinomas: An unfavorable and independent prognostic factor. *Oncology Reports, 26*(5), 1133–1142.

Sun, S., Ke, Z., Wang, F., Li, S., Chen, W., Han, A., et al. (2012). Overexpression of astrocyte-elevated gene-1 is closely correlated with poor prognosis in human nonsmall cell lung cancer and mediates its metastasis through up-regulation of matrix metalloproteinase-9 expression. *Human Pathology, 43*(7), 1051–1060.

Sutherland, H. G., Lam, Y. W., Briers, S., Lamond, A. I., & Bickmore, W. A. (2004). 3D3/lyric: A novel transmembrane protein of the endoplasmic reticulum and nuclear envelope, which is also present in the nucleolus. *Experimental Cell Research, 294*(1), 94–105.

Thirkettle, H. J., Girling, J., Warren, A. Y., Mills, I. G., Sahadevan, K., Leung, H., et al. (2009). LYRIC/AEG-1 is targeted to different subcellular compartments by ubiquitinylation and intrinsic nuclear localization signals. *Clinical Cancer Research, 15*(9), 3003–3013.

Thirkettle, H. J., Mills, I. G., Whitaker, H. C., & Neal, D. E. (2009). Nuclear LYRIC/AEG-1 interacts with PLZF and relieves PLZF-mediated repression. *Oncogene, 28*(41), 3663–3670.

Tokunaga, E., Nakashima, Y., Yamashita, N., Hisamatsu, Y., Okada, S., Akiyoshi, S., et al. (2012). Overexpression of metadherin/MTDH is associated with an aggressive phenotype and a poor prognosis in invasive breast cancer. *Breast Cancer*, http://dx.doi.org/10.1007/s12282-012-0398-2.

Vartak-Sharma, N., & Ghorpade, A. (2012). Astrocyte elevated gene-1 regulates astrocyte responses to neural injury: Implications for reactive astrogliosis and neurodegeneration. *Journal of Neuroinflammation*, *9*, 195.

Wang, N., Du, X., Zang, L., Song, N., Yang, T., Dong, R., et al. (2012). Prognostic impact of Metadherin-SND1 interaction in colon cancer. *Molecular Biology Reports*, *39*(12), 10497–10504.

Wang, F., Ke, Z. F., Sun, S. J., Chen, W. F., Yang, S. C., Li, S. H., et al. (2011). Oncogenic roles of astrocyte elevated gene-1 (AEG-1) in osteosarcoma progression and prognosis. *Cancer Biology & Therapy*, *12*(6), 539–548.

Wang, Z., Trillo-Pazos, G., Kim, S. Y., Canki, M., Morgello, S., Sharer, L. R., et al. (2004). Effects of human immunodeficiency virus type 1 on astrocyte gene expression and function: Potential role in neuropathogenesis. *Journal of Neurovirology*, *10*(Suppl. 1), 25–32.

Ward, A., Balwierz, A., Zhang, J. D., Kublbeck, M., Pawitan, Y., Hielscher, T., et al. (2012). Re-expression of microRNA-375 reverses both tamoxifen resistance and accompanying EMT-like properties in breast cancer. *Oncogene*, *32*(9), 1173–1182.

Watkins, S., & Sontheimer, H. (2012). Unique biology of gliomas: Challenges and opportunities. *Trends in Neurosciences*, *35*(9), 546–556.

Xia, Y., Dai, J., Lu, P., Huang, Y., Zhu, Y., & Zhang, X. (2008). Distinct effect of CD40 and TNF-signaling on the chemokine/chemokine receptor expression and function of the human monocyte-derived dendritic cells. *Cellular and Molecular Immunology*, *5*(2), 121–131.

Xia, Z., Zhang, N., Jin, H., Yu, Z., Xu, G., & Huang, Z. (2010). Clinical significance of astrocyte elevated gene-1 expression in human oligodendrogliomas. *Clinical Neurology and Neurosurgery*, *112*(5), 413–419.

Xue, Y., Ren, J., Gao, X., Jin, C., Wen, L., & Yao, X. (2008). GPS 2.0, a tool to predict kinase-specific phosphorylation sites in hierarchy. *Molecular & Cellular Proteomics*, 7(9), 1598–1608.

Yan, J., Zhang, M., Chen, Q., & Zhang, X. (2012). Expression of AEG-1 in human T-cell lymphoma enhances the risk of progression. *Oncology Reports*, *28*(6), 2107–2114.

Yang, Y., Wu, J., Guan, H., Cai, J., Fang, L., Li, J., et al. (2012). MiR-136 promotes apoptosis of glioma cells by targeting AEG-1 and Bcl-2. *FEBS Letters*, *586*(20), 3608–3612.

Ying, Z., Li, J., & Li, M. (2011). Astrocyte elevated gene 1: Biological functions and molecular mechanism in cancer and beyond. *Cell and Bioscience*, *1*(1), 36.

Yoo, B. K., Emdad, L., Lee, S. G., Su, Z. Z., Santhekadur, P., Chen, D., et al. (2011). Astrocyte elevated gene-1 (AEG-1): A multifunctional regulator of normal and abnormal physiology. *Pharmacology and Therapeutics*, *130*(1), 1–8.

Yoo, B. K., Emdad, L., Su, Z. Z., Villanueva, A., Chiang, D. Y., Mukhopadhyay, N. D., et al. (2009). Astrocyte elevated gene-1 regulates hepatocellular carcinoma development and progression. *The Journal of Clinical Investigation*, *119*(3), 465–477.

Yoo, B. K., Santhekadur, P. K., Gredler, R., Chen, D., Emdad, L., Bhutia, S., et al. (2011). Increased RNA-induced silencing complex (RISC) activity contributes to hepatocellular carcinoma. *Hepatology*, *53*(5), 1538–1548.

Yu, C., Chen, K., Zheng, H., Guo, X., Jia, W., Li, M., et al. (2009). Overexpression of astrocyte elevated gene-1 (AEG-1) is associated with esophageal squamous cell carcinoma (ESCC) progression and pathogenesis. *Carcinogenesis*, *30*(5), 894–901.

Zhang, B., Liu, X. X., He, J. R., Zhou, C. X., Guo, M., He, M., et al. (2011). Pathologically decreased miR-26a antagonizes apoptosis and facilitates carcinogenesis by targeting MTDH and EZH2 in breast cancer. *Carcinogenesis*, *32*(1), 2–9.

Zhang, F., Yang, Q., Meng, F., Shi, H., Li, H., Liang, Y., et al. (2012). Astrocyte elevated gene-1 interacts with beta-catenin and increases migration and invasion of colorectal carcinoma. *Molecular Carcinogenesis*, http://dx.doi.org/10.1002/mc.21894.

Zhao, Y., Kong, X., Li, X., Yan, S., Yuan, C., Hu, W., et al. (2011). Metadherin mediates lipopolysaccharide-induced migration and invasion of breast cancer cells. *PLoS One*, *6*(12), e29363.

Zhou, Z., Deng, H., Yan, W., Huang, H., Deng, Y., Li, Y., et al. (2012). Expression of metadherin/AEG-1 gene is positively related to orientation chemotaxis and adhesion of human hepatocellular carcinoma cell lines of different metastatic potentials. *Journal of Huazhong University of Science and Technology. Medical Sciences*, *32*(3), 353–357.

Zhou, J., Li, J., Wang, Z., Yin, C., & Zhang, W. (2012). Metadherin is a novel prognostic marker for bladder cancer progression and overall patient survival. *Asia-Pacific Journal of Clinical Oncology*, *8*(3), e42–e48.

Zhu, K., Dai, Z., Pan, Q., Wang, Z., Yang, G. H., Yu, L., et al. (2011). Metadherin promotes hepatocellular carcinoma metastasis through induction of epithelial-mesenchymal transition. *Clinical Cancer Research*, *17*(23), 7294–7302.

CHAPTER TWO

AEG-1/MTDH/LYRIC: Clinical Significance

Devanand Sarkar[*,†,‡,1], **Paul B. Fisher**[*,†,‡,1]
[*]Department of Human and Molecular Genetics, School of Medicine, Virginia Commonwealth University, School of Medicine, Richmond, Virginia, USA
[†]VCU Institute of Molecular Medicine, School of Medicine, Virginia Commonwealth University, School of Medicine, Richmond, Virginia, USA
[‡]VCU Massey Cancer Center, School of Medicine, Virginia Commonwealth University, School of Medicine, Richmond, Virginia, USA
[1]Corresponding authors: e-mail address: dsarkar@vcu.edu; pbfisher@vcu.edu

Contents

Advances in Cancer Research, Volume 120
ISSN 0065-230X
http://dx.doi.org/10.1016/B978-0-12-401676-7.00002-4

Abstract

"Gain-of-function" and "loss-of-function" studies in human cancer cells and analysis of a transgenic mouse model have convincingly established that AEG-1/MTDH/LYRIC performs a seminal role in regulating proliferation, invasion, angiogenesis, metastasis, and chemoresistance, the salient defining hallmarks of cancer. These observations are strongly buttressed by clinicopathologic correlations of AEG-1/MTDH/LYRIC expression in a diverse array of cancers distinguishing AEG-1/MTDH/LYRIC as an independent biomarker for highly aggressive metastatic disease with poor prognosis. AEG-1/MTDH/LYRIC has been shown to be a marker predicting response to chemotherapy, and serum anti-AEG-1/MTDH/LYRIC antibody titer also serves as a predictor of advanced stages of aggressive cancer. However, inconsistent findings have been reported regarding the localization of AEG-1/MTDH/LYRIC protein in the nucleus or cytoplasm of cancer cells and the utility of nuclear or cytoplasmic AEG-1/MTDH/LYRIC to predict the course and prognosis of disease. This chapter provides a comprehensive analysis of the existing literature to emphasize the common and conflicting findings relative to the clinical significance of AEG-1/MTDH/LYRIC in cancer.

1. INTRODUCTION

Astrocyte elevated gene-1 (AEG-1) was first cloned in 2002 as an HIV- and TNF-α-inducible gene in primary human fetal astrocytes (Kang et al., 2005; Su et al., 2002). Subsequently, *in vivo* phage screening allowed the cloning of the mouse gene as a protein mediating metastasis of breast cancer cells to lung and was named metadherin (MTDH) (Brown & Ruoslahti, 2004). The mouse/rat gene was also cloned as a tight junction protein named LYsine-RIch CEACAM1 coisolated (LYRIC) and by gene trapping techniques as an endoplasmic reticulum (ER)/nuclear envelop protein and was named 3D3/LYRIC (Britt et al., 2004; Sutherland, Lam, Briers, Lamond, & Bickmore, 2004). Human AEG-1/MTDH/LYRIC mRNA encodes a single-pass transmembrane protein of predicted molecular mass of ~64 kDa and p*I* of 9.3 (Kang et al., 2005). It is a highly basic protein rich in lysines. There is an N-terminal transmembrane domain and three putative nuclear localization signals in AEG-1/MTDH/LYRIC.

Expression analysis revealed that AEG-1/MTDH/LYRIC is a unique protein that is overexpressed in all cancers studied to date (Sarkar et al., 2009; Yoo, Emdad, et al., 2011). The spectrum of cancers analyzed includes all organs and tissues belonging to all biological systems. AEG-1/MTDH/

LYRIC expression gradually increases as the disease process progresses, and AEG-1/MTDH/LYRIC expression level clearly correlates with adverse patient prognosis. Overexpression of AEG-1/MTDH/LYRIC augments proliferation, migration, invasion, angiogenesis, chemoresistance, and metastasis, while inhibition of AEG-1/MTDH/LYRIC abrogates the above-mentioned phenotypes indicating a pivotal role of AEG-1/MTDH/LYRIC in regulating tumorigenesis (Sarkar et al., 2009; Yoo, Emdad, et al., 2011).

Multiple mechanisms underlie AEG-1/MTDH/LYRIC overexpression in cancers. AEG-1/MTDH/LYRIC is transcriptionally regulated by c-Myc, which is located downstream of Ha-ras and PI3K pathways (Lee, Su, Emdad, Sarkar, & Fisher, 2006). As such, activation or increase in any of these three components will lead to AEG-1/MTDH/LYRIC overexpression. The AEG-1/MTDH/LYRIC gene is located at chromosome 8q22, which is a center of activity for genomic amplification in multiple cancers. Indeed, genomic amplification of AEG-1/MTDH/LYRIC has been detected in breast and liver cancers (Hu et al., 2009; Yoo, Emdad, et al., 2009). AEG-1/MTDH/LYRIC is regulated by multiple tumor suppressor miRNAs, miR-375, miR-136, and miR-26a, which are downregulated in several cancers (He et al., 2012; Hui et al., 2011; Nohata et al., 2011; Yang et al., 2012; Zhang et al., 2011). Cytoplasmic polyadenylation element-binding protein-1 binds to the 3′-UTR of AEG-1/MTDH/LYRIC mRNA and increases its translation in glioma cells (Kochanek & Wells, 2013). Monoubiquitination of AEG-1/MTDH/LYRIC protein increases its stabilization and cytoplasmic accumulation in cancer cells (Srivastava et al., 2012; Thirkettle et al., 2009). These diverse mechanisms ensure that AEG-1/MTDH/LYRIC is overexpressed in all cancers (Fig. 2.1), thereby permitting AEG-1/MTDH/LYRIC to serve as an important participant in aggressive progression of cancers.

A literature search using AEG-1/MTDH/LYRIC as a key word identifies 114 papers, a large number of which analyze the clinical significance of AEG-1/MTDH/LYRIC overexpression in cancers. Indeed, more papers are devoted to analyzing AEG-1/MTDH/LYRIC expression profile and its clinicopathological significance rather than scrutinizing the molecular mechanism(s) of AEG-1/MTDH/LYRIC function. These studies have firmly established the importance of AEG-1/MTDH/LYRIC in regulating cancer progression and metastasis, which is reflected in the inclusion of AEG-1/MTDH/LYRIC in MammaPrint early metastasis risk assessment

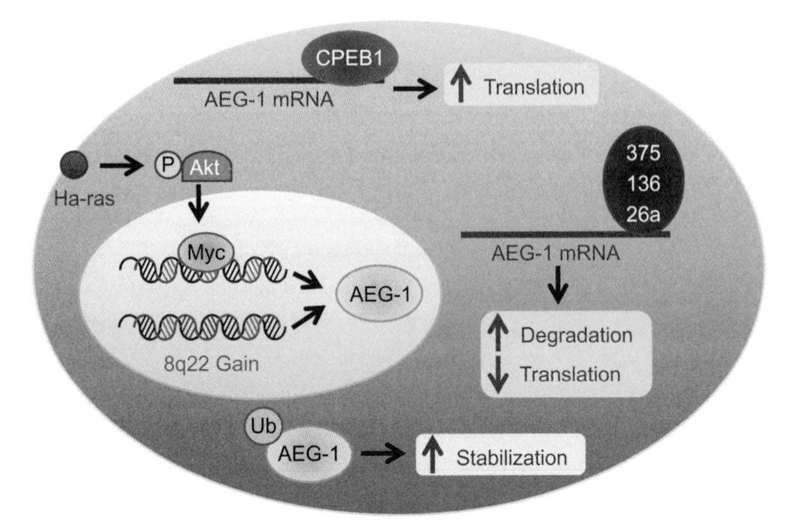

Figure 2.1 Molecular mechanism of AEG-1/MTDH/LYRIC overexpression in cancer. Genomic amplification (8q22 gain) leading to increased AEG-1/MTDH/LYRIC expression has been documented in breast and liver cancers. Activation of Ha-ras results in activation of PI3K/Akt and subsequent binding of c-Myc to AEG-1/MTDH/LYRIC promoter increasing transcription. Monoubiquitination of AEG-1/MTDH/LYRIC protein in cancer cells leads to increased stabilization and cytoplasmic accumulation. AEG-1/MTDH/LYRIC mRNA is posttranscriptionally regulated by several tumor suppressor mRNAs, such as miR-375, miR-136, and miR-26a, which are downregulated in multiple cancers. Cytoplasmic polyadenylation element-binding protein 1 (CEBP1) binds to the 3′-UTR of AEG-1/MTDH/LYRIC mRNA and promotes its translation in glioma cells. (See Page 2 in Color Section at the back of the book.)

assay (http://www.agendia.com/pages/mammaprint/21.php). MammaPrint is the first and only FDA-approved individualized metastasis risk assessment assay for breast cancer that includes a unique 70-gene signature, and AEG-1/MTDH/LYRIC is included in this gene signature. The present review provides a comprehensive synopsis of the clinicopathological correlation studies on AEG-1/MTDH/LYRIC stressing the relevant features in each cancer indication.

2. GASTROINTESTINAL SYSTEM

2.1. Colorectal cancer

Colorectal cancer (CRC) is the third most common cancer and the fourth most common cause of cancer-related deaths worldwide with about 600,000 estimated deaths annually (Ferlay et al., 2010). Several studies have been performed in CRC patients to analyze the clinical significance of AEG-1/MTDH/LYRIC. Song, Li, Li, and Geng (2010) analyzed by immunohistochemistry (IHC) Formalin-fixed paraffin-embedded (FFPE) sections of low-grade adenoma ($n=31$), high-grade adenoma ($n=15$), colorectal carcinoma ($n=146$), and normal colorectal mucosa ($n=45$). Additionally, tissues with hepatic ($n=10$), pulmonary ($n=2$), and lymph node ($n=250$) metastases were included. None of these patients received chemo- or radiotherapy prior to surgery. Based on the Union Internationale Centre le Cancer (UICC) classification, among the CRC patients, 42 were stage I, 38 were stage II, 54 were stage III, and 12 were stage IV. While weak to no AEG-1/MTDH/LYRIC expression was detected in normal mucosa, the expression of this protein gradually increased from low-grade adenoma to high-grade adenoma with the highest expression being detected in CRC. In normal mucosa and adenoma, AEG-1/MTDH/LYRIC expression was cytoplasmic, while in CRC samples as well as in pulmonary metastases, overexpressed AEG-1/MTDH/LYRIC was detected in the nucleus. In CRC patients, stage III/IV patients showed higher nuclear AEG-1/MTDH/LYRIC staining compared to stage I/II patients (46.67% vs. 24.32%, respectively; $p=0.037$). Statistical analysis revealed that in CRC patients, AEG-1/MTDH/LYRIC expression correlated significantly with UICC stage, TNM classification, Ki-67 expression, and histological differentiation. No statistical correlation was observed with age, gender, tumor location, and size. A significant correlation was observed between AEG-1/MTDH/LYRIC expression level and the shorter overall survival (OS) time of CRC patients ($p<0.001$) with a correlation coefficient of -0.380. The 5-year cumulative survival rate was 73.4% in low AEG-1/MTDH/LYRIC-expressing group, while it was 41.5% in high AEG-1/MTDH/LYRIC-expressing group. These findings indicate that AEG-1/MTDH/LYRIC might be a significant prognostic factor for CRC patients.

Zhang et al. (2012) performed a similar IHC-based study in 120 pairs of CRC and adjacent nontumor tissue (ANT) and 60 samples of lymph node metastases of CRC tissues. High AEG-1/MTDH/LYRIC expression was detected in 54 CRC samples (45%) and 13 ANT samples (10.8%). AEG-1/MTDH/LYRIC expression was significantly higher in CRC with lymph node metastasis versus CRC without lymph node metastasis ($p<0.001$). Nuclear accumulation of β-catenin, a downstream AEG-1/MTDH/LYRIC molecule, was observed in 62% CRC and 0% ANT samples and showed positive statistical correlation with high AEG-1/MTDH/LYRIC expression ($p<0.001$).

Jiang et al. analyzed tumor tissues, matched normal tissues, and liver metastasis specimens from 520 CRC cases for AEG-1/MTDH/LYRIC expression Jiang, Zhu, Zhu, and Piao (2012). Among the 520 cases, 37 (7.12%), 204 (39.23%), 262 (50.38%), and 17 (3.27%) were with Duke's A, B, C, and D clinical-stage tumors, respectively. AEG-1/MTDH/LYRIC mRNA and protein expressions were significantly higher in CRC compared to matched normal tissues and in patients with liver metastases compared to those without liver metastases. However, the overexpressed AEG-1/MTDH/LYRIC protein was detected in the cytoplasm and membrane but not in the nucleus. A positive correlation was observed between AEG-1/MTDH/LYRIC expression and age, Duke's stage, and distant metastasis ($p=0.001$, 0.001, and 0.016, respectively), but not with gender and histological grade. A separate set of 56 patients with postoperative distant metastases was also analyzed. Among these patients, 48 (88.89%) developed liver metastases in which 58.33% showed positive AEG-1/MTDH/LYRIC expression ($p=0.016$). High AEG-1/MTDH/LYRIC expression was an independent prognostic marker for shorter OS. The significant finding of this study is the identification of AEG-1/MTDH/LYRIC as a determinant of liver metastases of CRC.

Gnosa et al. (2012) analyzed AEG-1/MTDH/LYRIC mRNA expression by qPCR in primary tumor and adjacent normal mucosa from 156 CRC patients, and AEG-1/MTDH/LYRIC protein expression by IHC in 74 distant normal colorectal mucosa, 107 adjacent normal colorectal mucosa, 158 primary CRC, 35 lymph node metastases, and 9 liver metastases. The mean level of AEG-1/MTDH/LYRIC mRNA was 371.56 ± 348.37 in the primary tumor and 214.98 ± 156.39 in the adjacent normal mucosa in 156 patients ($p=0.0005$). AEG-1/MTDH/LYRIC protein expression was significantly higher in primary CRC when compared to either adjacent or distant normal mucosa. Interestingly, both high cytoplasmic and nuclear AEG-1/MTDH/LYRIC expressions were detected in CRC patients. Even though AEG-1/MTDH/LYRIC expression in lymph node metastases showed a higher trend compared to that in primary tumors, the difference was not statistically significant. However, both cytoplasmic and nuclear AEG-1/MTDH/LYRIC expressions in liver metastases were significantly higher than that in primary tumors and lymph node metastases. Interestingly, AEG-1/MTDH/LYRIC mRNA expression showed significantly higher expression in rectal cancer patients compared to colon cancer patients ($p=0.047$). However, this study did not find any correlation between AEG-1/MTDH/LYRIC protein expression and other clinicopathological variables, such as age, gender, location, differentiation, or patient survival. In primary tumors, both cytoplasmic and nuclear AEG-1/MTDH/LYRIC protein expression correlated with phosphorylated NF-κB (Ser 536), p73, and Rad50 with statistical significance. Cytoplasmic, but not nuclear, AEG-1/MTDH/LYRIC showed association with increased apoptosis as determined by TUNEL assay and it was speculated that AEG-1 might be involved in DNA damage-induced apoptosis which is mediated by p73. However, this hypothesis requires experimental validation to establish its significance in colorectal carcinogenesis.

Although these studies authenticate the importance of AEG-1/MTDH/LYRIC in promoting tumorigenesis and metastasis in CRC patients, one disconcerting observation is the varied localization of AEG-1/MTDH/LYRIC detected by different investigators. Use of different antibodies recognizing different epitopes of AEG-1/MTDH/LYRIC or experimental technique might be potential reasons as is the case with Jiang et al. and Gnosa et al. who used the same antibody, but reported dissimilar findings. Thus, an optimized IHC protocol needs to be developed that might be universally used by molecular diagnostic laboratories to employ cytoplasmic and/or nuclear AEG-1/MTDH/LYRIC as a diagnostic/prognostic marker for CRC (Table 2.1).

Multiple studies have identified staphylococcal nuclease domain-containing protein-1 (SND1), a multifunctional protein regulating transcription, mRNA splicing and stability, and miRNA function, as an AEG-1/MTDH/LYRIC-interacting protein (Blanco et al., 2011; Meng et al., 2012; Yoo, Santhekadur, et al., 2011). Coexpression of AEG-1/MTDH/LYRIC and SND1 was analyzed in tumor tissues and adjacent noncancerous tissues in 196 colon cancer patients by IHC (Wang et al., 2012). AEG-1/MTDH/LYRIC and SND1 expression was detected in 149 (76%) and 137 (69.9%) cases, respectively, specifically in the cancer tissue and not in the adjacent normal tissue. Among the 149 AEG-1/MTDH/LYRIC-positive cases, 132 showed positive staining for SND1. Among the 47 AEG-1/MTDH/LYRIC-negative cases, 42 were negative for SND1, suggesting a high consistency of expression of these two proteins in colon cancer, which was statistically significant ($r=0.86$, $p<0.001$). Significantly higher AEG-1/MTDH/LYRIC and SND1 expression was detected in cases with aggressive nodal status (N2), late pathological stage, and poor differentiation when compared to N0–N1 nodal status ($p=0.02$), early pathological stage ($p=0.006$), and moderate differentiation ($p=0.03$), respectively. No correlation was observed with age, sex, or tumor status.

Table 2.1 AEG-1/MTDH/LYRIC is a diagnostic/prognostic marker for multiple cancers

Type of cancer	No. of cases	References
Colorectal cancer	146	Song, Li, Li, et al. (2010)
	180	Zhang et al. (2012)
	520	Jiang et al. (2012)
	202	Gnosa et al. (2012)
	196	Wang et al. (2012)
Hepatocellular carcinoma	109	Yoo, Emdad, et al. (2009)
	323	Zhu et al. (2011)
	73 (HBV-HCC)	Gong et al. (2012)
Gastric cancer	105	Jian-Bo et al. (2011)
	30	Baygi and Nikpour (2012)
Esophageal cancer	168	Yu et al. (2009)
Gallbladder carcinoma	41	Sun et al. (2011)
	108	Liu and Yang (2011)
Breast cancer	225	Li et al. (2008)
	170	Hu et al. (2009)
	249	Su, Zhang, and Yang (2010)
	125 (Triple negative)	Li et al., 2011
Ovarian cancer	157 (epithelial)	Li et al. (2011)
	101 (serous)	Li et al. (2012)
Endometrial cancer	174	Song, Li, Lu, et al. (2010)
Prostate cancer	143	Thirkettle et al. (2009)
Renal cell carcinoma	102	Chen, Ke, Shi, Yang, and Wang (2010)
Bladder cancer	60	Zhou, Li, Wang, Yin, and Zhang (2012)
Glioblastoma multiforme	296	Liu et al. (2010)
Brain cancer	98	Emdad et al. (2010)

Table 2.1 AEG-1/MTDH/LYRIC is a diagnostic/prognostic marker for multiple cancers—cont'd

Type of cancer	No. of cases	References
Neuroblastoma	32	Liu, Liu, Han, Zhang, and Sun (2012)
Oligodendroglioma	75	Xia et al. (2010)
Nonsmall cell lung cancer	220	Song et al. (2009)
	67	Sun et al. (2012)
Salivary gland carcinoma	141	Liao et al. (2011)
Head and neck squamous cell carcinoma	20	Nohata et al. (2011)
Squamous cell carcinoma of the tongue	93	Ke et al. (2012)
Osteosarcoma	62	Wang et al. (2011)
Diffuse large B-cell lymphoma	30	Ge et al. (2012)
T-cell non-Hodgkin lymphoma	129	Yan, Zhang, Chen, and Zhang (2012)

The 196 patients were divided into four groups based on AEG-1/MTDH/LYRIC and SND1 coexpression: AEG-1-/SND1− (42 cases); AEG-1-/SND1+ (5 cases); AEG-1+/SND1− (17 cases); and AEG-1+/SND1+ (132 cases). AEG-1+/SND1+ status was significantly associated with aggressive nodal status ($p=0.02$), late pathological stage ($p=0.01$), poor differentiation ($p<0.001$), and shorter OS ($p=0.01$). Indeed, OS in AEG-1-/SND1-cases was significantly longer when compared to the other three groups ($p=0.006$). Cox multivariate analysis revealed that AEG-1/MTDH/LYRIC and SND1 coexpression negatively correlated with postoperative OS and positively correlated with mortality ($p=0.009$), indicating that AEG-1/MTDH/LYRIC and SND1 are potential prognostic factors for colon cancer.

2.2. Hepatocellular carcinoma

Hepatocellular carcinoma (HCC) which represents >80% of all primary liver cancers is the fifth most common cancer and the third most common cause of cancer-related deaths worldwide (El-Serag, 2011). Yoo, Emdad, et al. (2009) first analyzed AEG-1/MTDH/LYRIC expression profile at mRNA and protein levels in multiple cohorts of HCC patients. IHC was performed in tissue microarrays (TMA) containing 86 primary HCC, 23 metastatic HCC, and 9 normal adjacent liver samples. Very little to no AEG-1/MTDH/LYRIC immunostaining was detected in the nine normal liver samples, while significant AEG-1/MTDH/LYRIC staining was observed in HCC samples. AEG-1 expression was detected predominantly in the perinuclear region. Among the 109 HCC samples, only 7 scored negative for AEG-1/MTDH/LYRIC and the remaining 102 (93.58%) showed variable levels of AEG-1/MTDH/LYRIC. Expression of AEG-1/MTDH/LYRIC gradually increased with the stages from I to IV as well as with the grades of differentiation from well differentiated to poorly differentiated, and a statistically significant correlation ($p<0.0001$) was obtained between AEG-1/MTDH/LYRIC expression level and the stage of HCC. Gene expression data from Affymetrix microarray of a separate set of patients, including 132 human samples in various stages of human hepatocarcinogenesis: normal liver ($n=10$), cirrhotic tissue ($n=13$), low-grade dysplastic nodules ($n=10$), high-grade dysplastic nodules ($n=8$), and HCC ($n=91$), were analyzed to check AEG-1/MTDH/LYRIC mRNA expression level. Expression of AEG-1/MTDH/LYRIC in hepatitis C virus (HCV)-related HCC was significantly increased in comparison to normal liver and cirrhotic tissue. Mean upregulation in comparison to normal liver and cirrhosis were 1.7 (t-test, $p=0.04$)- and 1.65 (t-test, $p<0.001$)-fold increase, respectively. Analysis of DNA copy gain in the AEG-1 locus identified genomic amplification of AEG-1 in 26% of the patients. These studies, however, did not analyze the clinicopathological correlation of AEG-1/MTDH/LYRIC expression in HCC patients.

AEG-1/MTDH/LYRIC expression was assessed by IHC in TMA of 323 HCC patients demonstrating high AEG-1/MTDH/LYRIC expression in 54.2% (175 of 323) of all the patients (Zhu et al., 2011). AEG-1/MTDH/LYRIC expression was closely associated with microvascular invasion ($p<0.001$), pathologic satellites ($p=0.007$), tumor differentiation ($p=0.002$), and TNM stage ($p=0.001$). Expression of AEG-1/MTDH/LYRIC did not correlate with other clinicopathologic characteristics such as age, gender, liver cirrhosis, serum alpha-fetoprotein, tumor diameter, tumor encapsulation, or BCLC stage. The 1-, 3-, and 5-year OS and cumulative recurrence rates in the whole cohort were 85.4% and 25.4%, 62.2% and 50.2%, 50.7% and 59.7%, respectively. Further, the 1-, 3-, and 5-year OS rates in high AEG-1/MTDH/LYRIC-expressing group were significantly lower than those in low AEG-1/MTDH/LYRIC-expressing group (83.0% vs. 89.7%, 52.0% vs. 75.3%, 37.4% vs. 66.9%, respectively); the 1-, 3-, and 5-year cumulative recurrence rates were markedly higher in the high AEG-1/MTDH/LYRIC-expressing group than those in the low AEG-1/MTDH/LYRIC-expressing group (32.4% vs. 16.8%, 61.2% vs. 38.2%, 70.7% vs. 47.8%, respectively). Univariate and multivariate analyses revealed that along with tumor diameter, encapsulation, microvascular invasion, and TNM stage, AEG-1/MTDH/LYRIC was an independent prognostic factor for both OS (HR = 1.870, $p < 0.001$) and recurrence (HR = 1.695, $p < 0.001$).

In a separate study, AEG-1/MTDH/LYRIC expression levels were identified to be elevated in HBV-related HCC tissues ($n = 73$) compared to normal liver tissues ($n = 11$) and hepatitis samples ($n = 45$) (Gong et al., 2012). AEG-1/MTDH/LYRIC expression significantly correlated with the American Joint Committee on Cancer (AJCC, 7th edition) stage ($p = 0.020$), T classification ($p = 0.007$), N classification ($p = 0.044$), vascular invasion ($p = 0.006$), and histological differentiation ($p = 0.020$) in the HBV-related HCC patients. In addition, patients with high AEG-1/MTDH/LYRIC levels had shorter survival times compared to those with low AEG-1/MTDH/LYRIC expression ($p = 0.001$).

The gene expression profiles of HBV-HCC and HCV-HCC show significant differences. The identification of AEG-1/MTDH/LYRIC overexpression in both HBV-HCC and HCV-HCC cases suggests that AEG-1/MTDH/LYRIC might be a useful biomarker for HCC irrespective of its etiology.

2.3. Gastric cancer

Gastric cancer is the fourth most frequent malignancy in the world and is common in Asian countries, such as China, Japan, and Korea (Chan, Wong, & Lam, 2001). AEG-1/MTDH/LYRIC expression was analyzed in 105 cases of gastric cancer by IHC (Jian-Bo et al., 2011). Overexpression of AEG-1 was detected in 66 patients mainly in the cytoplasm. In cancerous tissue, AEG-1/MTDH/LYRIC expression intensity score was three or higher, while in the adjacent noncancerous tissue, the corresponding score was 0–2. AEG-1/MTDH/LYRIC expression strongly correlated with the clinical stages of the disease ($p < 0.01$) and T ($p < 0.01$), N ($p < 0.01$), and M ($p < 0.05$) classifications indicating that advanced clinical stage is associated with AEG-1/MTDH/LYRIC overexpression. The proliferation marker Ki-67 correlated with AEG-1/MTDH/LYRIC expression

($p < 0.01$) indicating that AEG-1/MTDH/LYRIC might promote tumor growth. The median overall 5-year survival rate was 23 months in patients with AEG-1/MTDH/LYRIC overexpression versus 38 months in AEG-1/MTDH/LYRIC-negative patients ($p < 0.001$). In multivariate analysis, TNM stage, lymph node metastasis, and AEG-1/MTDH/LYRIC overexpression were associated with poor OS thus suggesting that AEG-1/MTDH/LYRIC might be a prognostic marker for gastric cancer.

Another independent study analyzed AEG-1/MTDH/LYRIC expression in 30 paired gastric tumoral and nontumoral tissue samples of Iranian patients by qRT-PCR (Baygi & Nikpour, 2012). Although overall AEG-1/MTDH/LYRIC expression was higher in tumor tissue compared to normal tissue ($p = 0.05$), a significant heterogeneity was observed in the level of expression. In 46.6% cases, AEG-1/MTDH/LYRIC expression was higher, while in 36.6% cases, the expression was lower in the tumor tissue versus the nontumor tissue. No statistically significant association was observed between AEG-1/MTDH/LYRIC expression and the grades and types of tumor. This apparent lack of importance of AEG-1/MTDH/LYRIC in gastric cancer might be due to a number of reasons. AEG-1/MTDH/LYRIC expression at the protein level is much higher in tumor tissue versus normal tissue when compared to AEG-1/MTDH/LYRIC expression at the mRNA level suggesting that posttranscriptional regulation might be more important than genomic amplification or transcriptional regulation to confer AEG-1/MTDH/LYRIC overexpression. Regulation of AEG-1/MTDH/LYRIC by multiple tumor suppressor miRNAs and cancer-specific stabilization of AEG-1/MTDH/LYRIC protein by monoubiquitination further stress this notion (He et al., 2012; Hui et al., 2011; Nohata et al., 2011; Srivastava et al., 2012; Thirkettle et al., 2009; Yang et al., 2012; Zhang et al., 2011). As such, simultaneous analysis of AEG-1/MTDH/LYRIC at both mRNA and protein levels is necessary to get a comprehensive picture of the expression pattern. A racial difference in AEG-1/MTDH/LYRIC expression might also be an issue.

2.4. Esophageal cancer

Esophageal cancer is a highly aggressive cancer of the gastrointestinal tract and is the sixth most common cause of cancer-related deaths worldwide (Kamangar, Dores, & Anderson, 2006). Esophageal cancer is typically either esophageal squamous cell cancer (ESCC) or adenocarcinoma, the former more prevalent in Asian countries (Hiyama, Yoshihara, Tanaka, &

Chayama, 2007). AEG-1/MTDH/LYRIC was evaluated as a biomarker for ESCC by IHC of FFPE sections in 168 patients that include 9 cases of stage I, 73 cases of stage IIa, 14 cases of stage IIb, 62 cases of stage III, and 10 cases of stage IV cancer (Yu et al., 2009). AEG-1/MTDH/LYRIC protein expression was detected in 156 cases (92.9%) and was upregulated when compared to adjacent normal esophageal tissues. AEG-1/MTDH/LYRIC expression strongly correlated with clinical staging ($p=0.001$), T ($p=0.002$), N ($p=0.034$), M ($p=0.021$) classifications, and histological differentiation ($p=0.035$), which was confirmed by Spearman correlations. The cumulative 5-year survival rate in low AEG-1/MTDH/LYRIC-expressing patients was 40.7% (95% confidence interval, 0.5095–0.3044) versus 22.6% in high AEG-1/MTDH/LYRIC-expressing patients (95% confidence interval, 0.3177–0.1343). AEG-1/MTDH/LYRIC expression levels inversely correlated with OS time ($p=0.001$) establishing AEG-1/MTDH/LYRIC as a potential prognostic biomarker. One interesting finding in this study, which was not observed in other studies, is statistically significant ($p=0.041$) higher AEG-1/MTDH/LYRIC expression in male ESCC patients compared to female patients. The molecular mechanism of male-specific overexpression and the potential implications of this observation remain to be determined.

2.5. Gallbladder carcinoma

Gallbladder carcinoma (GBC) is the most common malignancy of the biliary tract and the fifth most common malignancy of the digestive tract (Miller & Jarnagin, 2008). The 5-year survival rate of GBC after surgery is 0–10%, thus making it a leading cause of cancer-related deaths (Malka et al., 2004). AEG-1/MTDH/LYRIC expression was detected by IHC in 41 GBC, 10 adenomas, and chronic or acute cholecystitis samples (Sun et al., 2011). Weak or negative AEG-1/MTDH/LYRIC expression was observed in normal gallbladder mucosa. The frequency and intensity of AEG-1/MTDH/LYRIC expression gradually increased from normal mucosa to adenoma to GBC. AEG-1/MTDH/LYRIC overexpression was detected in 26 of 41 GBC (63.4%) patients ($p=0.0003$ vs. cholecystitis). While no nuclear AEG-1/MTDH/LYRIC staining was observed in normal mucosa or adenomas, overexpressed AEG-1/MTDH/LYRIC was detected in the nucleus. Increased AEG-1/MTDH/LYRIC expression correlated with differentiation degree ($p=0.0259$), Nevin stage ($p=0.0339$), liver infiltration ($p=0.0328$), and Ki-67 expression ($p=0.0032$). However, no correlation was observed between AEG-1/MTDH/LYRIC level and age, gender, tumor location, tumor size, venous invasion, lymph node metastasis, or pathological type. AEG-1/MTDH/LYRIC was identified as an independent prognostic factor for OS rate of GBC patients by the Cox proportional hazards model. The mean survival time for high and low AEG-1/MTDH/LYRIC-expressing GBC patients was 21 and 37.1 months, respectively ($p=0.008$). The cumulative 1-, 3-, and 5-year OS rates were 57.7%, 19.2%, and 3.8% in the high AEG-1/MTDH/LYRIC-expressing group versus 80%, 53.3%, and 33.3% in low AEG-1/MTDH/LYRIC-expressing group. Although the clinicopathological findings of this study are similar to studies in other cancers, two interesting differences were observed. First, AEG-1/MTDH/LYRIC levels did not correlate with lymph node metastases, even though in other cancers AEG-1/MTDH/LYRIC is a primary determinant of metastasis. Second, the overexpressed AEG-1/MTDH/LYRIC was detected in the nucleus, while in the majority of other cancers, it is detected in the cytoplasm and in the cell membrane. Whether the nuclear AEG-1/MTDH/LYRIC preferentially regulates any signaling pathways or gene expression in GBC patients remains to be determined.

A second study analyzed 108 GBC patient samples, including 36 well-differentiated adenocarcinoma, 31 moderately differentiated adenocarcinomas, 30 poorly differentiated adenocarcinomas, and 11 mucinous adenocarcinomas, and 96 benign samples that included 46 peritumoral tissues from the 108 GBC patients, 15 gallbladder polyps, and 35 chronic cholecystitis cases (Liu & Yang, 2011). Among the 46 peritumoral tissues, 10 were normal, while 10, 12, and 14 cases showed mild, moderate, or severe dysplasia, respectively. Among the 35 chronic cholecystitis cases, 11 were considered as normal, while 12, 7, and 5 cases showed mild, moderate, or severe dysplasia, respectively. The expression of AEG-1/MTDH/LYRIC and another marker EphA7 was analyzed by IHC. Compared to the peritumoral tissues, polyps, and chronic cholecystitis, significantly increased AEG-1/MTDH/LYRIC and EphA7 expression was detected in GBC ($p<0.01$). Those benign cases showed that expression of AEG-1/MTDH/LYRIC and EphA7 also showed moderate to severe dysplasia. Among the 57 EphA7 positive GBC cases, 43 showed positive expression of AEG-1/MTDH/LYRIC, while among the 51 EphA7 negative cases, 32 were negative for AEG-1/MTDH/LYRIC ($\chi^2=13.11$, $p<0.001$), suggesting a concordant expression pattern of these two markers. AEG-1/MTDH/LYRIC and EphA7 expression was lower in well-differentiated cases with small tumor size (<2 cm), no lymph node metastasis, and no invasion compared to poorly differentiated cases with large tumor size (>2 cm), lymph node metastasis, and invasion into surrounding tissues and organs ($p < 0.05$). No correlation was observed with mucinous adenocarcinoma or with sex, age, or history of gallstones. Survival information was available for 67 out of 108 GBC patients. Among these 67 patients, 37 and 34 patients showed positive AEG-1/MTDH/LYRIC and EphA7 expression, respectively. The average survival time in EphA7 and AEG-1/MTDH/LYRIC positive cases was 8.1 months when compared to 13.2 months in EphA7 and AEG-1/MTDH/LYRIC-negative cases ($p < 0.001$). Cox multivariate analysis revealed that tumor size (>2 cm), lymph node metastasis, invasion as well as AEG-1/MTDH/LYRIC and EphA7 expression levels were negatively correlated with postoperative survival and positively correlated with mortality, suggesting that AEG-1/MTDH/LYRIC and EphA7 might be prognostic factors for GBC.

3. GENITOURINARY SYSTEM

3.1. Breast cancer

Breast cancer is the second most frequent cancer in the world and the first most commonly diagnosed cancer, and the first leading cause of cancer-related deaths in women in the United States (Ferlay et al., 2010; Siegel, Naishadham, & Jemal, 2012). Although the role of AEG-1/MTDH/LYRIC in breast cancer metastasis was first reported by Brown and Ruoslahti (2004) using mouse models, Li et al. (2008) first demonstrated the clinical significance of AEG-1/MTDH/LYRIC overexpression in breast cancer. The study was performed by IHC using FFPE sections of 225 breast cancer patients, including 28, 98, 66, and 33 cases of stage I, II, III, and IV, respectively, and 9 cases of matched lung or liver metastases. AEG-1/MTDH/LYRIC expression was detected in 210 of 225 (93.3%) cases. While AEG-1/MTDH/LYRIC expression was marginally detectable in normal breast tissue and in the adjacent noncancerous tissues in all tumor sections, high levels of expression were detected in areas containing cancer cells in primary breast tumors as well as metastatic tumors. More intense staining was detected in hepatic metastases derived from patients with late relapses than that in the paired primary breast tumors. In the primary breast cancer cells, AEG-1/MTDH/LYRIC staining was detected mostly in the cytoplasm, while metastatic tumors showed a high percentage of nuclear staining. AEG-1/MTDH/LYRIC expression level showed statistical correlation with advanced clinical staging ($p = 0.001$) as well as T ($p = 0.004$), N,

($p=0.026$), and M ($p=0.001$) classifications. Higher levels of AEG-1/MTDH/LYRIC expression were associated with shorter survival time ($p<0.001$ with a correlation coefficient of -0.304). The cumulative 5-year survival rate was 75.7% in the low AEG-1/MTDH/LYRIC expression group (95% confidence interval, 0.773–0.857), while it was only 45.1% in the high AEG-1/MTDH/LYRIC expression group (95% confidence interval, 0.377–0.581). AEG-1/MTDH/LYRIC was recognized as an independent prognostic marker for breast cancer using multivariate survival analysis. However, no correlation was observed between the level of AEG-1/MTDH/LYRIC and that of estrogen receptor, progesterone receptor, and ErbB2 in these patients. In a follow-up study, the authors demonstrated a significant correlation between AEG-1/MTDH/LYRIC and Ki-67 levels ($p=0.003$) in the same breast cancer patients indicating involvement of AEG-1/MTDH/LYRIC in highly proliferative breast cancers (Li et al., 2009).

Hu et al. (2009) who first identified 8q22 gain, resulting in AEG-1/MTDH/LYRIC expression, to confer chemoresistance and metastasis in breast cancer, analyzed 170 patient samples by IHC, out of which 47% showed moderate to high level of AEG-1/MTDH/LYRIC staining. Although AEG-1/MTDH/LYRIC overexpression did not show any correlation with any specific breast tumor subtype based on HER2 status, triple marker status (ER/PR/HER2), or the basal epithelial cell marker C5/6 status, a significant association was observed with a higher risk of metastasis ($p=0.0058$) and shorter survival time ($p=0.0008$). Univariate survival analysis employing the Cox proportional hazard model demonstrated strong association between high AEG-1/MTDH/LYRIC level with higher hazard ratio (HR) and worse clinical outcome (HR $=3.7$, $p=0.01$ for metastasis; HR $=8.3$, $p=0.005$ for cancer-related death). Interestingly, not all patients showing high levels of AEG-1/MTDH/LYRIC expression presented with corresponding gains in 8q22 with 12% of samples having normal DNA copy number of AEG-1/MTDH/LYRIC but still showing high-level expression at the protein level. However, the association of AEG-1/MTDH/LYRIC with poor survival outcome was observed in all samples with AEG-1/MTDH/LYRIC overexpression, independent of 8q22 gain or alternative mechanisms of activation. More importantly, a multivariate analysis confirmed that AEG-1/MTDH/LYRIC might be a prognostic marker, independent of other clinicopathologic parameters including ER, PR, HER2, p53 status, and size of the primary tumor at the time of diagnosis. A multivariate Cox analysis combining all these parameters with AEG-1/MTDH/LYRIC expression showed that the hazards of metastasis were significantly higher with AEG-1/MTDH/LYRIC expression ($p=0.023$) even when all the other factors were considered.

IHC was used to analyze AEG-1/MTDH/LYRIC expression in 249 patients including 29 with ductal hyperplasia (UDH) without atypia, 14 atypical ductal hyperplasia (ADH), 37 ductal carcinoma *in situ* (DCIS) including 15 low-grade, 7 intermediate-grade, and 15 high-grade patients, 162 invasive ductal carcinomas, and 7 normal breast tissues from reduction mammoplasty (Su et al., 2010). AEG-1/MTDH/LYRIC overexpression was detected in 24.14% cases of UDH, 28.57% cases of ADH, 72.97% cases of DCIS, and 55.56% cases of invasive breast cancer with no overexpression detected in normal breast tissue. The observation that AEG-1/MTDH/LYRIC expression is detected in higher percentage of DCIS patients compared to invasive cancer suggests that AEG-1/MTDH/LYRIC might be involved in initiation of ductal carcinoma which contradicts with other studies indicating higher AEG-1/MTDH/LYRIC expression in metastatic lesions compared to the primary tumors. In DCIS patients, no correlation was observed between AEG-1/MTDH/LYRIC expression with that of ER, PR, and HER2. However, a significant correlation was observed with Ki-67 expression ($p=0.008$) as well as histologically high-grade tumors ($p=0.035$). However, in invasive cancer patients, AEG-1/MTDH/LYRIC expression strongly correlated with the patients' age ($p=0.042$), Ki-67 status ($p=0.036$), ER status ($p=0.018$), and p53 status ($p=0.001$), which were not observed in previous studies.

Triple-negative breast cancer is characterized by a lack of expression or ER, PR, and HER2 and carries a poor prognosis (Li, Li, et al., 2011). AEG-1/MTDH/LYRIC expression analysis was carried out in 125 cases of triple-negative invasive breast cancer that included 25 stage I, 71 stage II, and 29 stage III cases. A major focus of this study was to analyze association of AEG-1/MTDH/LYRIC with angiogenesis, which was checked by staining for VEGF and for CD34 that denotes microvascular density (MVD). Among the 125 cases, 71 (56.8%) showed high AEG-1/MTDH/LYRIC expression out of which 54 cases showed high VEGF expression, thus exhibiting strong statistical correlation ($p<0.001$). When analyzing MVD, 59 cases out of 125 showed high MVD and 42 cases of these high-MVD patients also showed high AEG-1/MTDH/LYRIC expression again demonstrating significant statistical correlation ($p=0.002$). Kaplan–Meyer 5-year survival curve analysis revealed that poor disease-free survival and OS were associated with high AEG-1/MTDH/LYRIC and VEGF levels.

3.2. Ovarian and endometrial cancer

Epithelial ovarian cancer (EOC) is the most lethal gynecological cancer in Western countries (Siegel et al., 2012). The peritoneal cavity is the most common site of metastatic spread and recurrence of EOC (Kikkawa et al., 1994). Li, Liu, et al. (2011) studied AEG-1/MTDH/LYRIC expression in 157 patients with EOC using IHC in FFPE sections. These cases included 49 patients with lymph node metastasis and 128 patients with peritoneal dissemination. In addition, 25 normal ovaries from hysterectomy specimens resected for nonovarian disease were analyzed. Normal ovaries showed little to no AEG-1/MTDH/LYRIC expression. Low AEG-1/MTDH/LYRIC expression was detected in 25 EOC samples (35.7%) without peritoneal metastasis and 62 (88.6%) samples without lymph node metastasis. However, 83 samples (64.8%) with peritoneal metastasis and 41 samples (83.7%) with lymph node metastasis exhibited high AEG-1/MTDH/LYRIC expression. The intensity and frequency of AEG-1/MTDH/LYRIC staining gradually increased from primary lesions to peritoneal dissemination and lymph node metastases in the same patient (25 cases). AEG-1/MTDH/LYRIC expression correlated with FIGO stage ($p=0.0011$), histopathological differentiation ($p=0$), and residual tumor size ($p<0.0001$). A multivariate logistic regression analysis revealed that ascites ($p<0.0001$; odds ratio 12.613; 95% confidence interval 3.148–28.517) and AEG-1/MTDH/LYRIC expression ($p=0.0017$; odds ratio 8.541; 95% confidence interval 2.561–37.461) were associated with peritoneal dissemination, and a positive correlation was observed between lymph node metastasis and AEG-1/MTDH/LYRIC expression ($p<0.0001$; odds ratio 9.581; 95% confidence interval 2.613–23.214).

Ovarian serous carcinoma is the most common type of ovarian cancer (Siegel et al., 2012). AEG-1/MTDH/LYRIC expression was analyzed by IHC in FFPE sections of 101 patients with stages II–IV ovarian serous carcinomas and 25 normal ovarian tissues (Li et al., 2012). Weak to no AEG-1/MTDH/LYRIC staining was detected in normal ovarian tissues. The intensity and frequency of AEG-1/MTDH/LYRIC staining was significantly higher in chemoresistant (to cisplatin) patients than that in chemosensitive patients ($p<0.0001$). AEG-1/MTDH/LYRIC expression correlated with poor differentiation ($p=0.0182$), lymph node metastasis ($p=0.0021$), and higher residual disease volume ($p<0.001$). Median progression-free survival and OS were 30.4 months and 35.28 months, respectively, in high AEG-1/MTDH/LYRIC expression group versus 63.6 months and >50 months in low AEG-1/MTDH/LYRIC expression group ($p<0.001$). A multivariate logistic regression analysis identified AEG-1/MTDH/LYRIC expression status is an independent factor in predicting a poor likelihood of response to chemotherapy (cisplatin) treatment ($p=0.0001$). Multiple studies have shown that overexpression of AEG-1/MTDH/LYRIC confers marked chemoresistance to the cells. This study establishes those observations by correlating AEG-1 expression with sensitivity of patients to chemotherapy, thus providing another aspect of using AEG-1/MTDH/LYRIC as a biomarker.

Endometrial cancer is the most common cancer of the female genital tract and typically a disease of postmenopausal women (Siegel et al., 2012). The disease is usually diagnosed early with a 5-year survival rate of more than 80% (van Wijk, van der Burg, Burger, Vergote, & van Doorn, 2009). However, there are a group of patients presenting with recurrence and metastasis for which a biomarker needs to be developed for properly assessing the prognosis of the disease. AEG-1/MTDH/LYRIC expression was analyzed by IHC in FFPE sections in 174 endometrial cancer patients that included 161 cases of endometroid carcinoma, 8 cases of serous carcinoma, and 5 cases of clear cell carcinoma (Song, Li, Lu, Zhang, & Geng, 2010). In addition, tissues from 35 healthy patients were used as control. AEG-1/MTDH/LYRIC staining gradually increased from normal to atypical hyperplasia and was highest in endometrial cancer ($p<0.001$). Advanced and invasive tumors showed nuclear AEG-1/MTDH/LYRIC staining. AEG-1/MTDH/LYRIC expression correlated with FIGO stage ($p<0.001$), depth of myometrial invasion ($p=0.015$), lymph node metastasis ($p=0.005$), lymph vascular space invasion ($p<0.001$), recurrence ($p<0.001$), and Ki-67 expression ($p=0.032$). Mean OS and disease-free survival (DFS) were 74 months and 72 months, respectively, in low AEG-1/MTDH/LYRIC-expressing group compared to 58 months and 54 months, respectively, in high AEG-1/MTDH/LYRIC-expressing group ($p<0.001$). AEG-1/MTDH/LYRIC was identified as an independent prognostic factor for poor OS and DFS by multivariate analysis.

3.3. Prostate cancer

Prostate cancer is the most frequent cancer and the second most common cause of cancer-associated death in men in the United States (Siegel et al., 2012). AEG-1/MTDH/LYRIC expression was analyzed by IHC in TMA containing 63 benign prostatic hyperplasia (BPH) and 143 prostate cancer samples. Additionally, 11 bone metastasis of prostate cancer patients were analyzed (Thirkettle et al., 2009). All tissues showed some AEG-1/MTDH/LYRIC staining. In benign tissues, predominantly nuclear AEG-1/MTDH/LYRIC staining was observed. Interestingly, tumor tissues showed low-level nuclear staining; however, nucleolar staining was observed in tumor tissues, which was not observed in benign tissues. When compared to normal bone, 9 of 11 bone metastases showed an increased expression of AEG-1/MTDH/LYRIC, which was detected exclusively in the cytoplasm and in the cell membrane. The intensity of AEG-1/MTDH/LYRIC staining was significantly increased in prostate cancer compared to BPH ($p=0.037$). Interestingly, unlike other cancers, AEG-1/MTDH/LYRIC expression was high in patients with low Gleason score and decreased in patients with high Gleason score, although these changes were not statistically significant. In 52 of 63 BPH cases (82.5%), AEG-1/MTDH/LYRIC expression was detected in the nucleus of luminal cells and some staining of basal cells was detected in benign tissue, which was lost in cancer samples. In tumors, 38 out of 143 cases (26.6%) showed nuclear AEG-1/MTDH/LYRIC staining. The decrease in nuclear staining was associated with increased Gleason grade ($p<0.001$) with reciprocal increase in cytoplasmic staining. Patients with nuclear AEG-1/MTDH/LYRIC had a mean survival of 70 months compared with 39 months for patients without any nuclear AEG-1/MTDH/LYRIC staining ($p=0.0023$). Patients who did not receive any hormone treatment showed higher AEG-1/MTDH/LYRIC expression compared to the patients who received hormone treatment ($p=0.009$) regardless of whether the patient responded to the hormone treatment or not. Although no significant change in AEG-1/MTDH/LYRIC localization was observed with hormone therapy, hormone-sensitive patients demonstrated more nuclear AEG-1/MTDH/LYRIC staining. The observations from this study suggest a differential function between cytoplasmic and nuclear AEG-1/MTDH/LYRIC. While nuclear AEG-1/MTDH/LYRIC might function in normal cells and exert a protective role against tumor formation, cytoplasmic AEG-1/MTDH/LYRIC might exert functions contributing to tumorigenesis. Indeed, studies from other cancer indications demonstrate that cytoplasmic AEG-1/MTDH/LYRIC regulates miRNA function as part of the RNA-induced silencing complex and regulates translation of specific mRNAs, such as multidrug resistance protein-1 or coagulation factor XII, which contribute to the oncogenic function of AEG-1/MTDH/LYRIC (Srivastava et al., 2012; Yoo et al., 2010; Yoo, Santhekadur, et al., 2011). AEG-1/

MTDH/LYRIC staining intensity as well as its localization might therefore be used as a biomarker for prostate cancer.

3.4. Renal cell carcinoma

Renal cell carcinoma (RCC) is the most common malignancy of adult urinary tract and accounts for ~3% of all adult malignancies (Nelson, Evans, & Lara, 2007). AEG-1/MTDH/LYRIC mRNA and protein overexpression were documented by RT-PCR and Western blot analysis in tumor tissues of eight RCC patients compared to matched normal kidney tissue (Chen et al., 2010). Analyzing AEG-1/MTDH/LYRIC expression by IHC in FFPE sections of 102 RCC patients that included 86 clear cell carcinomas, 10 papillary carcinomas, 3 chromophobe cell types, and 3 cases of cancer of the collecting duct of Bellini extended these observations. These cases also included six matched lymph node metastases and seven cases of neoplastic embolus in the renal vein. Weak AEG-1/MTDH/LYRIC expression was detected in the tubular epithelium and no expression was detected in the glomeruli of the adjacent normal kidney. However, 96 out of the 102 RCC cases showed high AEG-1/MTDH/LYRIC expression. Expression was higher in the lymph node metastases and in the neoplastic emboli, and the staining was detected in the cytoplasm. AEG-1/MTDH/LYRIC expression showed positive correlation with poorly differentiated nuclear grade of clear cell-type and papillary-type RCC with sarcomatoid areas showing the highest expression. Significant correlation was observed between AEG-1/MTDH/LYRIC and clinical stage ($p=0.026$), T classification ($p=0.013$), and M classification ($p=0.032$), while no correlation was observed with age, gender, and N classification. The cumulative 5-year survival rate was 91.3% in the low AEG-1/MTDH/LYRIC expression group, while it was 52.4% in the high AEG-1/MTDH/LYRIC expression group. The mean survival time in the low AEG-1/MTDH/LYRIC expression group was 76.98 months (95% confidence interval, 72.94–81.02), while it was 60.94 months in high AEG-1/MTDH/LYRIC expression group (95% confidence interval, 53.83–68.06) indicating that AEG-1/MTDH/LYRIC regulates advanced progression of RCC.

3.5. Bladder cancer

Bladder cancer is the second most common cancer of the urinary system (Kiriluk, Prasad, Patel, Steinberg, & Smith, 2012). AEG-1/MTDH/LYRIC expression analysis was performed in 60 cases of primary bladder carcinoma and 15 specimens of normal urothelial tissue (Zhou et al., 2012). Quantitative RT-PCR showed significantly higher levels of AEG-1/MTDH/LYRIC mRNA in cancer tissues when compared to normal tissues. In IHC analysis, all the normal samples had a staining index (SI) of 3 or less indicating negative expression, while 65% of cancer samples had a SI of 4 or more. Based on the Union for International Cancer Control (UICC) staging, 77.8% patients with positive AEG-1/MTDH/LYRIC staining belonged to the invasive (T_2–T_4) stage, while 26.7% patients with positive AEG-1/MTDH/LYRIC staining belonged to the superficial (T_a–T_1) stage indicating a statistically significant correlation ($p<0.001$). According to WHO classification, 30%, 50%, and 86.7% AEG-1/MTDH/LYRIC positively stained samples were identified in G1, G2, and G3 stages, respectively, with a significant statistical correlation ($p=0.001$). AEG-1/MTDH/LYRIC expression correlated with tumor recurrence ($p=0.015$) and with multiple tumors. Positive AEG-1/MTDH/LYRIC expression was observed in 82.4% multiple tumors versus 42.3% single tumors ($p=0.026$). AEG-1/MTDH/LYRIC expression level correlated with Ki-67 expression level ($p<0.001$), the latter also showing correlation with UICC stage, tumor recurrence, and tumor multiplicity. In a multivariate survival analysis, patients with high SI (SI > 6) had a shorter OS compared to patients with low SI (SI < 6) ($p<0.001$). The 5-year cumulative survival rate was 92.3% in the low SI group versus 81.3% in the high SI group suggesting that AEG-1/MTDH/LYRIC might be an independent prognostic marker for bladder cancer.

4. NERVOUS SYSTEM

4.1. Malignant glioma

Cancers of the nervous system account for 1.7% of new disease, representing 189,000 cases annually resulting in 142,000 deaths globally (Stupp et al., 2005). Gliomas arising from glial cells comprise more than one-third of these cases and malignant gliomas, including anaplastic astrocytoma and glioblastoma multiforme (GBM), are the most common primary tumors (Butowski, Sneed, & Chang, 2006). AEG-1/MTDH/LYRIC expression was analyzed by IHC in FFPE sections from 296 glioma patients that include 39 cases of grade 1 (13.2%), 121 cases of grade 2 (40.9%), 88 cases of grade 3 (29.7%), and 48 cases of grade 4 (16.2%) gliomas (Liu et al., 2010). AEG-1/MTDH/LYRIC staining was detected in 265 out of 296 cases (89.5%) among which 143 (48.3%) were identified as low AEG-1/MTDH/LYRIC expression and

153 (51.7%) was considered as high AEG-1/MTDH/LYRIC expression. A statistically significant difference was observed in AEG-1/MTDH/LYRIC expression level between normal brain and glioma samples ($p<0.001$). AEG-1/MTDH/LYRIC expression correlated with age of the patient showing higher expression in patients >45 years age ($p<0.001$) and the clinicopathologic grade of the patients ($p<0.001$).

In a separate study, AEG-1/MTDH/LYRIC expression was analyzed by Western blot using frozen brain tissues from 9 normal individuals and tumor samples from 25 GBM, 18 astrocytomas, 18 meningiomas, 19 oligodendrogliomas, and 18 other types of brain cancers (Emdad et al., 2010). Compared to normal brain, AEG-1/MTDH/LYRIC protein expression was increased in >90% cases with a 3- to 10-fold increase, the highest changes being observed in GBM patients. These findings were also confirmed by IHC in FFPE sections.

As yet, no study analyzed AEG-1/MTDH/LYRIC expression level with the clinical outcome and survival time of glioma patients. Association of MMP-2 and MMP-9 with AEG-1/MTDH/LYRIC was observed in glioma patients but stringent statistical analysis was not performed (Emdad et al., 2010; Liu et al., 2010).

4.2. Neuroblastoma

Neuroblastoma, a tumor of peripheral neural crest origin, is the most common extracranial solid tumor in infancy and childhood and accounts for 7–10% of pediatric malignancies and approximately 15% of pediatric cancer-related deaths (Brodeur, 2003). AEG-1/MTDH/LYRIC expression was analyzed by IHC in FFPE sections of 32 neuroblastoma patients (Liu et al., 2012). Positive AEG-1/MTDH/LYRIC staining was observed in all cases with 75% showing high expression, which was detected in vascular endothelial cells and glandula. AEG-1/MTDH/LYRIC expression strongly correlated with age ($p=0.012$), clinical stage ($p=0.030$), and histological stage ($p=0.041$) and inversely correlated with reduced survival and poor prognosis ($p=0.031$) indicating that similar to other cancers AEG-1/MTDH/LYRIC might also be a potential biomarker for neuroblastoma.

4.3. Oligodendroglioma

Oligodendroglioma is the third most common intracranial glioma after GBM and anaplastic astrocytoma (Fuller & Scheithauer, 2007). AEG-1/MTDH/LYRIC expression was analyzed by IHC in FFPE sections of

75 oligodendroglioma patients including 52 grade 2 (well-differentiated) and 23 grade 3 (anaplastic) patients (Xia et al., 2010). While AEG-1/MTDH/LYRIC was hardly detected in normal brain, its expression was detected in 51 out of 75 cases (68%), which showed strong correlation with grade ($p=0$) but no correlation with age and sex. Ki-67 and AEG-1/MTDH/LYRIC expression showed positive correlation in these patients. The median survival time of high AEG-1/MTDH/LYRIC-expressing patients was 28 months (95% confidence interval, 25.54–30.46) while that of low AEG-1/MTDH/LYRIC-expressing patients was 57 months (95% confidence interval, 46.37–67.63) ($p=0$). The cumulative 3-year survival rate was 19.69% in high AEG-1/MTDH/LYRIC-expressing group, while it was 88.05% in the low AEG-1/MTDH/LYRIC-expressing group. Multivariate survival analysis identified AEG-1/MTDH/LYRIC as an independent prognostic factor for patient outcome.

5. RESPIRATORY SYSTEM: NONSMALL CELL LUNG CANCER

Lung cancer is the leading cause of cancer-related deaths in both men and women worldwide (Ferlay et al., 2010). Nonsmall cell lung cancer (NSCLC) accounts for 80% of all lung cancers. AEG-1/MTDH/LYRIC expression was detected by IHC in 95 cases of NSCLC with nonlymphatic metastases, 105 cases with lymphatic metastases, and 20 cases of matched distant metastases derived mostly from patients with relapse of lung cancer (Song et al., 2009). AEG-1/MTDH/LYRIC expression was undetectable or detectable at a low level in the adjacent noncancerous tissues while significantly high AEG-1/MTDH/LYRIC expression was detected in all the cancerous regions ($p<0.001$). AEG-1/MTDH/LYRIC expression strongly correlated with N classification ($p=0.015$), distant metastases ($p=0.004$), and pathological differentiation ($p=0.027$). An inverse correlation was observed between high AEG-1/MTDH/LYRIC expression and OS time ($p<0.001$) with a correlation coefficient of -0.341. The survival time between high and low AEG-1/MTDH/LYRIC-expressing groups was significantly different in poorly differentiated cases ($p<0.001$) but not in well-differentiated cases, and this difference was in observed both squamous cell carcinoma and adenocarcinoma. In univariate and multivariate analyses, AEG-1/MTDH/LYRIC was identified as an independent prognostic factor for patient outcome.

In another study, AEG-1/MTDH/LYRIC expression was analyzed by IHC in FFPE sections of 67 NSCLC patients that included 27 squamous cell carcinomas and 40 adenocarcinomas. Out of these 67 cases, 32 contained corresponding normal lung tissue (Sun et al., 2012). High AEG-1/MTDH/LYRIC expression was detected in 46 cases (68.7%) mainly in the cytoplasm. More intense staining was observed in patients with poor differentiation or lymph node metastases. AEG-1/MTDH/LYRIC expression correlated with clinical staging ($p=0.048$), degree of differentiation ($p=0.023$), and lymph node metastases ($p=0.032$). The sections were also stained for MMP-2 and MMP-9, and correlation was observed between AEG-1/MTDH/LYRIC and MMP-9 and not MMP-2. Spearman correlation coefficients between AEG-1/MTDH/LYRIC and MMP-2 or MMP-9 were −0.191 ($p=0.121$) and 0.449 ($p<0.001$), respectively. An inverse correlation was observed between high AEG-1/MTDH/LYRIC level and OS ($p<0.001$). The cumulative 5-year survival rates were 46.4% and 4.8% in low and high AEG-1/MTDH/LYRIC-expressing groups, respectively.

6. HEAD AND NECK CANCER

6.1. Salivary gland carcinoma

Salivary gland carcinoma (SGC) is a relatively rare cancer accounting for <5% of head and neck cancers (Milano, Longo, Basile, Iaffaioli, & Caponigro, 2007). However, SGC is a complicated tumor containing up to 24 histological subtypes (Thompson, 2006). AEG-1/MTDH/LYRIC expression was detected by IHC in FFPE sections of 141 SGC samples including 9 histological subtypes, mucoepidermoid carcinoma, adenoid cystic carcinoma, acinar cell carcinoma, adenocarcinoma, squamous cell carcinoma, salivary duct carcinoma, and basal cell carcinoma (Liao et al., 2011). In 136 out of 141 cases (96.5%), positive AEG-1/MTDH/LYRIC staining was detected while weak or negative signals were detected in control normal tissues. AEG-1/MTDH/LYRIC expression was upregulated in all the histological subtypes and was detected mainly in the cytoplasm. Quantitative IHC analysis documented that AEG-1/MTDH/LYRIC expression in all primary SGC was significantly higher than that in normal tissues and increased with the progression of the tumor grades I–IV ($p<0.001$). Spearman correlation and χ^2-tests confirmed correlation of AEG-1/MTDH/LYRIC level with clinical stage ($p=0.001$), T classification ($p=0.008$), N classification ($p=0.008$), distant metastases ($p=0.006$), and lymph node involvement ($p=0.008$). However, no correlation was observed with age, gender, histological subtypes, and history of drinking or smoking. High AEG-1/MTDH/LYRIC expression was associated with shorter survival time ($p < 0.001$) with a correlation coefficient of −0.383. The cumulative 5-year survival rate was 78.4% (95% confidence interval, 0.665–0.903) in the low AEG-1/MTDH/LYRIC group, while it was only 45% (95% confidence interval, 0.303–0.597) in the high AEG-1/MTDH/LYRIC group. In clinical stages III–IV, high AEG-1/MTDH/LYRIC expression correlated with poor survival. However, in clinical stage I–II, no significant difference was observed between low or high AEG-1/MTDH/LYRIC and OS. Shorter OS time with high AEG-1/MTDH/LYRIC expression was observed in patients with distant metastases, while no such correlation was observed in patients without distant metastases. These findings suggest that AEG-1/MTDH/LYRIC might be a prognostic marker for patients with late stage and aggressive SGC.

6.2. Head and neck squamous cell carcinoma

Head and neck squamous cell carcinoma (HNSCC), arising from the oral cavity, oropharynx, larynx, and hypopharynx, is the sixth most common malignancy worldwide (Ferlay et al., 2010). AEG-1/MTDH/LYRIC expression was analyzed in 20 primary HNSCC cases (oral cavity: 6, larynx: 3, oropharynx: 5, hypopharynx: 6) and corresponding normal epithelial samples (Nohata et al., 2011). Compared to normal tissue, AEG-1/MTDH/LYRIC expression was significantly higher and miR-375 expression was significantly lower in the cancer tissue ($p = 0.0154$ and $p = 0.008$, respectively). A direct regulation of AEG-1/MTDH/LYRIC by miR-375 expression was documented thereby unraveling a novel mechanism of AEG-1/MTDH/LYRIC regulation in these patients.

6.3. Squamous cell carcinoma of the tongue

Squamous cell carcinoma of the tongue (TSCC) comprises ~41% of squamous cell carcinoma of the oral cavity and oropharynx (Rusthoven, Ballonoff, Raben, & Chen, 2008). The incidence of TSCC is rising and TSCC patients often present with lymphatic metastasis. AEG-1/MTDH/LYRIC expression was analyzed by IHC in 93 TSCC patients of different stages, 30 of whom also contained corresponding normal tongue tissue (Ke et al., 2012). While in normal tongue tissue, weak or negative signals for AEG-1/MTDH/LYRIC were detected, positive AEG-1/MTDH/LYRIC staining was detected in 45 of 93 (48.39%) TSCC samples. The staining intensity was significantly high in the tumor tissue when compared to the normal tongue ($p < 0.001$), and AEG-1/MTDH/LYRIC expression gradually increased with the progression of tumor grades I–IV. Both by χ^2-test and by Spearman correlation analysis, AEG-1/MTDH/LYRIC expression significantly correlated with differentiation degree ($p < 0.001$), clinical stage ($p < 0.001$), T classification ($p = 0.007$), N classification ($p = 0.012$), and lymph node involvement ($p = 0.013$). No correlation was observed between AEG-1/MTDH/LYRIC expression and age, gender, and smoking. Low AEG-1/MTDH/LYRIC expression was significantly associated with higher OS ($p = 0.004$) and multivariate Cox regression analysis identified AEG-1 as an independent prognostic marker for early TSCC with well-differentiated stages, but not in moderately or poorly differentiated stages. AEG-1/MTDH/LYRIC expression was detected mainly in the cytoplasm of the cancer cells.

7. OSTEOSARCOMA

The most commonly diagnosed primary malignancy of bone is osteosarcoma which is prevalent in children and adolescents (Damron, Ward, & Stewart, 2007). AEG-1/MTDH/LYRIC expression was analyzed in FFPE sections of 62 osteosarcoma patients and 20 normal bone samples by IHC (Wang et al., 2011). While AEG-1/MTDH/LYRIC expression was barely detectable in normal bone tissue, 51 of 62 (82.3%) showed positive staining for AEG-1/MTDH/LYRIC out of which 32 cases showed high-level expression. AEG-1/MTDH/LYRIC expression demonstrated association with clinical stage ($r = 0.547$, $p < 0.001$), tumor classification ($r = 0.489$, $p < 0.001$), metastasis ($r = 0.373$, $p = 0.003$), and poor differentiation ($r = 0.520$, $p < 0.001$). Interestingly, AEG-1/MTDH/LYRIC expression significantly associated with gender exhibiting higher expression in female patients ($r = 0.300$, $p = 0.018$). The average survival time in low AEG-1/MTDH/LYRIC expression group was 91.73 months (95% confidence interval, 77.950–105.511) while that in high AEG-1/MTDH/LYRIC expression group was 57.188 months (95% confidence interval, 44.608–70.308). Multivariate survival analysis identified AEG-1/MTDH/LYRIC as an independent prognostic factor for osteosarcoma.

8. LYMPHOMA

Contribution of AEG-1/MTDH/LYRIC to both B- and T-cell lymphomas has been reported. Diffuse large B-cell lymphoma (DLBCL) is an aggressive malignancy of mature B lymphocytes and is the most common type of lymphoma in adults accounting for 25–50% of adult non-Hodgkin lymphoma in the Western countries (Muris et al., 2005). AEG-1/MTDH/LYRIC mRNA and protein expression were analyzed by real-time PCR and Western blot, respectively, using tissues from 21 DLBCL patients and 25 patients with reactive hyperplasia of lymph nodes. A significant increase ($p < 0.0001$) in AEG-1/MTDH/LYRIC expression at an mRNA and protein level was observed in DLBCL patients compared to controls (Ge et al., 2012). AEG-1/MTDH/LYRIC expression was further analyzed in samples from 30 DLBCL patients and 15 reactive hyperplasia of the lymph nodes by IHC in FFPE sections. Little or no AEG-1/MTDH/LYRIC expression was detected in the 15 patients with reactive hyperplasia. In the cases of DLBCL, 7 cases were negative for AEG-1/MTDH/LYRIC expression while the remaining 23 (76.67%) cases showed a variable level of AEG-1/MTDH/LYRIC expression. AEG-1/MTDH/LYRIC level significantly correlated ($p < 0.05$) with the clinical staging of DLBCL patients, which was confirmed by Spearman rank correlation analysis (0.507; $p = 0.004$). However, no correlation was observed between AEG-1/MTDH/LYRIC expression and age, gender, and B symptoms of the patients.

AEG-1/MTDH/LYRIC expression was analyzed in lymph node biopsies from 129 T-cell non-Hodgkin lymphoma (T-NHL) patients and 17 control individuals. While AEG-1/MTDH/LYRIC expression was barely detectable in normal lymph node tissues, high AEG-1/MTDH/LYRIC level was detected in T-NHL patients ($p < 0.01$) (Yan et al., 2012). However, no clinicopathological correlation study with AEG-1/MTDH/LYRIC was performed.

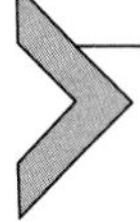

9. ANTI-AEG-1/MTDH/LYRIC ANTIBODY AS A BIOMARKER FOR CANCER

Autoantibodies against tumor-associated antigens are present in the blood of cancer patients (Caron, Choquet-Kastylevsky, & Joubert-Caron, 2007). As AEG-1/MTDH/LYRIC overexpression is detected on the membrane of cancer cells, it was hypothesized that cancer patients might have elevated levels of anti-AEG-1/MTDH/LYRIC antibody in the serum. Sera from 483 different cancer patients were analyzed by ELISA to detect anti-AEG-1/MTDH/LYRIC antibody using the lung-homing domain (aa 381–443) of human AEG-1/MTDH/LYRIC as the antigen (Chen et al., 2012). Antibody at titers of $\geq 1{:}50$ was detected in 238 of

483 (49%) cancer patients, including 44 out of 98 (45%) breast cancer patients, 48 out of 96 (50%) HCC patients, 43 out of 88 (49%) CRC patients, 51 out of 113 (45%) lung cancer patients, and 43 out of 88 (49%) gastric cancer patients. In contrast, 0 out of 230 normal individual showed serum anti-AEG-1/MTDH/LYRIC antibody ($p < 0.01$). Although no difference was observed between the antibody titer and sex or metastasis status of the patient, antibody at a titer of $\geq 1{:}50$ was detected in 168 of 287 (59%) patients <60 years old versus 70 out of 196 (36%) patients >60 years old ($p < 0.01$), suggesting that positive rate of anti-AEG-1/MTDH/LYRIC antibody decreases with age.

In stage I and II cancer patients, anti-AEG-1/MTDH/LYRIC antibody was detected in 40 out of 127 (31%) cancer patients at titers of $\geq 1{:}50$ that include 9 of 30 (30%) breast cancer patients, 8 of 31 (26%) HCC patients, 8 of 23 (35%) CRC patients, 6 of 20 (30%) lung cancer patients, and 9 of 23 (39%) gastric cancer patients. However, in stage III and IV cancer patients, anti-AEG-1/MTDH/LYRIC antibody was detected in 198 out of 356 (56%) cancer patients at titers of $\geq 1{:}50$ that include 35 of 68 (51%) breast cancer patients, 40 of 65 (62%) HCC patients, 43 of 90 (48%) CRC patients, 37 of 68 (54%) lung cancer patients, and 43 of 65 (64%) gastric cancer patients ($p < 0.01$). Thus, anti-AEG-1/MTDH/LYRIC antibody might be a marker for progression of cancer. Although exciting, this is the only study documenting the utility of anti-AEG-1/MTDH/LYRIC antibody as a biomarker and needs to be validated in other population-based studies of different geographic locations.

10. CLINICAL SIGNIFICANCE OF AEG-1/MTDH/LYRIC POLYMORPHISM

Two studies analyzed AEG-1/MTDH/LYRIC polymorphism and their link to clinicopathological status of the patients. In the first study, blood cells from 108 breast cancer patients, including 7 lobular carcinoma, 82 ductal carcinoma, and 19 classified as other subtypes, and 100 healthy women from the Han ethnic group in China were analyzed for single-nucleotide polymorphism in the AEG-1/MTDH/LYRIC gene (Liu et al., 2011). 13 variants in the control group and 11 variants in breast cancer patients were detected which were distributed throughout the gene but more frequently at the ends of the gene. Three novel variants were observed in breast cancer patients but not in the control group. Among them, one polymorphism in exon 6 (1333 C > G) causes an Asp to Glu conversion in MTDH protein, which might alter the structure and/or function of the protein. In another patient, a G/C variant (C > G) just one nucleotide prior to exon 8 changes the "GT/AG" rule for mRNA splicing and might lead to a variant form of the protein with altered functions. Among the high-frequency polymorphisms, untitled_3 occurring in intron 11 showed significant association with breast cancer patients. T/T genotype was observed in 89 control and 84 breast cancer patients, C/C genotype was observed in 2 controls and 0 cancer patients, and C/T genotype was observed in 9 control and 24 breast cancer patients accounting for a significant statistical difference ($p = 0.008$). Although the paper states that the homozygous T/T genotype increases the risk of breast cancer, based on the provided data there is more association of C/T genotype to breast cancer. The conclusion drawn by the authors thus is not clear and might reflect a flaw in data analysis or a simple typographic error.

A second polymorphism in exon 9 (1681 G > A, rs2331652) was also found to increase susceptibility to breast cancer. G/G genotype was observed in 64 control and 52 breast cancer patients, A/A genotype was observed in 5 controls and 10 cancer patients, and A/G genotype was observed in 31 control and 46 breast cancer patients. Although A/A or A/G genotype alone did not show any statistical difference, when A/A and A/G genotypes were combined, a statistically significant difference was observed ($p = 0.026$). However, neither of these polymorphisms exhibited any significant correlations with clinicopathologic parameters, such as age at diagnosis, pathological diagnosis, tumor size and grade, and nodal metastasis, indicating that these polymorphisms may not have any prognostic indication.

The same group performed a second study in 145 ovarian cancer patients and 245 age-matched healthy women of the Han ethnicity (Yuan et al., 2012). In this study, a polymorphism was identified in the promoter region of AEG-1/MTDH/LYRIC gene (−470 G > A). G/G genotype was observed in 55.5% control and 64.1% cancer patients, AA genotype was observed in 9.84% control and 3.45% cancer cases, and G/A genotype was observed in 34.6% control and 32.4% cancer patients. A statistically significant correlation ($p = 0.042$) was observed in ovarian cancer in the additive genetic model, GG versus GS versus AA. Using the dominant genetic model GG + GA versus AA, a statistically significant difference in ovarian cancer risk was observed ($p = 0.0198$). These findings indicate that the AA genotype might be protective for ovarian cancer. However, when the expression level of AEG-1/MTDH/LYRIC was analyzed by Western blot, none of these genotypes demonstrated a difference in level of expression either in the control or in the ovarian cancer patients. A polymorphism in the promoter region might indicate alteration of promoter function leading to changes in expression level. The lack of this observation raises concern regarding the functional contribution of the polymorphism in regulation of AEG-1/MTDH/LYRIC expression. Interestingly, a statistically significant correlation ($p = 0.038$) was observed between the polymorphism and the stage of the cancer although no other clinicopathologic parameters showed any significant correlation.

Those two studies were performed in a specific ethnic group in China. To establish the clinical relevance of AEG-1/MTDH/LYRIC polymorphism in disease etiology, diagnosis and prognosis studies need to be performed in large patient population of diverse genetic background.

11. CONCLUSION

The inclusion of AEG-1/MTDH/LYRIC in the 70-gene signature for risk assessment for breast cancer metastasis is the first step in the translational utilization of AEG-1/MTDH/LYRIC in clinical settings. However, AEG-1/MTDH/LYRIC has a much broader scope of application in patient care. A large number of studies in diverse cancer indications clearly demonstrate an inverse relationship between AEG-1/MTDH/LYRIC expression level and OS and recurrence of the disease. The results of these studies should pave the way for routinely analyzing AEG-1/MTDH/LYRIC expression level in tissue biopsies to determine the prognosis of the patients. One strong phenotype conferred by AEG-1/MTDH/LYRIC is chemoresistance. However, very few studies have been carried out to analyze AEG-1/MTDH/LYRIC expression status and sensitivity to chemotherapy in cancer patients. Given that bench-top laboratory analyses have elucidated the molecular mechanisms of AEG-1/MTDH/LYRIC-induced chemoresistance, stringent statistical clinical research needs to be performed to use AEG-1/MTDH/LYRIC as a screening biomarker before developing a chemotherapy-based treatment protocols (Bhutia et al., 2010; Hu et al., 2009; Yoo et al., 2010; Yoo, Gredler, et al., 2009). The identification of anti-AEG-1/MTDH/LYRIC antibody titer as a serum biomarker for aggressive cancer is an encouraging approach, and comparative studies need to be carried out to establish its specificity and selectivity versus other currently employed serum biomarkers (Chen et al., 2012). Additionally, studies need to be carried out to determine whether changes in anti-AEG-1/

MTDH/LYRIC antibody titer might be used as a marker for response to therapy and whether this titer might be periodically checked as a determinant for failure of therapy or disease recurrence. Overexpression of AEG-1/MTDH/LYRIC on the surface of cancer cells might be exploited to develop antibody-based diagnostic or therapeutic approaches. Indeed, *in vitro* proof-of-principle studies combining anti-AEG-1/MTDH/LYRIC antibody with gold nanoparticles have demonstrated the potential utility of this approach (Unak et al., 2012). More extensive future research might elevate AEG-1/MTDH/LYRIC as a central tool that could be routinely employed in a clinical diagnostic laboratory in diagnosing diverse cancers, monitoring cancer progression, defining efficacy of cancer therapy, and signaling cancer relapse after therapy.

ACKNOWLEDGMENTS

Research support is acknowledged from grants provided by the James S. McDonnell Foundation and National Cancer Institute Grant R01 CA138540 (DS), the Samuel Waxman Cancer Research Foundation (SWCRF) Grant (DS and PBF), the National Foundation for Cancer Research (PBF), and NIH Grant R01 CA134721 (PBF). PBF holds the Thelma Newmeyer Corman Chair in Cancer Research and is a SWCRF Investigator. DS is the Harrison Endowed Scholar in Cancer Research, a Blick scholar and a SWCRF Investigator. The authors declare no conflict of interest.

REFERENCES

Baygi, M. E., & Nikpour, P. (2012). Deregulation of MTDH gene expression in gastric cancer. *Asian Pacific Journal of Cancer Prevention, 13*, 2833–2836.

Bhutia, S. K., Kegelman, T. P., Das, S. K., Azab, B., Su, Z. Z., Lee, S. G., et al. (2010). Astrocyte elevated gene-1 induces protective autophagy. *Proceedings of the National Academy of Sciences of the United States of America, 107*, 22243–22248.

Blanco, M. A., Aleckovic, M., Hua, Y., Li, T., Wei, Y., Xu, Z., et al. (2011). Identification of staphylococcal nuclease domain-containing 1 (SND1) as a Metadherin-interacting protein with metastasis-promoting functions. *The Journal of Biological Chemistry, 286*, 19982–19992.

Britt, D. E., Yang, D. F., Yang, D. Q., Flanagan, D., Callanan, H., Lim, Y. P., et al. (2004). Identification of a novel protein, LYRIC, localized to tight junctions of polarized epithelial cells. *Experimental Cell Research, 300*, 134–148.

Brodeur, G. M. (2003). Neuroblastoma: Biological insights into a clinical enigma. *Nature Reviews. Cancer, 3*, 203–216.

Brown, D. M., & Ruoslahti, E. (2004). Metadherin, a cell surface protein in breast tumors that mediates lung metastasis. *Cancer Cell, 5*, 365–374.

Butowski, N. A., Sneed, P. K., & Chang, S. M. (2006). Diagnosis and treatment of recurrent high-grade astrocytoma. *Journal of Clinical Oncology, 24*, 1273–1280.

Caron, M., Choquet-Kastylevsky, G., & Joubert-Caron, R. (2007). Cancer immunomics using autoantibody signatures for biomarker discovery. *Molecular & Cellular Proteomics, 6*, 1115–1122.

Chan, A. O., Wong, B. C., & Lam, S. K. (2001). Gastric cancer: Past, present and future. *Canadian Journal of Gastroenterology, 15*, 469–474.

Chen, X., Dong, K., Long, M., Lin, F., Wang, X., Wei, J., et al. (2012). Serum anti-AEG-1 autoantibody is a potential novel biomarker for malignant tumors. *Oncology Letters*, *4*, 319–323.
Chen, W., Ke, Z., Shi, H., Yang, S., & Wang, L. (2010). Overexpression of AEG-1 in renal cell carcinoma and its correlation with tumor nuclear grade and progression. *Neoplasma*, *57*, 522–529.
Damron, T. A., Ward, W. G., & Stewart, A. (2007). Osteosarcoma, chondrosarcoma, and Ewing's sarcoma: National Cancer Data Base Report. *Clinical Orthopaedics and Related Research*, *459*, 40–47.
El-Serag, H. B. (2011). Hepatocellular carcinoma. *The New England Journal of Medicine*, *365*, 1118–1127.
Emdad, L., Sarkar, D., Lee, S. G., Su, Z. Z., Yoo, B. K., Dash, R., et al. (2010). Astrocyte elevated gene-1: A novel target for human glioma therapy. *Molecular Cancer Therapeutics*, *9*, 79–88.
Ferlay, J., Shin, H. R., Bray, F., Forman, D., Mathers, C., & Parkin, D. M. (2010). Estimates of worldwide burden of cancer in 2008: GLOBOCAN 2008. *International Journal of Cancer*, *127*, 2893–2917.
Fuller, G. N., & Scheithauer, B. W. (2007). The 2007 Revised World Health Organization (WHO) Classification of Tumours of the Central Nervous System: Newly codified entities. *Brain Pathology*, *17*, 304–307.
Ge, X., Lv, X., Feng, L., Liu, X., Gao, J., Chen, N., et al. (2012). Metadherin contributes to the pathogenesis of diffuse large B-cell lymphoma. *PLoS One*, 7, e39449.
Gnosa, S., Shen, Y. M., Wang, C. J., Zhang, H., Stratmann, J., Arbman, G., et al. (2012). Expression of AEG-1 mRNA and protein in colorectal cancer patients and colon cancer cell lines. *Journal of Translational Medicine*, *10*, 109.
Gong, Z., Liu, W., You, N., Wang, T., Wang, X., Lu, P., et al. (2012). Prognostic significance of metadherin overexpression in hepatitis B virus-related hepatocellular carcinoma. *Oncology Reports*, *27*, 2073–2079.
He, X. X., Chang, Y., Meng, F. Y., Wang, M. Y., Xie, Q. H., Tang, F., et al. (2012). MicroRNA-375 targets AEG-1 in hepatocellular carcinoma and suppresses liver cancer cell growth in vitro and in vivo. *Oncogene*, *31*, 3357–3369.
Hiyama, T., Yoshihara, M., Tanaka, S., & Chayama, K. (2007). Genetic polymorphisms and esophageal cancer risk. *International Journal of Cancer*, *121*, 1643–1658.
Hu, G., Chong, R. A., Yang, Q., Wei, Y., Blanco, M. A., Li, F., et al. (2009). MTDH activation by 8q22 genomic gain promotes chemoresistance and metastasis of poor-prognosis breast cancer. *Cancer Cell*, *15*, 9–20.
Hui, A. B., Bruce, J. P., Alajez, N. M., Shi, W., Yue, S., Perez-Ordonez, B., et al. (2011). Significance of dysregulated metadherin and microRNA-375 in head and neck cancer. *Clinical Cancer Research*, *17*, 7539–7550.
Jian-Bo, X., Hui, W., Yu-Long, H., Chang-Hua, Z., Long-Juan, Z., Shi-Rong, C., et al. (2011). Astrocyte-elevated gene-1 overexpression is associated with poor prognosis in gastric cancer. *Medical Oncology*, *28*, 455–462.
Jiang, T., Zhu, A., Zhu, Y., & Piao, D. (2012). Clinical implications of AEG-1 in liver metastasis of colorectal cancer. *Medical Oncology*, *29*, 2858–2863.
Kamangar, F., Dores, G. M., & Anderson, W. F. (2006). Patterns of cancer incidence, mortality, and prevalence across five continents: Defining priorities to reduce cancer disparities in different geographic regions of the world. *Journal of Clinical Oncology*, *24*, 2137–2150.
Kang, D. C., Su, Z. Z., Sarkar, D., Emdad, L., Volsky, D. J., & Fisher, P. B. (2005). Cloning and characterization of HIV-1-inducible astrocyte elevated gene-1, AEG-1. *Gene*, *353*, 8–15.
Ke, Z. F., He, S., Li, S., Luo, D., Feng, C., & Zhou, W. (2012). Expression characteristics of astrocyte elevated gene-1 (AEG-1) in tongue carcinoma and its correlation with poor prognosis. *Cancer Epidemiology*, *37*, 179–185.

Kikkawa, F., Kawai, M., Mizuno, K., Ishikawa, H., Kojima, M., Maeda, O., et al. (1994). Recurrence of epithelial ovarian carcinoma after clinical remission. *Gynecologic and Obstetric Investigation, 38*, 65–69.

Kiriluk, K. J., Prasad, S. M., Patel, A. R., Steinberg, G. D., & Smith, N. D. (2012). Bladder cancer risk from occupational and environmental exposures. *Urologic Oncology, 30*, 199–211.

Kochanek, D. M., & Wells, D. G. (2013). CPEB1 regulates the expression of MTDH/AEG-1 and glioblastoma cell migration. *Molecular Cancer Research, 11*(2), 149–160.

Lee, S. G., Su, Z. Z., Emdad, L., Sarkar, D., & Fisher, P. B. (2006). Astrocyte elevated gene-1 (AEG-1) is a target gene of oncogenic Ha-ras requiring phosphatidylinositol 3-kinase and c-Myc. *Proceedings of the National Academy of Sciences of the United States of America, 103*, 17390–17395.

Li, C., Li, R., Song, H., Wang, D., Feng, T., Yu, X., et al. (2011). Significance of AEG-1 expression in correlation with VEGF, microvessel density and clinicopathological characteristics in triple-negative breast cancer. *Journal of Surgical Oncology, 103*, 184–192.

Li, C., Li, Y., Wang, X., Wang, Z., Cai, J., Wang, L., et al. (2012). Elevated expression of astrocyte elevated gene-1 (AEG-1) is correlated with cisplatin-based chemoresistance and shortened outcome in patients with stages III-IV serous ovarian carcinoma. *Histopathology, 60*, 953–963.

Li, C., Liu, J., Lu, R., Yu, G., Wang, X., Zhao, Y., et al. (2011). AEG-1 overexpression: A novel indicator for peritoneal dissemination and lymph node metastasis in epithelial ovarian cancers. *International Journal of Gynecological Cancer, 21*, 602–608.

Li, J., Yang, L., Song, L., Xiong, H., Wang, L., Yan, X., et al. (2009). Astrocyte elevated gene-1 is a proliferation promoter in breast cancer via suppressing transcriptional factor FOXO1. *Oncogene, 28*, 3188–3196.

Li, J., Zhang, N., Song, L. B., Liao, W. T., Jiang, L. L., Gong, L. Y., et al. (2008). Astrocyte elevated gene-1 is a novel prognostic marker for breast cancer progression and overall patient survival. *Clinical Cancer Research, 14*, 3319–3326.

Liao, W. T., Guo, L., Zhong, Y., Wu, Y. H., Li, J., & Song, L. B. (2011). Astrocyte elevated gene-1 (AEG-1) is a marker for aggressive salivary gland carcinoma. *Journal of Translational Medicine, 9*, 205.

Liu, H. Y., Liu, C. X., Han, B., Zhang, X. Y., & Sun, R. P. (2012). AEG-1 is associated with clinical outcome in neuroblastoma patients. *Cancer Biomarkers, 11*, 115–121.

Liu, L., Wu, J., Ying, Z., Chen, B., Han, A., Liang, Y., et al. (2010). Astrocyte elevated gene-1 upregulates matrix metalloproteinase-9 and induces human glioma invasion. *Cancer Research, 70*, 3750–3759.

Liu, D. C., & Yang, Z. L. (2011). MTDH and EphA7 are markers for metastasis and poor prognosis of gallbladder adenocarcinoma. *Diagnostic Cytopathology, 41*(3), 199–205.

Liu, X., Zhang, N., Li, X., Moran, M. S., Yuan, C., Yan, S., et al. (2011). Identification of novel variants of metadherin in breast cancer. *PLoS One, 6*, e17582.

Malka, D., Boige, V., Dromain, C., Debaere, T., Pocard, M., & Ducreux, M. (2004). Biliary tract neoplasms: Update 2003. *Current Opinion in Oncology, 16*, 364–371.

Meng, X., Zhu, D., Yang, S., Wang, X., Xiong, Z., Zhang, Y., et al. (2012). Cytoplasmic Metadherin (MTDH) provides survival advantage under conditions of stress by acting as RNA-binding protein. *The Journal of Biological Chemistry, 287*, 4485–4491.

Milano, A., Longo, F., Basile, M., Iaffaioli, R. V., & Caponigro, F. (2007). Recent advances in the treatment of salivary gland cancers: Emphasis on molecular targeted therapy. *Oral Oncology, 43*, 729–734.

Miller, G., & Jarnagin, W. R. (2008). Gallbladder carcinoma. *European Journal of Surgical Oncology, 34*, 306–312.

Muris, J. J., Cillessen, S. A., Vos, W., van Houdt, I. S., Kummer, J. A., van Krieken, J. H., et al. (2005). Immunohistochemical profiling of caspase signaling pathways predicts

clinical response to chemotherapy in primary nodal diffuse large B-cell lymphomas. *Blood, 105*, 2916–2923.
Nelson, E. C., Evans, C. P., & Lara, P. N., Jr. (2007). Renal cell carcinoma: Current status and emerging therapies. *Cancer Treatment Reviews, 33*, 299–313.
Nohata, N., Hanazawa, T., Kikkawa, N., Mutallip, M., Sakurai, D., Fujimura, L., et al. (2011). Tumor suppressive microRNA-375 regulates oncogene AEG-1/MTDH in head and neck squamous cell carcinoma (HNSCC). *Journal of Human Genetics, 56*, 595–601.
Rusthoven, K., Ballonoff, A., Raben, D., & Chen, C. (2008). Poor prognosis in patients with stage I and II oral tongue squamous cell carcinoma. *Cancer, 112*, 345–351.
Sarkar, D., Emdad, L., Lee, S. G., Yoo, B. K., Su, Z. Z., & Fisher, P. B. (2009). Astrocyte elevated gene-1: Far more than just a gene regulated in astrocytes. *Cancer Research, 69*, 8529–8535.
Siegel, R., Naishadham, D., & Jemal, A. (2012). Cancer statistics, 2012. *CA: A Cancer Journal for Clinicians, 62*, 10–29.
Song, H., Li, C., Li, R., & Geng, J. (2010). Prognostic significance of AEG-1 expression in colorectal carcinoma. *International Journal of Colorectal Disease, 25*, 1201–1209.
Song, H., Li, C., Lu, R., Zhang, Y., & Geng, J. (2010). Expression of astrocyte elevated gene-1: A novel marker of the pathogenesis, progression, and poor prognosis for endometrial cancer. *International Journal of Gynecological Cancer, 20*, 1188–1196.
Song, L., Li, W., Zhang, H., Liao, W., Dai, T., Yu, C., et al. (2009). Over-expression of AEG-1 significantly associates with tumour aggressiveness and poor prognosis in human non-small cell lung cancer. *The Journal of Pathology, 219*, 317–326.
Srivastava, J., Siddiq, A., Emdad, L., Santhekadur, P., Chen, D., Gredler, R., et al. (2012). Astrocyte elevated gene-1 (AEG-1) promotes hepatocarcinogenesis: Novel insights from a mouse model. *Hepatology, 56*(5), 1782–1791.
Stupp, R., Mason, W. P., van den Bent, M. J., Weller, M., Fisher, B., Taphoorn, M. J., et al. (2005). Radiotherapy plus concomitant and adjuvant temozolomide for glioblastoma. *The New England Journal of Medicine, 352*, 987–996.
Su, Z. Z., Kang, D. C., Chen, Y., Pekarskaya, O., Chao, W., Volsky, D. J., et al. (2002). Identification and cloning of human astrocyte genes displaying elevated expression after infection with HIV-1 or exposure to HIV-1 envelope glycoprotein by rapid subtraction hybridization, RaSH. *Oncogene, 21*, 3592–3602.
Su, P., Zhang, Q., & Yang, Q. (2010). Immunohistochemical analysis of Metadherin in proliferative and cancerous breast tissue. *Diagnostic Pathology, 5*, 38.
Sun, W., Fan, Y. Z., Xi, H., Lu, X. S., Ye, C., & Zhang, J. T. (2011). Astrocyte elevated gene-1 overexpression in human primary gallbladder carcinomas: An unfavorable and independent prognostic factor. *Oncology Reports, 26*, 1133–1142.
Sun, S., Ke, Z., Wang, F., Li, S., Chen, W., Han, A., et al. (2012). Overexpression of astrocyte-elevated gene-1 is closely correlated with poor prognosis in human non-small cell lung cancer and mediates its metastasis through up-regulation of matrix metalloproteinase-9 expression. *Human Pathology, 43*, 1051–1060.
Sutherland, H. G., Lam, Y. W., Briers, S., Lamond, A. I., & Bickmore, W. A. (2004). 3D3/lyric: A novel transmembrane protein of the endoplasmic reticulum and nuclear envelope, which is also present in the nucleolus. *Experimental Cell Research, 294*, 94–105.
Thirkettle, H. J., Girling, J., Warren, A. Y., Mills, I. G., Sahadevan, K., Leung, H., et al. (2009). LYRIC/AEG-1 is targeted to different subcellular compartments by ubiquitinylation and intrinsic nuclear localization signals. *Clinical Cancer Research, 15*, 3003–3013.
Thompson, L. (2006). World Health Organization classification of tumours: Pathology and genetics of head and neck tumours. *Ear, Nose, & Throat Journal, 85*, 74.

Unak, G., Ozkaya, F., Medine, E. I., Kozgus, O., Sakarya, S., Bekis, R., et al. (2012). Gold nanoparticle probes: Design and in vitro applications in cancer cell culture. *Colloids and Surfaces. B, Biointerfaces*, *90*, 217–226.

van Wijk, F. H., van der Burg, M. E., Burger, C. W., Vergote, I., & van Doorn, H. C. (2009). Management of recurrent endometrioid endometrial carcinoma: An overview. *International Journal of Gynecological Cancer*, *19*, 314–320.

Wang, N., Du, X., Zang, L., Song, N., Yang, T., Dong, R., et al. (2012). Prognostic impact of Metadherin-SND1 interaction in colon cancer. *Molecular Biology Reports*, *39*, 10497–10504.

Wang, F., Ke, Z. F., Sun, S. J., Chen, W. F., Yang, S. C., Li, S. H., et al. (2011). Oncogenic roles of astrocyte elevated gene-1 (AEG-1) in osteosarcoma progression and prognosis. *Cancer Biology & Therapy*, *12*, 539–548.

Xia, Z., Zhang, N., Jin, H., Yu, Z., Xu, G., & Huang, Z. (2010). Clinical significance of astrocyte elevated gene-1 expression in human oligodendrogliomas. *Clinical Neurology and Neurosurgery*, *112*, 413–419.

Yan, J., Zhang, M., Chen, Q., & Zhang, X. (2012). Expression of AEG-1 in human T-cell lymphoma enhances the risk of progression. *Oncology Reports*, *28*, 2107–2114.

Yang, Y., Wu, J., Guan, H., Cai, J., Fang, L., Li, J., et al. (2012). MiR-136 promotes apoptosis of glioma cells by targeting AEG-1 and Bcl-2. *FEBS Letters*, *586*, 3608–3612.

Yoo, B. K., Chen, D., Su, Z.-Z., Gredler, R., Yoo, J., Shah, K., et al. (2010). Molecular mechanism of chemoresistance by Astrocyte Elevated Gene-1 (AEG-1). *Cancer Research*, *70*, 3249–3258.

Yoo, B. K., Emdad, L., Lee, S.-G., Su, Z.-Z., Santhekadur, P. K., Chen, D., et al. (2011). Astrocyte elevated gene-1 (AEG-1): A multifunctional regulator of normal and abnormal physiology. *Pharmacology & Therapeutics*, *130*, 1–8.

Yoo, B. K., Emdad, L., Su, Z. Z., Villanueva, A., Chiang, D. Y., Mukhopadhyay, N. D., et al. (2009). Astrocyte elevated gene-1 regulates hepatocellular carcinoma development and progression. *The Journal of Clinical Investigation*, *119*, 465–477.

Yoo, B. K., Gredler, R., Vozhilla, N., Su, Z. Z., Chen, D., Forcier, T., et al. (2009). Identification of genes conferring resistance to 5-fluorouracil. *Proceedings of the National Academy of Sciences of the United States of America*, *106*, 12938–12943.

Yoo, B. K., Santhekadur, P. K., Gredler, R., Chen, D., Emdad, L., Bhutia, S. K., et al. (2011). Increased RNA-induced silencing complex (RISC) activity contributes to hepatocelllular carcinoma. *Hepatology*, *53*, 1538–1548.

Yu, C., Chen, K., Zheng, H., Guo, X., Jia, W., Li, M., et al. (2009). Overexpression of astrocyte elevated gene-1 (AEG-1) is associated with esophageal squamous cell carcinoma (ESCC) progression and pathogenesis. *Carcinogenesis*, *30*, 894–901.

Yuan, C., Li, X., Yan, S., Yang, Q., Liu, X., & Kong, B. (2012). The MTDH (−470G>A) polymorphism is associated with ovarian cancer susceptibility. *PLoS One*, 7, e51561.

Zhang, B., Liu, X. X., He, J. R., Zhou, C. X., Guo, M., He, M., et al. (2011). Pathologically decreased miR-26a antagonizes apoptosis and facilitates carcinogenesis by targeting MTDH and EZH2 in breast cancer. *Carcinogenesis*, *32*, 2–9.

Zhang, F., Yang, Q., Meng, F., Shi, H., Li, H., Liang, Y., & Han, A. (2012). Astrocyte elevated gene-1 interacts with beta-catenin and increases migration and invasion of colorectal carcinoma. *Mol Carcinog*, in press.

Zhou, J., Li, J., Wang, Z., Yin, C., & Zhang, W. (2012). Metadherin is a novel prognostic marker for bladder cancer progression and overall patient survival. *Asia-Pacific Journal of Clinical Oncology*, *8*, e42–e48.

Zhu, K., Dai, Z., Pan, Q., Wang, Z., Yang, G. H., Yu, L., et al. (2011). Metadherin promotes hepatocellular carcinoma metastasis through induction of epithelial-mesenchymal transition. *Clinical Cancer Research*, 17, 7294–7302.

CHAPTER THREE

AEG-1/MTDH/LYRIC: Signaling Pathways, Downstream Genes, Interacting Proteins, and Regulation of Tumor Angiogenesis

Luni Emdad[*,†,1], **Swadesh K. Das**[*], **Santanu Dasgupta**[*,†], **Bin Hu**[*], **Devanand Sarkar**[*,†,‡], **Paul B. Fisher**[*,†,‡,1]

[*]Department of Human and Molecular Genetics, Virginia Commonwealth University, School of Medicine, Richmond, Virginia, USA
[†]VCU Massey Cancer Center, Virginia Commonwealth University, School of Medicine, Richmond, Virginia, USA
[‡]VCU Institute of Molecular Medicine, Virginia Commonwealth University, School of Medicine, Richmond, Virginia, USA
[1]Corresponding authors: e-mail address: lemdad@vcu.edu; pbfisher@vcu.edu

Contents

Advances in Cancer Research, Volume 120
ISSN 0065-230X
http://dx.doi.org/10.1016/B978-0-12-401676-7.00003-6

Abstract

Astrocyte elevated gene-1 (AEG-1), also known as metadherin (MTDH) and lysine-rich CEACAM1 coisolated (LYRIC), was initially cloned in 2002. AEG-1/MTDH/LYRIC has emerged as an important oncogene that is overexpressed in multiple types of human cancer. Expanded research on AEG-1/MTDH/LYRIC has established a functional role of this molecule in several crucial aspects of tumor progression, including transformation, proliferation, cell survival, evasion of apoptosis, migration and invasion, metastasis, angiogenesis, and chemoresistance. The multifunctional role of AEG-1/MTDH/LYRIC in tumor development and progression is associated with a number of signaling cascades, and recent studies identified several important interacting partners of AEG-1/MTDH/LYRIC in regulating cancer promotion and other biological functions. This review evaluates the current literature on AEG-1/MTDH/LYRIC function relative to signaling changes, interacting partners, and angiogenesis and highlights new perspectives of this molecule, indicating its potential as a significant target for the clinical treatment of various cancers and other diseases.

1. INTRODUCTION

Cancer progression is driven by uncontrolled cell growth and is invariably associated with genetic alterations linked to the regulation of proliferation, cell death, apoptosis, and genetic stability, such as tumor suppressor genes, oncogenes, growth factors, and cell adhesion molecules, which vary among different cancer types (Fisher, 1984; Gupta & Massague, 2006; Hanahan & Weinberg, 2000, 2011). Over the past several decades, significant advances in cancer research have led to the identification of a comprehensive list of oncogenes, tumor suppressors, and signaling pathways that are potential targets for anticancer therapeutics. Astrocyte elevated gene-1 (*AEG-1*), also known as metadherin (MTDH) and lysine-rich CEACAM1 coisolated (LYRIC), a novel gene that was cloned only a decade ago, is now firmly established as an oncogene in a wide array of cancer indications, reviewed in Hu, Wei, and Kang (2009) and Yoo, Emdad, et al. (2011). The increased mRNA levels of AEG-1/MTDH/LYRIC may be a direct result of the copy gain (or amplification) of the locus covering the gene, or the Ha-ras-driven activation of the PI3K/Akt pathway, which eventually leads to the recruitment of c-Myc to the AEG-1/MTDH/LYRIC promoter region (Hu, Wei, et al., 2009; Yoo, Emdad, et al., 2011). In essence, AEG-1/MTDH/LYRIC is oncogenic because it plays important roles in the activation of diverse signaling pathways (Akt, NF-κB, and Wnt) involved in cancer proliferation, survival, and invasion (overviewed in Figs. 3.1 and 3.2) in association with turning on epithelial to mesenchymal transition (EMT) and angiogenesis. Through overexpression (by vectors) or repression (by RNAi technology) of AEG-1/MTDH/LYRIC in cancer cell lines, studies have demonstrated that the gene is involved in the increased expression of oncogenes (e.g., CCND1, CTNNB1), metastasis-promoting genes (e.g., MMP9 and MMP2), and drug resistance genes (e.g., MDR1, DPYD, ABCC11, AKR1C2, ALDH3A1), or decreased expression of tumor suppressor genes (adenomatous polyposis coli (APC), CDKN1A, FOXO1, FOXO3; overviewed in Fig. 3.1). This review summarizes the role of the AEG-1/MTDH/LYRIC effectors as well as the crucial signaling pathways involved

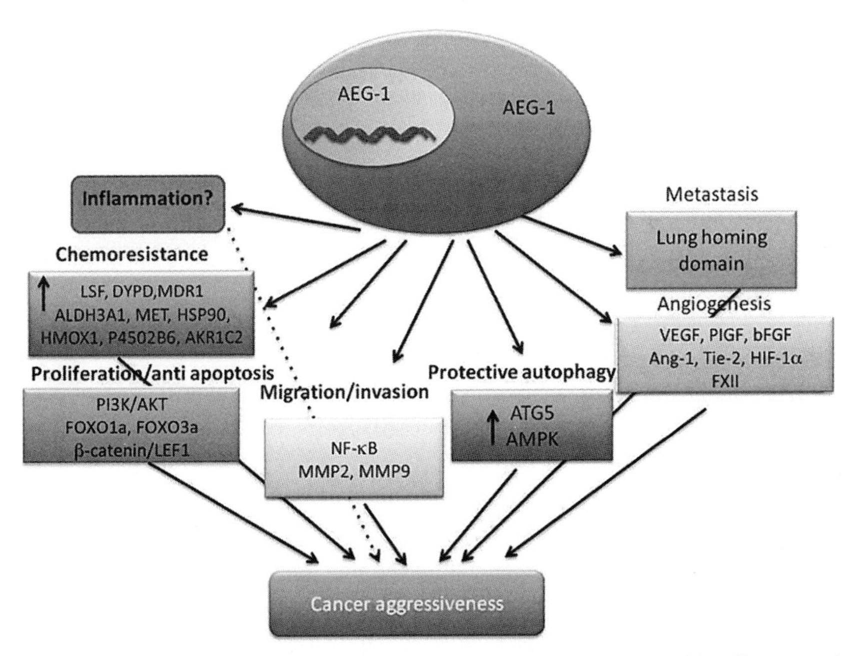

Figure 3.1 A model illustrating diverse biological functions and possible effector molecules of AEG-1/MTDH/LYRIC. Dotted line indicates an area requiring further validation. (See Page 3 in Color Section at the back of the book.)

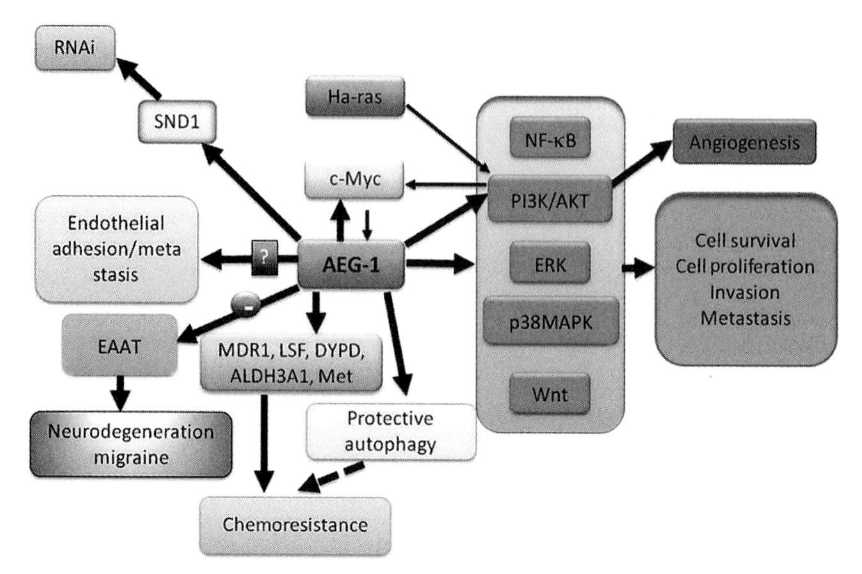

Figure 3.2 AEG-1/MTDH/LYRIC facilitates tumor progression and other functions via the integration of multiple signaling networks. Thick arrows indicate regulation by AEG-1/MTDH/LYRIC while thin arrows indicate molecular factors that regulate AEG-1/MTDH/LYRIC. Arrow with "–" indicates negative regulation. (See Page 3 in Color Section at the back of the book.)

in AEG-1/MTDH/LYRIC-mediated transformation and tumor progression. In addition, we discuss the impact of tumor angiogenesis in AEG-1/MTDH/LYRIC-mediated transformation and metastasis.

2. CLONING OF AEG-1/MTDH/LYRIC

AEG-1 was originally identified as a novel transcript induced in primary human fetal astrocytes (PHFAs) after human immunodeficiency virus (HIV)-1 infection or treatment with viral glycoprotein gp120 or tumor necrosis factor (TNF)-α (Su et al., 2003, 2002). Subsequently, using *in vivo* phage screening, Brown and Ruoslahti (2004) cloned mouse AEG-1 as a protein that allowed the specific adhesion of mouse 4T1 mammary tumor cells to lung endothelium and named it MTDH. The mouse/rat orthologue of AEG-1 was also cloned as a tight junction (TJ) protein in polarized rat epithelial cells, named LYRIC (Britt et al., 2004), and as a novel transmembrane protein that is present in endoplasmic reticulum (ER), nuclear envelope, and nucleolus, named 3D3/LYRIC (Sutherland, Lam, Briers, Lamond, & Bickmore, 2004).

3. AEG-1/MTDH/LYRIC: HIGH EXPRESSION IS LINKED TO CANCER AGGRESSIVENESS

Numerous recent studies have demonstrated that AEG-1/MTDH/LYRIC might play a pivotal role in the pathogenesis, progression, apoptosis regulation, angiogenesis, invasion, metastasis, and overall patient survival in diverse human cancers. Breast cancer, glioma, and prostate cancer were among the first tumor types in which AEG-1/MTDH/LYRIC upregulation was documented (Emdad et al., 2010; Kikuno et al., 2007; Li et al., 2008; Liu et al., 2010). Further evidence points to the fact that AEG-1/MTDH/LYRIC is located at chromosome 8q22, which is known to be a hot spot for genomic alterations in several cancer cells, including breast cancer, hepatocellular carcinoma (HCC), and malignant glioma (Hu, Chong, et al., 2009; Warr et al., 2001; Yoo, Emdad, et al., 2009). Several subsequent studies indicate that overexpression of AEG-1/MTDH/LYRIC is frequently observed in a diverse array of cancer indications studied so far including colorectal cancer (CRC), gastric and gall bladder carcinoma, ovarian and endometrial carcinoma, renal cancer, non small cell lung cancer (NSCLC) and laryngeal squamous cell carcinoma, esophageal squamous cell carcinoma, neuroblastoma, oligodendroglioma, pediatric solid tumors, T-cell lymphoma, diffuse large B cell lymphoma, head and neck cancer, salivary gland tumor, carcinoma of tongue, bladder cancer, and osteosarcoma compared to normal cells and matched nonneoplastic regions (Baygi & Nikpour, 2012; Chen, Ke, Shi, Yang, & Wang, 2010; Ge et al., 2012; Gnosa et al., 2012; Hui et al., 2011; Ke et al., 2012, 2013; Li, Liu, Lu, et al., 2011; Liao et al., 2011; Liu, Liu, Han, Zhang, & Sun, 2012; Liu et al., 2013; Liu & Yang, 2013; Orentas et al., 2012; Sarkar & Fisher, 2013; Song, Li, Lu, Zhang, & Geng, 2010; Wang et al., 2011; Xia et al., 2010; Yan, Zhang, Chen, & Zhang, 2012; Yu et al., 2009; Yuan et al., 2012; Zhou, Li, Wang, Yin, & Zhang, 2012). Moreover, recent clinical studies using a large cohort of patient samples representing diverse cancer indications have convincingly linked AEG-1/MTDH/LYRIC with tumor progression and poor clinical outcomes (Gong et al., 2012; Jiang, Zhu, Zhu, & Piao, 2012; Li et al., 2008, 2012; Sarkar & Fisher, 2013; Song, Li, Li, & Geng, 2010; Song et al., 2009; Tokunaga et al., 2012). These findings suggest that AEG-1/MTDH/LYRIC may be developed as a powerful autonomous poor-prognosis marker and a molecular target for anticancer therapeutics (Sarkar & Fisher, 2013).

4. AEG-1/MTDH/LYRIC: SIGNALING PATHWAYS

The past few years have witnessed increased interest and extensive studies on AEG-1/MTDH/LYRIC functions firmly linking major cellular signaling cascades with the ability of AEG-1/MTDH/LYRIC to implement diverse biological processes in multiple disease contexts. As a multifunctional protein, AEG-1/MTDH/LYRIC profoundly modulates a diverse array of signaling networks and effector molecules involved in tumor progression and implicated in additional disease processes such as neurodegenerative diseases (overviewed in Fig. 3.2).

4.1. Nuclear factor-κB

The nuclear factor-κB (NF-κB) transcription factor family is recognized as a fundamental mediator of the inflammatory process and a major participant in innate and adaptive immune responses (Ben-Neriah & Karin, 2011). Deregulated NF-κB signaling occurs in many pathological conditions such as autoimmune disease, chronic inflammation, and oncogenesis (Ben-Neriah & Karin, 2011). The NF-κB proteins found in most cells are p50/NF-κB1, p65/RelA, c-Rel, NF-κB2/p52, and Rel B (Ghosh, May, & Kopp, 1998). The cloning of NF-κB1/p105/p50 and p65/RelA and establishing their association with the avian viral oncoprotein v-Rel was the first evidence linking NF-κB to cancer (Gilmore, 2003; Karin, Cao, Greten, & Li, 2002; Nolan, Ghosh, Liou, Tempst, & Baltimore, 1991). Several well-established oncogenes, such as Ras, Rac, and Bcr-Abl, are inducers of NF-κB activity, and constitutive activation of NF-κB has been observed in a diverse array of human cancers, including but not restricted to pancreatic, gastric, colon, hepatocellular, breast, ovarian, and head and neck carcinomas, melanoma, leukemias, lymphomas, and Hodgkin's disease (Karin et al., 2002). However, oncogenic mutations that provide RelA, c-Rel, or other NF-κB proteins with transforming potential were found to be relatively rare and mainly occur in malignancies of lymphoid cells (Gilmore, 2003). It has been hypothesized that NF-κB activation in cancer may be the result of either exposure to proinflammatory stimuli in the tumor microenvironment or mutational activation of upstream components of IKK–NF-κB signaling pathways. NF-κB is regulated by two main pathways: the canonical and noncanonical pathways (Hayden & Ghosh, 2008). In resting cells, NF-κB is sequestered in the cytoplasm in complex with its inhibitor, IκB. These IκB proteins include IκBα, IκBβ, IκBε,

IκB ζ, and BCL-3, among others, and are defined by their ankyrin repeat domains (Hayden & Ghosh, 2004, 2011, 2012; Prasad, Ravindran, & Aggarwal, 2010). Ligand–receptor interactions lead to sequential activation of upstream signaling molecules, transmitting activation signals to the IκB kinase (IKK) complex. This multimeric protein complex comprises two catalytic subunits, IKKα and IKKβ, and one regulatory subunit, IKKγ, together with the Hsp90 chaperone complex (Salminen, Paimela, Suuronen, & Kaarniranta, 2008). Activated IKKα and IKKβ phosphorylate IκB on its N-terminal serine residues, resulting in its ubiquitination and subsequent proteasomal degradation (Kanarek & Ben-Neriah, 2012). Consequently, the released NF-κB heterodimer migrates into the nucleus. Following nuclear translocation, NF-κB binds to consensus NF-κB sequences in the promoter of diverse target genes, thereby augmenting their transcription.

Activated NF-κB increases the expression of numerous antiapoptotic genes including c-IAP1, c-IAP2, Bcl-xl, Survivin, XIAP, and Bcl-2. NF-κB activation also plays a critical role in the cell cycle by regulating certain genes such as cyclin D1, c-Myc, CDK2, and cyclin E (Ling & Kumar, 2012). Several genes involved in cancer invasion and metastasis are also controlled by NF-κB such as MMP2, VCAM-1, ICAM-1, uPA, iNOS, etc. (Ling & Kumar, 2012). Recent research has further revealed the link between NF-κB and many proinflammatory genes, such as TNF-α, COX2, MCP1, and E-selectin (Marrogi et al., 2000; Noguchi et al., 1996; van der Saag, Caldenhoven, & van de Stolpe, 1996). NF-κB is also crucially involved in tumor angiogenesis by regulation of proangiogenic molecules such as vascular endothelial growth factor (VEGF), hypoxia-inducible factor (HIF-1) α, CXCL1/8, IL-8, and TNF (Kumar, Takada, Boriek, & Aggarwal, 2004; Noguchi et al., 1996; Prasad et al., 2010; van der Saag et al., 1996).

NF-κB is the first signaling pathway showing activation by AEG-1/MTDH/LYRIC (Emdad et al., 2006; Sarkar et al., 2008). AEG-1/MTDH/LYRIC activates the NF-κB pathway by facilitating IκB degradation and by increasing binding of the transcriptional activator p50/p65 complex in the nucleus. NF-κB complex transcriptionally regulates several genes involved in adhesion, invasion, metastasis, and angiogenesis. A gene array analysis revealed that ectopic expression of AEG-1/MTDH/LYRIC by Ad.*AEG*-1 infection resulted in marked upregulation of NF-κB-responsive cell adhesion molecules (ICAM-2 and ICAM-3, selectin E, selectin L, and selectin P ligand), TLR4 and TLR5, FOS, JUN, and cytokines, such as IL-8 (Emdad et al., 2006). These proteins are important in mediating

NF-κB-induced proliferation, angiogenesis, and inflammation, all required for the carcinogenic process, indicating that activation of NF-κB also mediates multiple aspects of AEG-1/MTDH/LYRIC function. In HeLa cells and human malignant glioma cells, treatment with TNF-α results in AEG-1/MTDH/LYRIC translocation into the nucleus where it interacts with the p65 subunit of NF-κB and augments NF-κB-induced gene expression (Emdad et al., 2006; Sarkar et al., 2008). Inhibition of NF-κB pathway using an IκB super-repressor (Ad.*IκBα-mt32*) significantly reverted AEG-1/MTDH/LYRIC-induced agar cloning efficiency and invasion in human glioma cells (Sarkar et al., 2008). Functional analysis revealed a role of AEG-1/MTDH/LYRIC in invasion, migration, and NF-κB-activating properties that were mediated by the NH_2-terminal 71 amino acids. Activation of NF-κB by AEG-1/MTDH/LYRIC has also been reported in human prostate and liver cancer cells, as well as in NSCLC (Kikuno et al., 2007; Yoo, Emdad, et al., 2009). AEG-1/MTDH/LYRIC also promotes EMT in breast cancer cells in an NF-κB-dependent manner (Li, Kong, et al., 2011).

A potential role of AEG-1/MTDH/LYRIC in inflammatory processes is suggested by the observation that this gene activates NF-κB, which is a key regulator of proinflammatory cytokine activation. A recent study demonstrated that AEG-1/MTDH/LYRIC was induced via the NF-κB pathway in U937 human promonocytic cells following lipopolysaccharide (LPS) stimulation (Khuda et al., 2009). Inhibition of AEG-1/MTDH/LYRIC expression abrogated LPS-induced TNF-α and prostaglandin E2 production. Another recent study showed that LPS also upregulates the expression of AEG-1/MTDH/LYRIC in a number of breast cancer lines (Zhao et al., 2011). Stable knockdown of AEG-1/MTDH/LYRIC by shRNA in the human breast cancer cell line MDA-MB-231 abolished LPS-induced cell migration and invasion and also diminished NF-κB activation by LPS and inhibited LPS-induced IL-8 and MMP9 production. These findings suggest that AEG-1/MTDH/LYRIC might be a target molecule for the therapy of LPS-related diseases such as septic shock and systemic inflammatory response syndrome as well as in cancer-related inflammation.

4.2. PI3K/Akt pathway

The phosphatidylinositol 3-kinase (PI3K)/Akt signaling pathway is a key regulator of physiological cell processes, which include proliferation, differentiation, apoptosis, motility, metabolism, and autophagy (Engelman, Luo, & Cantley,

2006; Hennessy, Smith, Ram, Lu, & Mills, 2005). PI3K/Akt signaling is aberrantly upregulated in many cancers where it negatively influences prognosis (Cantley, 2002). The PI3K enzymes are involved in the phosphorylation of membrane inositol lipids (Vivanco & Sawyers, 2002). The activation of PI3K generates the second messenger phosphatidylinositol-3,4,5-trisphosphate (PIP3) from phosphatidylinositol 4,5-bisphosphate. This recruits proteins to the cell membrane, including the Akt/PKB kinases, resulting in their phosphorylation by phosphoinositide-dependent kinase 1 (PDK1) and PDK2 (Engelman et al., 2006; Yap et al., 2008). Activated Akt translocates to the cytoplasm and nucleus and activates downstream targets involved in survival, proliferation, cell cycle progression, growth, migration, and angiogenesis (Bellacosa, Kumar, Di Cristofano, & Testa, 2005; Yang et al., 2004). Akt is negatively regulated by the phosphatase and tensin homolog deleted on chromosome 10 (PTEN), a tumor suppressor gene that dephosphorylates PIP3 (Bartholomeusz & Gonzalez-Angulo, 2012).

Following activation, Akt modulates the function of numerous substrates involved in the regulation of cell survival, cell cycle progression, and cellular growth (Testa & Bellacosa, 2001). Akt phosphorylates and inactivates Bad, a proapoptotic member of the Bcl-2 family, thereby promoting survival (Vivanco & Sawyers, 2002). Akt inhibits the catalytic function of caspase-9 by phosphorylation and reduces its proapoptotic activity (Datta, Brunet, & Greenberg, 1999). Forkhead family of transcription factors (FOXO) is also well-established substrates of Akt that induce the expression of proapoptotic factors such as Fas ligand (Brunet et al., 1999). Akt also phosphorylates Ikk, which in turn degrades IκB. This eventually leads to nuclear translocation of NF-κB heterodimers and activation of NF-κB-regulated target genes. Glycogen synthase kinase-3 (GSK3), mammalian target of rapamycin (mTOR), insulin receptor substrate-1, the Forkhead family member FKHR, cyclin-dependent kinase inhibitors $p21^{Cip1/Waf-1/mda-6}$ and $p27^{KIP1}$, and, possibly, Raf1 are all Akt targets involved in protein synthesis, glycogen metabolism, and cell cycle regulation (Blume-Jensen & Hunter, 2001). Akt phosphorylates the CDK inhibitors, p21 and p27, and inhibits their antiproliferative effects (Liang et al., 2002; Zhou et al., 2001). Additionally, phosphorylation of MDM2 by Akt promotes the degradation of p53, leading to enhanced cell cycle activity in the G1/S phase (Sherr & Weber, 2000).

The second major signaling pathway activated by AEG-1/MTDH/LYRIC is the PI3K/Akt pathway. Lee, Su, Emdad, Sarkar, and Fisher (2006) first demonstrated that AEG-1/MTDH/LYRIC is a downstream target gene of Ha-ras, and this induction was attenuated by treatment with LY294002 or PTEN overexpression, indicating that activation of the PI3K signaling pathway regulates Ha-ras-mediated AEG-1/MTDH/LYRIC induction. Subsequent promoter mapping data indicate that Ha-ras increases binding of c-Myc to the E-box elements in the AEG-1/MTDH/LYRIC promoter through the PI3K/Akt/GSK3β/c-Myc pathway to contribute to Ha-ras-mediated oncogenesis through AEG-1/MTDH/LYRIC (Lee et al., 2006). Intriguingly, AEG-1/MTDH/LYRIC overexpression also increases phosphorylation of Akt and GSK3β, with subsequent c-Myc stabilization and MDM2 phosphorylation, leading to modulation of a number of additional Akt downstream factors that are crucial for cellular proliferation and survival (Lee et al., 2008). In neuroblastoma cells, overexpression of AEG-1/MTDH/LYRIC activates the PI3K/Akt pathways and stabilizes N-myc (Lee, Jeon, et al., 2009). Inhibition of AEG-1/MTDH/LYRIC by knockdown approach was shown to induce apoptosis through the upregulation of FOXO3a activity in prostate cancer cells (Kikuno et al., 2007) and FOXO1 activity in breast cancer cells (Li et al., 2009) via Akt signaling, respectively. In esophageal cancer cells, activation of Akt by AEG-1/MTDH/LYRIC leads to upregulation of cyclin D1 and downregulation of p27 (Yu et al., 2009). Activation of the PI3K/Akt pathway by AEG-1/MTDH/LYRIC led to an increase in multidrug resistance gene 1 (MDR1) levels by increased association of MDR1 mRNA to polysomes in drug-resistant human HCC cells (Yoo et al., 2010). In NSCLC, AEG-1/MTDH/LYRIC significantly increased the levels of PI3K, p110, and Akt phosphorylation and inhibited apoptosis by modulating caspase-3 and Bcl-2 (Ke et al., 2013). AEG-1/MTDH/LYRIC also mediates invasive ability of NSCLC via the NF-κB and PI3K/Akt pathways (Song et al., 2009). The activation of PI3K/Akt prosurvival pathways via upregulation of PIP3 has also been shown in endometrial cancer cells, and inhibition of AEG-1/MTDH/LYRIC reversed this signaling event and made cells more sensitive to TRAIL- and HDAC inhibitor-induced cell death (Meng et al., 2011).

In addition to growth-promoting activity, the PI3K/Akt signaling pathway is also involved in the regulation of angiogenesis through the control of VEGF, HIF-1, and thrombospondin (Blancher, Moore, Robertson, & Harris, 2001; Jiang et al., 2001). Once activated by the upstream growth factor/oncogene or by the loss of PTEN function, the PI3K/Akt signaling leads to upregulation of VEGF expression either directly or indirectly through increased expression of HIF-1α. The PI3K/Akt pathway also regulates AEG-1/MTDH/LYRIC-induced angiogenesis (Emdad et al., 2009). Akt activation is essential for AEG-1/MTDH/LYRIC-mediated endothelial cell tube formation, upregulation of HIF-1α in human umbilical vein endothelial cell (HUVEC), and activation of the VEGF promoter in human glioma cells. Additionally, Noch, Bookland, and Khalili (2011) demonstrated that AEG-1/MTDH/LYRIC was induced by hypoxia via the PI3K/Akt pathway and then AEG-1/MTDH/LYRIC feeds back to activate PI3K and create a positive feedback loop. Thus, PI3K/Akt activation mediates multiple aspects of AEG-1/MTDH/LYRIC function. However, the molecular mechanism of PI3K/Akt pathway activation by AEG-1/MTDH/LYRIC remains to be determined.

4.3. Wnt and mitogen-activated protein kinase pathways

The Wnt/β-catenin signaling pathway is another significant pathway that regulates cell proliferation, migration, and differentiation, thus making it an influential regulator of embryonic development and tumorigenesis (Clevers & Nusse, 2012). The Wnt/β-catenin pathway is an important protumorigenic pathway for many cancers (Clevers, 2006). Wnt proteins, which are secreted glycoproteins, bind to the low-density lipoprotein receptor-related protein5/6 (LRP5/6) and Frizzled (FZD), a seven-pass transmembrane receptor protein, to activate the Wnt/β-catenin signaling pathway (MacDonald, Tamai, & He, 2009; Polakis, 2007). In the absence of Wnts, β-catenin is sequestered in a complex that consists of the APC tumor suppressor, axin, GSK3β, and casein kinase 1 (CK1). This complex formation induces the phosphorylation of β-catenin by CK1 and GSK3β, which results in the ubiquitination and the subsequent degradation of β-catenin by the 26S proteasome. Conversely, when Wnt proteins are secreted from cells, they can form a ternary complex with FZD and LRP5/6 receptors, which results in the inhibition of GSK3β and the stabilization of cytosolic β-catenin. β-Catenin then translocates into the nucleus where it interacts with T-cell factor/lymphoid-enhancing factor (TCF/LEF). Binding of β-catenin to LEF/TCF proteins leads to transcription of a number of Wnt-responsive genes that regulate cell proliferation and migration such as c-Myc, cyclin D1, and members of the WISP family (Tanaka et al., 2003). In addition, mice demonstrating activated β-catenin signaling upregulate epidermal growth factor receptor (EGFR)

(Tan et al., 2005) and genes involved in glutamine metabolism, including glutamine synthetase, ornithine aminotransferase, and glutamate transporter-1 (GLT-1) (Cadoret et al., 2002).

AEG-1/MTDH/LYRIC has also been associated with the Wnt/β-catenin pathway in several cancer indications (Yoo, Emdad, et al., 2009). In HCC cells, AEG-1/MTDH/LYRIC was found to activate Wnt/β-catenin signaling by activating ERK42/44 and upregulating LEF1/TCF1, the ultimate executor of the Wnt pathway (Yoo, Emdad, et al., 2009). The importance of Wnt signaling in mediating AEG-1/MTDH/LYRIC action was demonstrated by inhibiting LEF1 that resulted in significant attenuation of AEG-1/MTDH/LYRIC-induced invasion of HCC cells. In gastric carcinoma, inhibition of AEG-1/MTDH/LYRIC downregulates LEF1 and cyclin D1 proteins, two critical downstream effectors in Wnt/β-catenin pathway involved in cell survival (Jian-bo et al., 2011). A recent study by Zhang et al. suggests that AEG-1/MTDH/LYRIC promotes invasion and metastasis of CRC by activation of β-catenin signaling pathway. This study documented a positive correlation between high AEG-1/MTDH/LYRIC expression and β-catenin nuclear expression in CRC, and overexpression of AEG-1/MTDH/LYRIC dramatically increased nuclear β-catenin accumulation in CRC cell lines (Zhang et al., 2012). The role of AEG-1/MTDH/LYRIC in activation of the Wnt/β-catenin pathway has also been documented in diffuse B cell lymphoma (Ge et al., 2012).

Aberrant activation of mitogen-activated protein kinase (MAPK) pathway is frequently observed in cancers and usually activated by growth factors, hormones, and chemokines. Once activated, MAPK is phosphorylated by MAPK kinases, MEK1 and MEK2, that recognize and phosphorylate tyrosine and threonine residues in the Thr-X-Tyr activation loop of the MAPKs, also known as ERK1 and ERK2 (Downward, 2003; Katz, Amit, & Yarden, 2007). A recent study indicates that specific inhibitors of the MAPK pathway are able to abolish the oncogenic effect of AEG-1/MTDH/LYRIC, namely, invasion and anchorage-independent growth, indicating that AEG-1/MTDH/LYRIC also plays an oncogenic role in tumor development and progression through activation of the MAPK signal pathway (Yoo, Emdad, et al., 2009). Another recent study showed that knockdown of AEG-1/MTDH/LYRIC can enhance the sensitivity of breast cancer cells to a novel ATP-noncompetitive inhibitor of MAP/ERK kinase via regulating FOXO3 activity and expression (Kong, Moran, Zhao, & Yang, 2012).

4.4. Regulation of glutamate signaling

Glutamate is the main excitatory amino acid transmitter in the mammalian central nervous system, and excessive glutamate exposure is toxic to neurons (Choi, 1988). Accumulation of excess extracellular glutamate in the synaptic cleft results in increased production of reactive and excitotoxic oxygen/ nitrogen species, which induce oxidative stress leading to neuronal death. Five excitatory amino acid transporters have been identified and cloned, which include EAAT1 (GLAST), EAAT2 (GLT-1), EAAT3, EAAT4, and EAAT5. EAAT2 is the most abundant glutamate transporter in brain, which plays an important role in keeping the glutamate concentration low by removing the glutamate released at the synapse. Excitotoxicity caused by impaired glutamate uptake by glial cells has been implicated in various neurodegenerative disease conditions such as ischemia/stroke, epilepsy, amyotrophic lateral sclerosis, traumatic brain injury (TBI), and HIV-associated dementia, and also in psychiatric disorders such as depression and schizophrenia (Choi, 1988; Doble, 1999). AEG-1/MTDH/LYRIC was originally isolated as a novel HIV-1- and TNF-α-induced transcript from PHFA (Su et al., 2003, 2002). However, the role of AEG-1/ MTDH/LYRIC in HIV pathogenesis still remains unknown. Previously, an interesting inverse correlation between expression levels of AEG-1/ MTDH/LYRIC and EAAT2 was documented (Kang et al., 2005). AEG-1/MTDH/LYRIC expression is elevated following HIV-1 and TNF-α treatment of astrocytes, whereas EAAT2 expression is down-regulated. Additionally, a recent study by Lee et al. (2011) suggests that AEG-1/MTDH/LYRIC also contributes to glioma-induced neurodegeneration. A strong negative correlation between expression of AEG-1/MTDH/LYRIC and EAAT2 was noticed in normal brain tissues and glioma patient samples. Gain and loss of function studies in PHFA and T98G cells revealed that AEG-1/MTDH/LYRIC repressed EAAT2 expression at a transcriptional level by inducing YY1 activity to inhibit CBP function as a coactivator on the EAAT2 promoter. Moreover, AEG-1/MTDH/LYRIC overexpression in glioma impairs glutamate uptake by reducing EAAT2 expression, culminating in glioma-induced neurodegeneration (Lee et al., 2011).

Reactive astrogliosis is frequently linked to CNS insults such as ischemia, trauma, infection, neurodegeneration, and postneurosurgical healing, commonly associated with the management of brain tumors (Eddleston & Mucke, 1993; Maragakis & Rothstein, 2006). EAAT2-positive cells were shown to be significantly decreased for a prolonged survival period following traumatic injury of human brain (van Landeghem, Weiss, Oehmichen, & von Deimling, 2006). This molecular phenotype was also confirmed in a mouse model of TBI (Wei et al., 2012). Interestingly, one recent study identified a novel role of AEG-1/MTDH/LYRIC in mediating reactive astrogliosis and in regulating astrocyte responses to injury in an *in vivo* brain injury model (Vartak-Sharma & Ghorpade, 2012). This potential cross talk between AEG-1/MTDH/LYRIC and EAAT2 may play a significant role in TBI, which needs further validation.

Migraine is a common neurological disorder usually presented by severe attacks of headache associated with other symptoms like nausea, vomiting, and photophobia. Two main types of migraine are distinguished based on the presence of an aura that can precede the headache: migraine with aura (MA) or migraine without aura. One recently published genome-wide association study (GWAS) using data from migraine patients found evidence for a role of the AEG-1/MTDH/LYRIC gene in common migraine. In this study, a single nucleotide polymorphism (SNP) (rs1835740) was identified that was located between two interesting candidate genes: AEG-1/MTDH/LYRIC and the plasma glutamate carboxypeptidase (PGCP) gene (Anttila, Wessman, Kallela, & Palotie, 2011). An expression quantitative locus (eQTL) analysis revealed that rs1835740 most likely affects migraine through *cis*-regulation of AEG-1/MTDH/LYRIC, which in turn downregulates EAAT2. Another recent study presented GWA meta-analysis for common migraine by the Dutch Icelandic migraine genetics consortium (DICE) (Ligthart et al., 2011). In this study, they also investigated SNP rs1835740 that was found to be significantly associated with MA in the first GWAS of clinic-based populations. The SNP itself was not associated with migraine in GWA-DICE study, but the gene-based analyses provided modest support for an association of AEG-1/MTDH/LYRIC with migraine.

4.5. Autophagy signaling

Autophagy is crucial for maintaining cellular homeostasis as well as remodeling during normal development, and dysfunctions in autophagy have been associated with a variety of diseases such as cancer, inflammatory bowel diseases, and neurodegenerative diseases (Levine & Klionsky, 2004; Mathew, Karantza-Wadsworth, & White, 2007). Autophagy can be either tumor suppressive or protective. During cellular or metabolic stress, a reduced cellular ATP level is sensed by AMPK (5′-AMP-activated protein kinase). Active AMPK leads to phosphorylation and activation of the TSC1/2 complex, which inhibits mTOR activity. The inhibition of mTORC1 activity leads to the activation of a set of evolutionarily conserved autophagy-regulating proteins (Atg proteins) and starts the cascades of events of autophagy (Levine & Klionsky, 2004). Autophagy can be induced by the canonical or noncanonical pathway depending on the involvement of Beclin 1 and the class III PI3K (hVps34).

A recent study (Bhutia et al., 2010) documents that AEG-1/MTDH/LYRIC mediates protective autophagy in various normal cell types, an important regulator of cancer survival under metabolic stress and apoptosis-deficient conditions, which may underlie its significant cancer-promoting properties. AEG-1/MTDH/LYRIC protects normal cells from serum starvation-induced death through protective autophagy, and inhibition of AEG-1/MTDH/LYRIC-induced autophagy results in serum starvation-induced cell death. AEG-1/MTDH/LYRIC induces non-canonical autophagy as evidenced by an increase in expression of ATG5, and knockdown of ATG5 significantly inhibited the autophagy-related phenotypes. AEG-1/MTDH/LYRIC-induced autophagy is activated in response to a decrease in the cellular ATP/AMP ratio. Ectopic expression of AEG-1/MTDH/LYRIC resulted in diminished cellular metabolism. AEG-1/MTDH/LYRIC causes a significant increase of AMPK phosphorylation at Thr-172, which in turn phosphorylates and activates TSC2, further inhibiting the activation of downstream targets, such as mTOR and S6 kinase. Phosphorylation of mTOR and S6 proteins is substantially decreased in AEG-1/MTDH/LYRIC-overexpressing cells. Inhibition of AMPK by siAMPK or compound C decreases expression of ATG5, ultimately attenuating AEG-1/MTDH/LYRIC-induced autophagy. Autophagy also provides resistance to therapy-mediated tumor cell death. When tumor cells induce protective autophagy, inhibition of autophagy could provide a way of sensitizing tumor cells to therapy by activating apoptosis. Indeed, protective autophagy plays a crucial role in AEG-1/MTDH/LYRIC-mediated chemoresistance in cancer cells, and inhibition of AEG-1/MTDH/LYRIC results in a decrease in protective autophagy and chemosensitization of cancer cells.

5. AEG-1/MTDH/LYRIC: DOWNSTREAM MOLECULES AND INTERACTING PROTEINS

AEG-1/MTDH/LYRIC exerts its diverse function by interacting with other proteins in multiprotein complexes. Research in the past several years has identified several intracellular interacting partners of AEG-1/MTDH/LYRIC

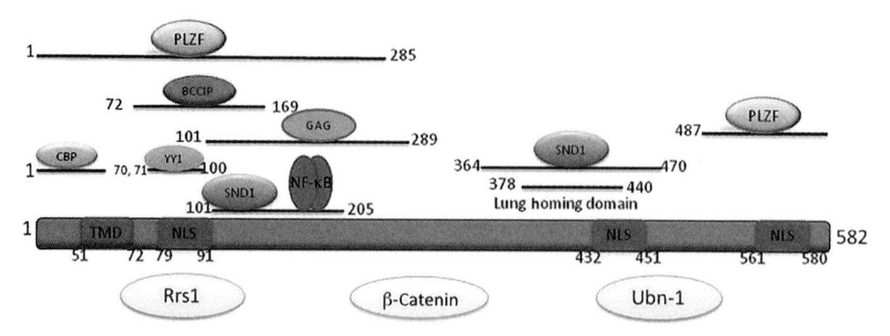

Figure 3.3 Schematic diagram of AEG-1/MTDH/LYRIC protein and the region that interacts with various proteins. Proteins interact with AEG-1/MTDH/LYRIC, but the precise regions of interaction are not defined yet are shown in light green. NLS, nuclear localization signal; TMD, transmembrane domain. (See Page 4 in Color Section at the back of the book.)

that provide significant insights into the mechanism of action of AEG-1/MTDH/LYRIC (reviewed in Fig. 3.3). In each intracellular location, AEG-1/MTDH/LYRIC interacts with specific protein(s) thus influencing diverse intracellular events.

5.1. NF-κB p65 and CBP

Activation of the NF-κB pathway by AEG-1/MTDH/LYRIC significantly contributes to AEG-1/MTDH/LYRIC-mediated oncogenic events such as invasion, migration, and anchorage-independent growth of cancer cells (Emdad et al., 2006; Sarkar et al., 2008).

In HeLa and malignant glioma cells, following infection with Ad.*AEG*-1 or treatment with TNF-α, AEG-1/MTDH/LYRIC was found to translocate into the nucleus where it interacted with the p65 subunit of NF-κB and enhanced NF-κB-induced gene expression. Activation of transcription by NF-κB requires transcriptional coactivator proteins, such as those possessing histone acetyltransferase (HAT) activity (Furia et al., 2002; Vanden Berghe, De Bosscher, Boone, Plaisance, & Haegeman, 1999; Zhong, May, Jimi, & Ghosh, 2002). HAT plays a key role in altering chromatin structure, allowing recruitment of the basal transcription factors and RNA polymerase II to initiate transcription (Huang, Ju, Hung, & Chen, 2007). NF-κB interacts with several HATs, such as CBP and its homologue p300, p300/CBP-associated factor, and members of the SRC/p160 family. These

HATs acetylate core histone proteins as well as p50–p65 NF-κB to stimulate NF-κB-dependent gene expression (Chen & Greene, 2003). Chromatin immunoprecipitation (ChIP) assays document that AEG-1/MTDH/LYRIC is located on the consensus NF-κB binding element in the IL-8 promoter together with p50–p65. Upon TNF-α treatment, AEG-1/MTDH/LYRIC interacts with p65 and cyclic AMP-responsive element-binding protein-binding protein (CBP) on the IL-8 promoter and increases IL-8 transcription. Inhibition of AEG-1/MTDH/LYRIC by siRNA does not affect recruitment of NF-κB but impedes recruitment of CBP to the IL-8 promoter indicating that AEG-1/MTDH/LYRIC functions as a bridging factor among NF-κB, CBP, and the basal transcription machinery to induce IL-8 transcription (Sarkar et al., 2008). Deletion of the NH_2-terminal 71 a.a. residues blocks AEG-1/MTDH/LYRIC-induced NF-κB activation, IL-8 expression, as well as invasion and anchorage-independent growth in malignant glioma cells. However, this region does not interact with NF-κB, rather a.a. 101–205 region was identified as the p65-interacting domain of AEG-1/MTDH/LYRIC. It might be possible that the proximal 71 a.a. region of AEG-1/MTDH/LYRIC interacts with CBP, thereby mediating NF-κB activation. Another recent study by Lee et al. (2011) documented that AEG-1/MTDH/LYRIC plays a critical role as a link between YY1 and CBP on the *EAAT2* promoter, causing YY1 to function as a negative regulator of *EAAT2* expression by inhibiting CBP. These results indicate that interactions among AEG-1/MTDH/LYRIC, YY1, and CBP are crucial for AEG-1/MTDH/LYRIC-mediated *EAAT2* repression and also suggest that AEG-1/MTDH/LYRIC functions in the nucleus as an authentic transcriptional cofactor. A further domain analysis study indicated that a.a. 1–70 and 71–100 of AEG-1/MTDH/LYRIC are responsible for interaction with CBP and YY1, respectively.

5.2. Staphylococcal nuclease domain-containing 1

Staphylococcal nuclease domain-containing 1 (SND1) has been identified as an interaction partner of AEG-1/MTDH/LYRIC by two independent approaches, including yeast two-hybrid screening using a human liver cDNA library and isolation of AEG-1/MTDH/LYRIC-interacting proteins by coimmunoprecipitation followed by mass spectrometry (MS) (Yoo, Santhekadur, et al., 2011). SND1 is localized both in the nucleus and in the cytoplasm where it functions differently. In the nucleus, it facilitates transcription as a coactivator and mRNA splicing by interacting with the spliceosome machinery (Yang et al., 2007). In the cytoplasm, SND1 functions as a nuclease in the RNA-induced silencing complex (RISC) in which small RNAs (such as siRNAs or miRNAs) are complexed with ribonucleoproteins resulting in RNAi-mediated gene silencing (Caudy et al., 2003). It has been demonstrated that AEG-1/MTDH/LYRIC interacts with SND1 via the region 101–205 a.a. in the cytoplasm but not in the nucleus. Both SND1 and AEG-1/MTDH/LYRIC functionally facilitate RISC activity (Yoo, Santhekadur, et al., 2011). Studies in HCC found that both AEG-1/MTDH/LYRIC and SND1 are overexpressed in HCC and RISC activity was found to be higher in human HCC cells compared to normal immortal hepatocytes. Moreover, inhibition of either AEG-1/MTDH/LYRIC or SND1 increased, while overexpression of AEG-1/MTDH/LYRIC or SND1 decreased the mRNA level of the tumor suppressor PTEN, a target of miRNA-221, which is overexpressed in HCC. In another study, Blanco et al. (2011) employed MS-based screen and identified SND1 as a candidate AEG-1/MTDH/LYRIC-interacting protein in breast cancer. They further mapped the region of interaction to amino acids 364–470 of the AEG-1/MTDH/LYRIC coding sequence, which is very similar to the AEG-1/MTDH/LYRIC lung-homing domain. By "gain-of-function" and "loss-of-function" studies, a role of SND1 in breast cancer metastasis to the lungs was demonstrated (Blanco et al., 2011). Additionally, Wang et al. (2012) recently reported that the expression pattern of SND1 significantly and positively correlated with that of AEG-1/MTDH/LYRIC in CRC. AEG-1/MTDH/LYRIC+/SND1+ coexpression was more significantly associated with aggressive nodal status, high pathological stage, and poor differentiation than either AEG-1/MTDH/LYRIC expression or SND1 expression.

5.3. BRCA2- and CDKN1A-interacting protein α

Using yeast two-hybrid screening, Ash, Yang, and Britt (2008) identified BCCIPα (a BRCA2- and CDKN1A ($p21^{Cip1/Waf-1/mda-6}$)-interacting protein α), a tumor suppressor gene, as an AEG-1/MTDH/LYRIC-interacting protein. Reduced BCCIPα expression has been observed in breast cancer and glioma cell lines and ectopic expression of the protein causes growth delay (Liu, Yuan, Huan, & Shen, 2001). BCCIPα binds to p21 (Cip1/Waf-1/mda-6) and enhances p21-mediated Cdk2 kinase inhibition. Coexpression of AEG-1/MTDH/LYRIC with BCCIPα in prostate tumor cells resulted in decreased BCCIPα protein levels, relative to control, suggesting that AEG-1/MTDH/LYRIC may serve as a possible negative regulator of BCCIPα. The interaction region between AEG-1/MTDH/LYRIC and BCCIPα was defined, and it was demonstrated that a.a. 72–169 in AEG-1/MTDH/LYRIC is the critical region for this interaction. However, the functional role of BCCIPα degradation in mediating AEG-1/MTDH/LYRIC function remains to be determined.

5.4. Promyelocytic leukemia zinc finger protein

Thirkettle, Mills, Whitaker, and Neal (2009) identified promyelocytic leukemia zinc finger (PLZF) protein as an interacting protein of AEG-1/MTDH/LYRIC using the yeast two-hybrid approach. PLZF is a transcriptional repressor associated with growth suppression and apoptosis (Bernardo, Yelo, Gimeno, Campillo, & Parrado, 2007). AEG-1/MTDH/LYRIC–PLZF interaction was detected in nuclear bodies and the NH_2-terminal 1–285 a.a. and COOH-terminal 487–582 a.a. of AEG-1/MTDH/LYRIC interacted with PLZF as shown by coimmunoprecipitation assays. The authors further showed that a.a. 322–404 in PLZF were sufficient to interact with AEG-1/MTDH/LYRIC, which contains two lysine residues required for activation of PLZF by SUMOylation. By using a point mutation construct where the key lysine residue for PLZF SUMOylation was mutated (K242R), the authors demonstrated that AEG-1/MTDH/LYRIC was not able to interact with PLZF, suggesting that AEG-1/MTDH/LYRIC only interacts with the active form of PLZF. Coexpression of AEG-1/MTDH/LYRIC and PLZF resulted in a significant increase in c-Myc transcript level. ChIP assay further revealed that AEG-1/MTDH/LYRIC suppressed the binding of the PLZF to the c-Myc promoter, thus causing increased transcription of c-Myc. PLZF neutralization thus might be another way, along with effects on the PI3K/Akt and Wnt signaling pathways, by which AEG-1/MTDH/LYRIC activates c-Myc. In these contexts, AEG-1/MTDH/LYRIC and c-Myc might have an intimate and supplementary role in mediating tumorigenesis.

5.5. HIV-1 Gag

HIV-1 Gag is the main structural protein driving assembly and release of HIV-1 virions from infected cells. Using a series of affinity purification experiments, Engeland et al. (2011) identified AEG-1/MTDH/LYRIC to be an HIV-1 Gag-interacting protein. Gag interacts with endogenous AEG-1/MTDH/LYRIC through its matrix (MA) and nucleocapsid domains. This interaction requires Gag multimerization and AEG-1/MTDH/LYRIC amino acids 101–289. Gag–AEG-1/MTDH/LYRIC interaction was also observed for murine leukemia virus and equine infectious anemia virus, suggesting that it represents a conserved feature among retroviruses. Expression of the Gag-binding domain of AEG-1/MTDH/LYRIC increased Gag expression levels and viral infectivity, whereas expression of an AEG-1/MTDH/LYRIC mutant lacking the Gag-binding site resulted in lower Gag expression and decreased viral infectivity. These data indicate a potential role of AEG-1/MTDH/LYRIC in regulating infectivity. Further experiments are needed to elucidate the precise purpose of this interaction.

5.6. Other interacting molecules: Rrs1, β-catenin, ubinuclein

In a knock-in mouse model of Huntington disease, regulator of ribosome synthesis 1 (Rrs1) is upregulated and might play an important role in HD pathogenesis by initiating ER stress (Carnemolla et al., 2009). In a recent study, Carnemolla et al. demonstrated that Rrs1 is localized both in the nucleolus and in the ER of neurons. By yeast two-hybrid screening, they identified AEG-1/MTDH/LYRIC as Rrs1 interactor that shares its dual subcellular localization in the ER and nucleolus (Carnemolla et al., 2009). The interaction between Rrs1 and AEG-1/MTDH/LYRIC was validated by coimmunoprecipitation experiments. These data suggest that both Rrs1 and AEG-1/MTDH/LYRIC might function as an ER stress sensor in HD and participate in transducing these signals to the nucleolus.

Another study showed that there was a positive correlation between AEG-1/MTDH/LYRIC high expression and β-catenin nuclear expression in CRC (Zhang et al., 2012). AEG-1/MTDH/LYRIC overexpression dramatically increased nuclear β-catenin accumulation in CRC cell lines. Furthermore, AEG-1/MTDH/LYRIC interacted with β-catenin in SW480 cells by immunoprecipitation assay (Zhang et al., 2012).

Ubinuclein (Ubn-1) is a cellular protein described as a nuclear protein interacting with the EBV transcription factor EB1 (also called ZEBRA or Zta) and other cellular transcription factor of the leucine-zipper family such as C/EBP or Jun (Lupo et al., 2012). One recent study identified AEG-1/MTDH/LYRIC as a new binding partner of Ubn-1 in epithelial cells. The authors used MS-based identifications of proteins eluted from pull-down assay and AEG-1/MTDH/LYRIC was validated as true partners of Ubn-1 through coimmunoprecipitation and confocal microscopy analyses (Lupo et al., 2012). In HT29 cells, the colocalization of AEG-1/MTDH/LYRIC and Ubn-1 was shown at the TJ level. It would be interesting to further identify the domain of this interaction and potential role of this interaction in tumorigenesis and/or other processes.

5.7. Chemoresistance-related genes

Recent studies discovered a crucial role of AEG-1/MTDH/LYRIC in chemoresistance in various cancer indications (Hu, Chong, et al., 2009; Liu et al., 2009; Yoo et al., 2009; Yoo, Gredler, et al., 2009). Pharmacogenomic analysis of the NCI-60 panel of cancer cell lines revealed a significant correlation of AEG-1/MTDH/LYRIC overexpression and resistance of cancer cells to a broad spectrum of chemotherapeutics (Hu, Wei, et al., 2009). Gene expression profiles comparing AEG-1/MTDH/LYRIC overexpressed HCC cells versus control cells identified a cluster of genes associated with chemoresistance, including drug-metabolizing enzymes for various chemotherapeutic agents, such as dihydropyrimidine dehydrogenase (DPYD), cytochrome P4502B6 (CYP2B6), and dihydrodiol dehydrogenase (AKR1C2), ATP-binding cassette transporter ABCC11 for drug efflux, and the transcription factor LSF (Yoo, Gredler, et al., 2009). LSF regulates the expression of thymidylate synthase, a target of 5-fluorouracil (5-FU). In addition, AEG-1/MTDH/LYRIC enhanced the expression of DPYD, which catalyzes the initial and rate-limiting step in the catabolism of 5-FU. The increase in LSF and DPYD confers AEG-1/MTDH/LYRIC-induced 5-FU resistance. A recent study in HCC reported another novel mechanism through which AEG-1/MTDH/LYRIC plays a role in chemoresistance (Yoo et al., 2010). In this study, AEG-1/MTDH/LYRIC was shown to increase the expression of MDR1 protein, resulting in increased efflux and decreased accumulation of doxorubicin, thereby promoting doxorubicin resistance (Yoo et al., 2010). In breast cancer, inhibition of AEG-1/MTDH/LYRIC sensitizes different cancer cell lines to paclitaxel, cisplatin, doxorubicin, 4-hydroxycylcophosphamide, hydrogen peroxide, and UV radiation (Hu, Chong, et al., 2009). Microarray analysis in breast cancer cells revealed that knockdown of AEG-1/MTDH/LYRIC led to decreased expression of chemoresistance genes ALDH3A1, MET, HSP90, and HMOX1 and increased expression of proapoptotic genes BNIP3 and TRAIL (Hu, Chong, et al., 2009). In neuroblastoma cells, a significant enhancement in chemosensitivity to cisplatin and doxorubicin was observed following knockdown of AEG-1/MTDH/LYRIC; however, no molecular characterization of downstream effectors was delineated (Liu et al., 2009). AEG-1/MTDH/LYRIC does not affect the uptake or retention of chemotherapy drugs. Instead, AEG-1/MTDH/LYRIC may increase chemoresistance by promoting cell survival after chemotherapeutic stress. This could be mediated by the prosurvival pathways such as PI3K and NF-κB or through other downstream genes affected by AEG-1/MTDH/LYRIC that directly regulate chemoresistance.

5.8. Invasion-associated molecules

The invasion-promoting ability of AEG-1/MTDH/LYRIC has been confirmed in diverse cancers including glioma, prostate cancer, HCC, neuroblastoma, osteosarcoma, CRC, and NSCLC (Emdad et al., 2010; Kikuno et al., 2007; Lee, Jeon, et al., 2009; Wang et al., 2011; Yoo, Emdad, et al., 2009; Zhang et al., 2012). Matrix metalloproteinases (MMPs) play a critical role in cancer invasion and metastasis. The MMPs are a family of zinc-dependent endopeptidases that remodel and degrade extracellular matrix, which are considered to play important roles in matrix degradation for tumor growth, invasion, and tumor-induced angiogenesis. AEG-1/MTDH/LYRIC has previously been shown to increase expression of adhesion molecules by activating NF-κB. Meanwhile, the promoter of MMP2 and MMP9 contains multiple functional elements, including NF-κB and AP-1. Among the MMPs, MMP2 has been shown to be modulated by AEG-1/MTDH/LYRIC in glioma and osteosarcoma (Emdad et al., 2010; Wang et al., 2011), while MMP9 was shown in glioma, prostate cancer, and NSCLC (Emdad et al., 2010; Kikuno et al., 2007; Liu et al., 2010; Sun et al., 2012).

5.9. Other Downstream Molecules

A recent study indicates that AEG-1/MTDH/LYRIC plays a seminal role in hepatocarcinogenesis and profoundly downregulates insulin-like growth factor-binding protein-7 (IGFBP7) (Chen et al., 2011). This study further suggested that IGFBP7 expression is significantly downregulated in human HCC samples and cell lines compared with normal liver and hepatocytes, respectively, and inversely correlates with the stages and grades of HCC. Forced overexpression of IGFBP7 in AEG-1/MTDH/LYRIC-overexpressing HCC cells inhibited *in vitro* growth and induced senescence and strongly suppressed *in vivo* growth in nude mice that might be an end result of inhibition of angiogenesis by IGFBP7. This study by Chen et al. provides evidence that IGFBP7 functions as a novel putative tumor suppressor for HCC and establishes the corollary that IGFBP7 downregulation can effectively modify AEG-1/MTDH/LYRIC function.

To further understand the role of AEG-1/MTDH/LYRIC in HCC, Srivastava et al. (2012) developed a transgenic mouse with hepatocyte-specific expression of AEG-1/MTDH/LYRIC (Alb/AEG-1). Treating Alb/AEG-1, but not wild-type (WT) mice, with *N*-nitrosodiethylamine (DEN) resulted in multinodular HCC with steatotic features. Using oligonucleotide microarray, the authors investigated the global gene expression changes in Alb/AEG-1 mice by comparing DEN-treated WT and Alb/AEG-1 livers. Apart from already known AEG-1/MTDH/LYRIC downstream effector molecules, a large number of novel molecules regulated by AEG-1/MTDH/LYRIC were discovered, including HCC marker alpha-fetoprotein; invasion and metastasis-associated genes tetraspanin 8 and lipocalin 2; several genes associated with fat metabolism, such as stearoyl coenzyme A desaturase-2, lipoprotein lipase, apolipoprotein A-IV, and apolipoprotein C-II; and genes regulating angiogenesis, such as trefoil factor 3 and mesenchyme homeobox 2.

6. AEG-1/MTDH/LYRIC AND ANGIOGENESIS

Cancer development is a multifactorial and multistep process and angiogenesis is a crucial component in this process. Angiogenesis, the process of forming neovascularization from existing vascular networks, is a fundamental event in the development and maintenance of solid tumors and their metastases. Increased intratumoral microvascular density relative to normal tissue is observed in tumors of different tissues including the brain, colon, and breast (Emdad et al., 2009). The process of angiogenesis is regulated by a continuous interplay of proangiogenic factors such as VEGF, basic fibroblast growth factor, epidermal growth factor, interleukins, nitric oxide synthase, transforming growth factor beta, TNF-α, platelet-derived growth factor (PDGF), and MMPs and angiogenic inhibitors such as endostatin, platelet factor-4, tumastin, thrombospondin-1, plasminogen activator inhibitor-1, and angiostatin (Carmeliet & Jain, 2011; Folkman, 2007; Weis & Cheresh, 2011). In many disorders including cancer, the balance between pro- and antiangiogenic factors is tilted to favor proangiogenic one, resulting in an "angiogenic switch."

Tumor angiogenesis is induced by hypoxic conditions through the regulation of several proangiogenic factors. Hypoxia leads to stabilization of VEGF mRNA and induces expression of HIF-1, which upregulates expression of VEGF through binding to a HIF-1 consensus sequence in the 5′ flanking region of the VEGF gene. VEGF is an established potent endothelial-specific mitogen and has a central role in tumor angiogenesis (An, Matsuda, Fujii, & Matsumoto, 2000; Dvorak, 2002; Semenza, 2010). Increased expression of VEGF has been known to be associated with a wide range of solid tumors and is considered as negative predictor of tumor prognosis.

Another important component of angiogenesis in normal and tumor tissue is the angiopoietin (Ang) pathway. Two Ang family members have been identified, Ang-1 and 2. Ang1 is the ligand for the Ang receptor Tie2. Recent studies demonstrated that Ang1 and Tie2 are expressed in human glioma cell lines and a transgenic mouse astrocytoma model (Davis et al., 1996; Stratmann, Risau, & Plate, 1998).

Another important regulatory molecule of tumor angiogenesis is HIF induced by hypoxia as indicated above. HIF-1 is a transcription factor that regulates the expression of several genes involved in various cellular functions including angiogenesis (Majmundar, Wong, & Simon, 2010), invasion, and metastasis, resistance to chemotherapy, metabolic reprogramming, EMT, and stem cell maintenance (Semenza, 2012). HIF-1 transcriptional activity requires formation of a heterodimer consisting of HIF-1α and HIF-1β. The heterodimer binds to hypoxia response elements in the promoter regions of its target genes, where it activates transcription. HIF-1β is ubiquitously expressed, whereas, HIF-1α expression is tightly regulated by oxygen tension (Majmundar et al., 2010). Among the growth/survival factors that are encoded by HIF-regulated genes are transforming growth factor-α, insulin-like growth factor-2, VEGF, endothelin 1, adrenomedullin, and erythropoietin. HIF-1 controls the expression of multiple proangiogenic molecules, including VEGF, stromal-derived factor 1, placental growth factor, PDGF, and Ang 1 and 2 (Semenza, 2012). In mouse models, inhibition of HIF-1 activity dramatically inhibits tumor vascularization (Lee, Zhang, et al., 2009). Additionally, HIF-1 activates transcription of genes involved in invasion and metastasis such as MMP2, MMP9, MMP14, MET, VEGF, and ANGPT2. Interestingly, one recent study demonstrated that AEG-1/MTDH/LYRIC is induced by hypoxia and glucose deprivation in glioblastoma and that hypoxic induction of AEG-1/MTDH/LYRIC depends on the stabilization of HIF-1α (Noch et al., 2011). AEG-1/

MTDH/LYRIC was induced by hypoxia via the PI3K/Akt pathway and then AEG-1/MTDH/LYRIC feeds back to activate PI3K, thereby creating a positive feedback loop and consequently might facilitate angiogenesis.

AEG-1/MTDH/LYRIC has also been shown to be proangiogenic both *in vitro* and *in vivo* and can also augment expression of key angiogenesis molecules, such as Ang1, MMP2, and HIF-1α. Tumors formed by AEG-1/MTDH/LYRIC-overexpressing cloned rat embryo fibroblasts (CREF-AEG-1) are highly aggressive and angiogenic. *In vitro* angiogenesis studies also demonstrated that AEG-1/MTDH/LYRIC promotes tube formation in Matrigel and increases invasion of HUVECs via the PI3K/Akt signaling pathway (Emdad et al., 2009). The proangiogenic property of AEG-1/MTDH/LYRIC was also reinforced by the chicken chorioallantoic membrane assay. The angiogenesis-promoting function of AEG-1/MTDH/LYRIC was also noted in human HCC (Yoo, Emdad, et al., 2009). AEG-1/MTDH/LYRIC overexpression in human HCC cells resulted in increased production of angiogenic factors, such as VEGF, PIGF, and fibroblast growth factor-α (FGF-α), and AEG-1/MTDH/LYRIC overexpression in human HCC cells resulted in highly aggressive, angiogenic, and metastatic tumors. Recent study (Srivastava et al., 2012) using a transgenic mouse with hepatocyte-specific expression of AEG-1/MTDH/LYRIC (Alb/AEG-1) has further supported the proangiogenic role of AEG-1/MTDH/LYRIC. Conditioned media from Alb/AEG-1 hepatocytes induced marked angiogenesis with elevation in several coagulation factors. Among these factors, AEG-1/MTDH/LYRIC facilitated the association of Factor XII (FXII) messenger RNA with polysomes, resulting in increased translation. FXII displays angiogenic activity, which is independent of its function in coagulation. FXII binds to urokinase plasminogen activator receptor or cross talks with EGFR on HUVEC membranes, leading to activation of MAPK and Akt with subsequent proliferation and differentiation. siRNA-mediated knockdown of FXII resulted in a profound inhibition of AEG-1/MTDH/LYRIC-induced angiogenesis, indicating a central role of FXII in this process.

In one recent study, analysis of clinical samples taken from large cohorts of patients of breast cancer showed that AEG-1/MTDH/LYRIC correlated positively with VEGF levels and microvascular density (Li, Li, Song, et al., 2011). VEGF and VEGFR play very important roles in angiogenesis (Carmeliet & Jain, 2011). IL-8 and COX2 are also suggested as proangiogenic factors regulated by NF-κB signaling pathway. IL-8 and COX2 are overexpressed in breast cancer (Lerebours et al., 2008), and the expression of IL-8 can be regulated by AEG-1/MTDH/LYRIC (Emdad et al., 2006), indicating that AEG-1/MTDH/LYRIC may promote the angiogenesis of breast cancer by activating the NF-κB and/or VEGF signaling pathway. Further experiments should be performed to clarify the mechanism.

Angiogenesis is classically triggered by hypoxia and inflammation, conditions which can also induce EMT (Sanchez-Tillo et al., 2012). EMT-activating transcription factors ZEB, Snail, and Twist are often overexpressed by endothelial cells in the peritumoral stroma and are shown to promote tumor angiogenesis *in vivo*. EMT and hypoxic conditions also increase the population of cancer cells with stem-like phenotype (Bao et al., 2012). Two recent studies, one in breast cancer and another in HCC, documented a pivotal role of AEG-1/MTDH/LYRIC in inducing EMT. In breast cancer, overexpression of AEG-1/MTDH/LYRIC led to upregulation of the mesenchymal marker fibronectin and downregulation of the epithelial marker E-cadherin, and also, transcription factors Snail and Slug were upregulated in AEG-1/MTDH/LYRIC-overexpressing breast cancer cells. Interestingly, overexpression of AEG-1/MTDH/LYRIC also led to increased acquisition of CD44+/CD24−/low markers that are characteristic of breast cancer stem cells (Li, Kong, et al., 2011). Similarly, knockdown of AEG-1/MTDH/LYRIC expression in HCC cell lines resulted in downregulation of N-cadherin and snail, upregulation of E-cadherin, and translocation of β-catenin (Zhu et al., 2011). An overview of the role of AEG-1/MTDH/LYRIC in modulating angiogenesis is presented in Fig. 3.4. The present findings indicating a potential cross talk between AEG-1/MTDH/LYRIC-induced EMT and AEG-1/MTDH/LYRIC-induced enhanced tumor angiogenesis deserve further in-depth exploration.

7. CONCLUSION AND FUTURE PERSPECTIVES

In recent years, AEG-1/MTDH/LYRIC has emerged as an important mediator of cancer development and progression through regulation of essential processes of tumorigenesis such as transformation, invasion, metastasis, angiogenesis, and chemoresistance. Although numerous contributions to our literature base since its discovery in 2002 by Su et al. through subtraction hybridization have documented a diverse array of AEG-1/MTDH/LYRIC functions, much work is still required to unravel the fundamental roles of AEG-1/MTDH/LYRIC in physiological and pathological

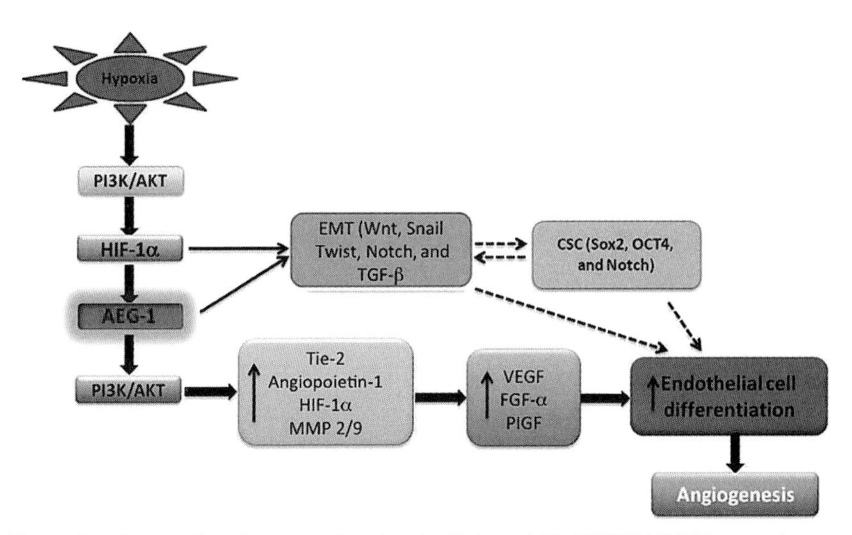

Figure 3.4 Potential pathways and molecules linking AEG-1/MTDH/LYRIC to angiogenesis. Dotted lines indicate potential pathways involved in AEG-1/MTDH/LYRIC-regulated angiogenesis that require further validation. (See Page 4 in Color Section at the back of the book.)

conditions. Evolutionally, AEG-1/MTDH/LYRIC is only found in vertebrates that have developed complete immune systems, and AEG-1/MTDH/LYRIC has recently been shown to be associated with the regulation of inflammation and immune responses. It would be worthwhile investigating further the significance of dysregulated AEG-1/MTDH/LYRIC expression in pathogenesis of cancer-related inflammation, autoimmune diseases, and other immunopathological conditions. Additionally, the intriguing novel role of AEG-1/MTDH/LYRIC in neurodegeneration and trauma also requires further exploration.

The multiple functions of AEG-1/MTDH/LYRIC highlight several important clinical ramifications. AEG-1/MTDH/LYRIC might be a ubiquitous biomarker for aggressive cancers with potential for routine screening of patients. However, to establish AEG-1/MTDH/LYRIC as a viable candidate biomarker, a larger cohort of patient studies and correlative investigations quantifying and comparing AEG-1/MTDH/LYRIC mRNA or protein in blood, urine, and biopsy samples with clinical characteristics are mandatory. AEG-1/MTDH/LYRIC expression might also help stratify patients receiving chemotherapeutic regimens based on the broad-spectrum chemoresistance conferred by AEG-1/MTDH/LYRIC. Cell surface expression of AEG-1/MTDH/LYRIC might also be useful for diagnostic and therapeutic purposes. Similarly, therapeutic toxins or radioactive isotopes might be conjugated with anti-AEG-1/MTDH/LYRIC antibody for targeted destruction of AEG-1/MTDH/LYRIC overexpressing cells. Qian et al. (2011) recently used an AEG-1/MTDH/LYRIC-based DNA vaccine, delivered by attenuated *Salmonella typhimurium*, and showed that this AEG-1/MTDH/LYRIC vaccine increased chemosensitivity to doxorubicin and inhibited breast cancer lung metastasis. The potential of AEG-1/MTDH/LYRIC-based vaccine requires further evaluation in other cancers.

The observation that AEG-1/MTDH/LYRIC is able to interact with a plethora of proteins and assist in the formation of multiprotein complexes indicates that AEG-1/MTDH/LYRIC might be a new member of scaffolding proteins. Moreover, the biochemical modifications occurring to AEG-1/MTDH/LYRIC itself are also of great interest and worthy of further investigation, as such modification(s), if proved to be functionally important for AEG-1/MTDH/LYRIC, would represent potentially promising targets for optimal inhibitors or therapeutic drugs against diseases potentially mediated by aberrant AEG-1/MTDH/LYRIC expression, including cancer. Solving the crystal structure of the novel protein–protein interaction domains might also help in the design of small molecule inhibitors of AEG-1/MTDH/LYRIC. High-throughput screening combined with NMR with small molecules that disrupt interactions of AEG-1/MTDH/LYRIC with its known interacting proteins might be yet another approach to identify AEG-1/MTDH/LYRIC inhibitors.

Experimental evidence derived from AEG-1/MTDH/LYRIC knockout or transgenic animal models would be important for further understanding the functional significance of AEG-1/MTDH/LYRIC in embryonic development and disease progression. As a positive regulator of cancer progression, transgenic mice expressing AEG-1/MTDH/LYRIC as well as conditional targeted AEG-1/MTDH/LYRIC knockout mice would prove extremely valuable for analyzing tumor progression *in vivo*. In this context, the transgenic mouse with hepatocyte-specific expression of AEG-1/MTDH/LYRIC explored several novel aspects of AEG-1/MTDH/LYRIC function that might not have been possible using *in vitro* models and nude mice xenograft studies. This mouse model might also be valuable in evaluating novel therapeutic approaches targeted toward nonalcoholic fatty liver disease and HCC. Additionally, developing more conditional transgenic animal model to express AEG-1/MTDH/LYRIC specifically in target tissue using a tissue-specific promoter would be beneficial in more precisely defining AEG-1/MTDH/LYRIC functions in additional cancer contexts. Moreover, crossing these animals with other tumor models to determine potential cross talk between AEG-1/MTDH/LYRIC and other tumor-promoting pathways, as well as with other disease models, would also prove very informative.

In summary, this review has tried to capture the varied and important functions of AEG-1/MTDH/LYRIC in diverse disease contexts. This is a continuing story and we are optimistic that new seminal information will be uncovered in the future as more investigators work on this intriguing molecule. Understanding the role of AEG-1/MTDH/LYRIC in normal physiological functions is also a worthwhile endeavor that clearly requires further research (Jeon et al., 2010). The observation that AEG-1/MTDH/LYRIC may be an important contributor to many diverse cancers highlights the significance of developing approaches for selectively inhibiting its expression. These efforts are extremely valuable and hold potential for heralding in a new way of developing rationale target-based therapies for cancer, inflammatory diseases, and neurodegeneration.

ACKNOWLEDGMENTS

The research that served as the basis of this review were provided in part by the National Institutes of Health, National Cancer Institute Grants CA134721 (P. B. F.) and CA138540 (D. S.), the Trauma Fund, Virginia Commonwealth University, School of Medicine (L. E. and P. B. F.), and the VCU Massey Cancer Center (MCC). D. S. is a Harrison Scholar in the VCU MCC, and P. B. F. holds the Thelma Newmeyer Chair in Cancer Research in the VCU MCC.

REFERENCES

An, F. Q., Matsuda, M., Fujii, H., & Matsumoto, Y. (2000). Expression of vascular endothelial growth factor in surgical specimens of hepatocellular carcinoma. *Journal of Cancer Research and Clinical Oncology*, *126*, 153–160.

Anttila, V., Wessman, M., Kallela, M., & Palotie, A. (2011). Towards an understanding of genetic predisposition to migraine. *Genome Medicine*, *3*, 17.

Ash, S. C., Yang, D. Q., & Britt, D. E. (2008). LYRIC/AEG-1 overexpression modulates BCCIPalpha protein levels in prostate tumor cells. *Biochemical and Biophysical Research Communications*, *371*, 333–338.

Bao, B., Azmi, A. S., Ali, S., Ahmad, A., Li, Y., Banerjee, S., et al. (2012). The biological kinship of hypoxia with CSC and EMT and their relationship with deregulated expression of miRNAs and tumor aggressiveness. *Biochimica et Biophysica Acta*, *1826*, 272–296.

Bartholomeusz, C., & Gonzalez-Angulo, A. M. (2012). Targeting the PI3K signaling pathway in cancer therapy. *Expert Opinion on Therapeutic Targets*, *16*, 121–130.

Baygi, M. E., & Nikpour, P. (2012). Deregulation of MTDH gene expression in gastric cancer. *Asian Pacific Journal of Cancer Prevention*, *13*, 2833–2836.

Bellacosa, A., Kumar, C. C., Di Cristofano, A., & Testa, J. R. (2005). Activation of AKT kinases in cancer: Implications for therapeutic targeting. *Advances in Cancer Research*, *94*, 29–86.

Ben-Neriah, Y., & Karin, M. (2011). Inflammation meets cancer, with NF-kappaB as the matchmaker. *Nature Immunology, 12*, 715–723.
Bernardo, M. V., Yelo, E., Gimeno, L., Campillo, J. A., & Parrado, A. (2007). Identification of apoptosis-related PLZF target genes. *Biochemical and Biophysical Research Communications, 359*, 317–322.
Bhutia, S. K., Kegelman, T. P., Das, S. K., Azab, B., Su, Z. Z., Lee, S. G., et al. (2010). Astrocyte elevated gene-1 induces protective autophagy. *Proceedings of the National Academy of Sciences of the United States of America, 107*, 22243–22248.
Blancher, C., Moore, J. W., Robertson, N., & Harris, A. L. (2001). Effects of ras and von Hippel-Lindau (VHL) gene mutations on hypoxia-inducible factor (HIF)-1alpha, HIF-2alpha, and vascular endothelial growth factor expression and their regulation by the phosphatidylinositol 3′-kinase/Akt signaling pathway. *Cancer Research, 61*, 7349–7355.
Blanco, M. A., Aleckovic, M., Hua, Y., Li, T., Wei, Y., Xu, Z., et al. (2011). Identification of staphylococcal nuclease domain-containing 1 (SND1) as a Metadherin-interacting protein with metastasis-promoting functions. *The Journal of Biological Chemistry, 286*, 19982–19992.
Blume-Jensen, P., & Hunter, T. (2001). Oncogenic kinase signalling. *Nature, 411*, 355–365.
Britt, D. E., Yang, D. F., Yang, D. Q., Flanagan, D., Callanan, H., Lim, Y. P., et al. (2004). Identification of a novel protein, LYRIC, localized to tight junctions of polarized epithelial cells. *Experimental Cell Research, 300*, 134–148.
Brown, D. M., & Ruoslahti, E. (2004). Metadherin, a cell surface protein in breast tumors that mediates lung metastasis. *Cancer Cell, 5*, 365–374.
Brunet, A., Bonni, A., Zigmond, M. J., Lin, M. Z., Juo, P., Hu, L. S., et al. (1999). Akt promotes cell survival by phosphorylating and inhibiting a Forkhead transcription factor. *Cell, 96*, 857–868.
Cadoret, A., Ovejero, C., Terris, B., Souil, E., Levy, L., Lamers, W. H., et al. (2002). New targets of beta-catenin signaling in the liver are involved in the glutamine metabolism. *Oncogene, 21*, 8293–8301.
Cantley, L. C. (2002). The phosphoinositide 3-kinase pathway. *Science, 296*, 1655–1657.
Carmeliet, P., & Jain, R. K. (2011). Molecular mechanisms and clinical applications of angiogenesis. *Nature, 473*, 298–307.
Carnemolla, A., Fossale, E., Agostoni, E., Michelazzi, S., Calligaris, R., De Maso, L., et al. (2009). Rrs1 is involved in endoplasmic reticulum stress response in Huntington disease. *The Journal of Biological Chemistry, 284*, 18167–18173.
Caudy, A. A., Ketting, R. F., Hammond, S. M., Denli, A. M., Bathoorn, A. M., Tops, B. B., et al. (2003). A micrococcal nuclease homologue in RNAi effector complexes. *Nature, 425*, 411–414.
Chen, L. F., & Greene, W. C. (2003). Regulation of distinct biological activities of the NF-kappaB transcription factor complex by acetylation. *Journal of Molecular Medicine (Berlin), 81*, 549–557.
Chen, W., Ke, Z., Shi, H., Yang, S., & Wang, L. (2010). Overexpression of AEG-1 in renal cell carcinoma and its correlation with tumor nuclear grade and progression. *Neoplasma, 57*, 522–529.
Chen, D., Yoo, B. K., Santhekadur, P. K., Gredler, R., Bhutia, S. K., Das, S. K., et al. (2011). Insulin-like growth factor-binding protein-7 functions as a potential tumor suppressor in hepatocellular carcinoma. *Clinical Cancer Research, 17*, 6693–6701.
Choi, D. W. (1988). Glutamate neurotoxicity and diseases of the nervous system. *Neuron, 1*, 623–634.
Clevers, H. (2006). Wnt/beta-catenin signaling in development and disease. *Cell, 127*, 469–480.
Clevers, H., & Nusse, R. (2012). Wnt/beta-catenin signaling and disease. *Cell, 149*, 1192–1205.

Datta, S. R., Brunet, A., & Greenberg, M. E. (1999). Cellular survival: A play in three Akts. *Genes & Development, 13*, 2905–2927.

Davis, S., Aldrich, T. H., Jones, P. F., Acheson, A., Compton, D. L., Jain, V., et al. (1996). Isolation of angiopoietin-1, a ligand for the TIE2 receptor, by secretion-trap expression cloning. *Cell, 87*, 1161–1169.

Doble, A. (1999). The role of excitotoxicity in neurodegenerative disease: Implications for therapy. *Pharmacology & Therapeutics, 81*, 163–221.

Downward, J. (2003). Targeting RAS signalling pathways in cancer therapy. *Nature Reviews. Cancer, 3*, 11–22.

Dvorak, H. F. (2002). Vascular permeability factor/vascular endothelial growth factor: A critical cytokine in tumor angiogenesis and a potential target for diagnosis and therapy. *Journal of Clinical Oncology, 20*, 4368–4380.

Eddleston, M., & Mucke, L. (1993). Molecular profile of reactive astrocytes—Implications for their role in neurologic disease. *Neuroscience, 54*, 15–36.

Emdad, L., Lee, S. G., Su, Z. Z., Jeon, H. Y., Boukerche, H., Sarkar, D., et al. (2009). Astrocyte elevated gene-1 (AEG-1) functions as an oncogene and regulates angiogenesis. *Proceedings of the National Academy of Sciences of the United States of America, 106*, 21300–21305.

Emdad, L., Sarkar, D., Lee, S. G., Su, Z. Z., Yoo, B. K., Dash, R., et al. (2010). Astrocyte elevated gene-1: A novel target for human glioma therapy. *Molecular Cancer Therapeutics, 9*, 79–88.

Emdad, L., Sarkar, D., Su, Z. Z., Randolph, A., Boukerche, H., Valerie, K., et al. (2006). Activation of the nuclear factor kappaB pathway by astrocyte elevated gene-1: Implications for tumor progression and metastasis. *Cancer Research, 66*, 1509–1516.

Engeland, C. E., Oberwinkler, H., Schumann, M., Krause, E., Muller, G. A., & Krausslich, H. G. (2011). The cellular protein lyric interacts with HIV-1 Gag. *Journal of Virology, 85*, 13322–13332.

Engelman, J. A., Luo, J., & Cantley, L. C. (2006). The evolution of phosphatidylinositol 3-kinases as regulators of growth and metabolism. *Nature Reviews. Genetics, 7*, 606–619.

Fisher, P. B. (1984). Enhancement of viral transformation and expression of the transformed phenotype by tumor promoters. In T. J. Slaga (Ed.), *Tumor promotion and cocarcinogenesis in vitro, mechanisms of tumor promotion* (pp. 57–123). Florida: CRC Press, Inc.

Folkman, J. (2007). Angiogenesis: An organizing principle for drug discovery? *Nature Reviews. Drug Discovery, 6*, 273–286.

Furia, B., Deng, L., Wu, K., Baylor, S., Kehn, K., Li, H., et al. (2002). Enhancement of nuclear factor-kappa B acetylation by coactivator p300 and HIV-1 Tat proteins. *The Journal of Biological Chemistry, 277*, 4973–4980.

Ge, X., Lv, X., Feng, L., Liu, X., Gao, J., Chen, N., et al. (2012). Metadherin contributes to the pathogenesis of diffuse large B-cell lymphoma. *PLoS One*, 7, e39449.

Ghosh, S., May, M. J., & Kopp, E. B. (1998). NF-kappa B and Rel proteins: Evolutionarily conserved mediators of immune responses. *Annual Review of Immunology, 16*, 225–260.

Gilmore, T. D. (2003). The Rel/NF-kappa B/I kappa B signal transduction pathway and cancer. *Cancer Treatment and Research, 115*, 241–265.

Gnosa, S., Shen, Y. M., Wang, C. J., Zhang, H., Stratmann, J., Arbman, G., et al. (2012). Expression of AEG-1 mRNA and protein in colorectal cancer patients and colon cancer cell lines. *Journal of Translational Medicine, 10*, 109.

Gong, Z., Liu, W., You, N., Wang, T., Wang, X., Lu, P., et al. (2012). Prognostic significance of metadherin overexpression in hepatitis B virus-related hepatocellular carcinoma. *Oncology Reports, 27*, 2073–2079.

Gupta, G. P., & Massague, J. (2006). Cancer metastasis: Building a framework. *Cell, 127*, 679–695.

Hanahan, D., & Weinberg, R. A. (2000). The hallmarks of cancer. *Cell, 100*, 57–70.

Hanahan, D., & Weinberg, R. A. (2011). Hallmarks of cancer: The next generation. *Cell, 144*, 646–674.
Hayden, M. S., & Ghosh, S. (2004). Signaling to NF-kappaB. *Genes & Development, 18*, 2195–2224.
Hayden, M. S., & Ghosh, S. (2008). Shared principles in NF-kappaB signaling. *Cell, 132*, 344–362.
Hayden, M. S., & Ghosh, S. (2011). NF-kappaB in immunobiology. *Cell Research, 21*, 223–244.
Hayden, M. S., & Ghosh, S. (2012). NF-kappaB, the first quarter-century: Remarkable progress and outstanding questions. *Genes & Development, 26*, 203–234.
Hennessy, B. T., Smith, D. L., Ram, P. T., Lu, Y., & Mills, G. B. (2005). Exploiting the PI3K/AKT pathway for cancer drug discovery. *Nature Reviews. Drug Discovery, 4*, 988–1004.
Hu, G., Chong, R. A., Yang, Q., Wei, Y., Blanco, M. A., Li, F., et al. (2009). MTDH activation by 8q22 genomic gain promotes chemoresistance and metastasis of poor-prognosis breast cancer. *Cancer Cell, 15*, 9–20.
Hu, G., Wei, Y., & Kang, Y. (2009). The multifaceted role of MTDH/AEG-1 in cancer progression. *Clinical Cancer Research, 15*, 5615–5620.
Huang, W. C., Ju, T. K., Hung, M. C., & Chen, C. C. (2007). Phosphorylation of CBP by IKKalpha promotes cell growth by switching the binding preference of CBP from p53 to NF-kappaB. *Molecular Cell, 26*, 75–87.
Hui, A. B., Bruce, J. P., Alajez, N. M., Shi, W., Yue, S., Perez-Ordonez, B., et al. (2011). Significance of dysregulated metadherin and microRNA-375 in head and neck cancer. *Clinical Cancer Research, 17*, 7539–7550.
Jeon, H. Y., Choi, M., Howlett, E. L., Vozhilla, N., Yoo, B. K., Lloyd, J. A., et al. (2010). Expression patterns of astrocyte elevated gene-1 (AEG-1) during development of the mouse embryo. *Gene Expression Patterns, 10*, 361–367.
Jian-bo, X., Hui, W., Yu-long, H., Chang-hua, Z., Long-juan, Z., Shi-rong, C., et al. (2011). Astrocyte-elevated gene-1 overexpression is associated with poor prognosis in gastric cancer. *Medical Oncology, 28*, 455–462.
Jiang, B. H., Jiang, G., Zheng, J. Z., Lu, Z., Hunter, T., & Vogt, P. K. (2001). Phosphatidylinositol 3-kinase signaling controls levels of hypoxia-inducible factor 1. *Cell Growth & Differentiation, 12*, 363–369.
Jiang, T., Zhu, A., Zhu, Y., & Piao, D. (2012). Clinical implications of AEG-1 in liver metastasis of colorectal cancer. *Medical Oncology, 29*, 2858–2863.
Kanarek, N., & Ben-Neriah, Y. (2012). Regulation of NF-kappaB by ubiquitination and degradation of the IkappaBs. *Immunological Reviews, 246*, 77–94.
Kang, D. C., Su, Z. Z., Sarkar, D., Emdad, L., Volsky, D. J., & Fisher, P. B. (2005). Cloning and characterization of HIV-1-inducible astrocyte elevated gene-1, AEG-1. *Gene, 353*, 8–15.
Karin, M., Cao, Y., Greten, F. R., & Li, Z. W. (2002). NF-kappaB in cancer: From innocent bystander to major culprit. *Nature Reviews. Cancer, 2*, 301–310.
Katz, M., Amit, I., & Yarden, Y. (2007). Regulation of MAPKs by growth factors and receptor tyrosine kinases. *Biochimica et Biophysica Acta, 1773*, 1161–1176.
Ke, Z. F., He, S., Li, S., Luo, D., Feng, C., & Zhou, W. (2012). Expression characteristics of astrocyte elevated gene-1 (AEG-1) in tongue carcinoma and its correlation with poor prognosis. *Cancer Epidemiology, 37*(2), 179–185.
Ke, Z. F., Mao, X., Zeng, C., He, S., Li, S., & Wang, L. T. (2013). AEG-1 expression characteristics in human non-small cell lung cancer and its relationship with apoptosis. *Medical Oncology, 30*, 383.
Khuda, I. I., Koide, N., Noman, A. S., Dagvadorj, J., Tumurkhuu, G., Naiki, Y., et al. (2009). Astrocyte elevated gene-1 (AEG-1) is induced by lipopolysaccharide as toll-like receptor 4 (TLR4) ligand and regulates TLR4 signalling. *Immunology, 128*, e700–e706.

Kikuno, N., Shiina, H., Urakami, S., Kawamoto, K., Hirata, H., Tanaka, Y., et al. (2007). Knockdown of astrocyte-elevated gene-1 inhibits prostate cancer progression through upregulation of FOXO3a activity. *Oncogene, 26*, 7647–7655.

Kong, X., Moran, M. S., Zhao, Y., & Yang, Q. (2012). Inhibition of metadherin sensitizes breast cancer cells to AZD6244. *Cancer Biology & Therapy, 13*, 43–49.

Kumar, A., Takada, Y., Boriek, A. M., & Aggarwal, B. B. (2004). Nuclear factor-kappaB: Its role in health and disease. *Journal of Molecular Medicine (Berlin), 82*, 434–448.

Lee, S. G., Jeon, H. Y., Su, Z. Z., Richards, J. E., Vozhilla, N., Sarkar, D., et al. (2009). Astrocyte elevated gene-1 contributes to the pathogenesis of neuroblastoma. *Oncogene, 28*, 2476–2484.

Lee, S. G., Kim, K., Kegelman, T. P., Dash, R., Das, S. K., Choi, J. K., et al. (2011). Oncogene AEG-1 promotes glioma-induced neurodegeneration by increasing glutamate excitotoxicity. *Cancer Research, 71*, 6514–6523.

Lee, S. G., Su, Z. Z., Emdad, L., Sarkar, D., & Fisher, P. B. (2006). Astrocyte elevated gene-1 (AEG-1) is a target gene of oncogenic Ha-ras requiring phosphatidylinositol 3-kinase and c-Myc. *Proceedings of the National Academy of Sciences of the United States of America, 103*, 17390–17395.

Lee, S. G., Su, Z. Z., Emdad, L., Sarkar, D., Franke, T. F., & Fisher, P. B. (2008). Astrocyte elevated gene-1 activates cell survival pathways through PI3K-Akt signaling. *Oncogene, 27*, 1114–1121.

Lee, K., Zhang, H., Qian, D. Z., Rey, S., Liu, J. O., & Semenza, G. L. (2009). Acriflavine inhibits HIF-1 dimerization, tumor growth, and vascularization. *Proceedings of the National Academy of Sciences of the United States of America, 106*, 17910–17915.

Lerebours, F., Vacher, S., Andrieu, C., Espie, M., Marty, M., Lidereau, R., et al. (2008). NF-kappa B genes have a major role in inflammatory breast cancer. *BMC Cancer, 8*, 41.

Levine, B., & Klionsky, D. J. (2004). Development by self-digestion: Molecular mechanisms and biological functions of autophagy. *Developmental Cell, 6*, 463–477.

Li, X., Kong, X., Huo, Q., Guo, H., Yan, S., Yuan, C., et al. (2011). Metadherin enhances the invasiveness of breast cancer cells by inducing epithelial to mesenchymal transition. *Cancer Science, 102*, 1151–1157.

Li, C., Li, R., Song, H., Wang, D., Feng, T., Yu, X., et al. (2011). Significance of AEG-1 expression in correlation with VEGF, microvessel density and clinicopathological characteristics in triple-negative breast cancer. *Journal of Surgical Oncology, 103*, 184–192.

Li, C., Li, Y., Wang, X., Wang, Z., Cai, J., Wang, L., et al. (2012). Elevated expression of astrocyte elevated gene-1 (AEG-1) is correlated with cisplatin-based chemoresistance and shortened outcome in patients with stages III–IV serous ovarian carcinoma. *Histopathology, 60*, 953–963.

Li, C., Liu, J., Lu, R., Yu, G., Wang, X., Zhao, Y., et al. (2011). AEG-1 overexpression: A novel indicator for peritoneal dissemination and lymph node metastasis in epithelial ovarian cancers. *International Journal of Gynecological Cancer, 21*, 602–608.

Li, J., Yang, L., Song, L., Xiong, H., Wang, L., Yan, X., et al. (2009). Astrocyte elevated gene-1 is a proliferation promoter in breast cancer via suppressing transcriptional factor FOXO1. *Oncogene, 28*, 3188–3196.

Li, J., Zhang, N., Song, L. B., Liao, W. T., Jiang, L. L., Gong, L. Y., et al. (2008). Astrocyte elevated gene-1 is a novel prognostic marker for breast cancer progression and overall patient survival. *Clinical Cancer Research, 14*, 3319–3326.

Liang, J., Zubovitz, J., Petrocelli, T., Kotchetkov, R., Connor, M. K., Han, K., et al. (2002). PKB/Akt phosphorylates p27, impairs nuclear import of p27 and opposes p27-mediated G1 arrest. *Nature Medicine, 8*, 1153–1160.

Liao, W. T., Guo, L., Zhong, Y., Wu, Y. H., Li, J., & Song, L. B. (2011). Astrocyte elevated gene-1 (AEG-1) is a marker for aggressive salivary gland carcinoma. *Journal of Translational Medicine, 9*, 205.

Ligthart, L., de Vries, B., Smith, A. V., Ikram, M. A., Amin, N., Hottenga, J. J., et al. (2011). Meta-analysis of genome-wide association for migraine in six population-based European cohorts. *European Journal of Human Genetics, 19*, 901–907.
Ling, J., & Kumar, R. (2012). Crosstalk between NFkB and glucocorticoid signaling: A potential target of breast cancer therapy. *Cancer Letters, 322*, 119–126.
Liu, H. Y., Liu, C. X., Han, B., Zhang, X. Y., & Sun, R. P. (2012). AEG-1 is associated with clinical outcome in neuroblastoma patients. *Cancer Biomarkers: Section A of Disease Markers, 11*, 115–121.
Liu, H., Song, X., Liu, C., Xie, L., Wei, L., & Sun, R. (2009). Knockdown of astrocyte elevated gene-1 inhibits proliferation and enhancing chemo-sensitivity to cisplatin or doxorubicin in neuroblastoma cells. *Journal of Experimental & Clinical Cancer Research, 28*, 19.
Liu, Y., Su, Z., Li, G., Yu, C., Ren, S., Huang, D., Fan, S., Tian, Y., Zhang, X., & Qiu, Y. (2013). Increased expression of metadherin protein predicts worse disease-free and overall survival in laryngeal squamous cell carcinoma. *International Journal of Cancer, 133*(3), 671–679.
Liu, L., Wu, J., Ying, Z., Chen, B., Han, A., Liang, Y., et al. (2010). Astrocyte elevated gene-1 upregulates matrix metalloproteinase-9 and induces human glioma invasion. *Cancer Research, 70*, 3750–3759.
Liu, D. C., & Yang, Z. L. (2013). MTDH and EphA7 are markers for metastasis and poor prognosis of gallbladder adenocarcinoma. *Diagnostic Cytopathology, 41*(3), 199–205.
Liu, J., Yuan, Y., Huan, J., & Shen, Z. (2001). Inhibition of breast and brain cancer cell growth by BCCIPalpha, an evolutionarily conserved nuclear protein that interacts with BRCA2. *Oncogene, 20*, 336–345.
Lupo, J., Conti, A., Sueur, C., Coly, P. A., Coute, Y., Hunziker, W., et al. (2012). Identification of new interacting partners of the shuttling protein ubinuclein (Ubn-1). *Experimental Cell Research, 318*, 509–520.
MacDonald, B. T., Tamai, K., & He, X. (2009). Wnt/beta-catenin signaling: Components, mechanisms, and diseases. *Developmental Cell, 17*, 9–26.
Majmundar, A. J., Wong, W. J., & Simon, M. C. (2010). Hypoxia-inducible factors and the response to hypoxic stress. *Molecular Cell, 40*, 294–309.
Maragakis, N. J., & Rothstein, J. D. (2006). Mechanisms of disease: Astrocytes in neurodegenerative disease. *Nature Clinical Practice. Neurology, 2*, 679–689.
Marrogi, A., Pass, H. I., Khan, M., Metheny-Barlow, L. J., Harris, C. C., & Gerwin, B. I. (2000). Human mesothelioma samples overexpress both cyclooxygenase-2 (COX-2) and inducible nitric oxide synthase (NOS2): In vitro antiproliferative effects of a COX-2 inhibitor. *Cancer Research, 60*, 3696–3700.
Mathew, R., Karantza-Wadsworth, V., & White, E. (2007). Role of autophagy in cancer. *Nature Reviews. Cancer, 7*, 961–967.
Meng, X., Brachova, P., Yang, S., Xiong, Z., Zhang, Y., Thiel, K. W., et al. (2011). Knockdown of MTDH sensitizes endometrial cancer cells to cell death induction by death receptor ligand TRAIL and HDAC inhibitor LBH589 co-treatment. *PLoS One, 6*, e20920.
Noch, E., Bookland, M., & Khalili, K. (2011). Astrocyte-elevated gene-1 (AEG-1) induction by hypoxia and glucose deprivation in glioblastoma. *Cancer Biology & Therapy, 11*, 32–39.
Noguchi, Y., Makino, T., Yoshikawa, T., Nomura, K., Fukuzawa, K., Matsumoto, A., et al. (1996). The possible role of TNF-alpha and IL-2 in inducing tumor-associated metabolic alterations. *Surgery Today, 26*, 36–41.
Nolan, G. P., Ghosh, S., Liou, H. C., Tempst, P., & Baltimore, D. (1991). DNA binding and I kappa B inhibition of the cloned p65 subunit of NF-kappa B, a rel-related polypeptide. *Cell, 64*, 961–969.
Orentas, R. J., Yang, J. J., Wen, X., Wei, J. S., Mackall, C. L., & Khan, J. (2012). Identification of cell surface proteins as potential immunotherapy targets in 12 pediatric cancers. *Frontiers in Oncology, 2*, 194.

Polakis, P. (2007). The many ways of Wnt in cancer. *Current Opinion in Genetics & Development, 17*, 45–51.
Prasad, S., Ravindran, J., & Aggarwal, B. B. (2010). NF-kappaB and cancer: How intimate is this relationship. *Molecular and Cellular Biochemistry, 336*, 25–37.
Qian, B. J., Yan, F., Li, N., Liu, Q. L., Lin, Y. H., Liu, C. M., et al. (2011). MTDH/AEG-1-based DNA vaccine suppresses lung metastasis and enhances chemosensitivity to doxorubicin in breast cancer. *Cancer Immunology, Immunotherapy, 60*, 883–893.
Salminen, A., Paimela, T., Suuronen, T., & Kaarniranta, K. (2008). Innate immunity meets with cellular stress at the IKK complex: Regulation of the IKK complex by HSP70 and HSP90. *Immunology Letters, 117*, 9–15.
Sanchez-Tillo, E., Liu, Y., de Barrios, O., Siles, L., Fanlo, L., Cuatrecasas, M., et al. (2012). EMT-activating transcription factors in cancer: Beyond EMT and tumor invasiveness. *Cellular and Molecular Life Sciences, 69*, 3429–3456.
Sarkar, D., & Fisher, P. B. (2013). AEG-1/MTDH/LYRIC: Clinical significance. *Advances in Cancer Research,* in press.
Sarkar, D., Park, E. S., Emdad, L., Lee, S. G., Su, Z. Z., & Fisher, P. B. (2008). Molecular basis of nuclear factor-kappaB activation by astrocyte elevated gene-1. *Cancer Research, 68*, 1478–1484.
Semenza, G. L. (2010). Defining the role of hypoxia-inducible factor 1 in cancer biology and therapeutics. *Oncogene, 29*, 625–634.
Semenza, G. L. (2012). Hypoxia-inducible factors: Mediators of cancer progression and targets for cancer therapy. *Trends in Pharmacological Sciences, 33*, 207–214.
Sherr, C. J., & Weber, J. D. (2000). The ARF/p53 pathway. *Current Opinion in Genetics & Development, 10*, 94–99.
Song, H., Li, C., Li, R., & Geng, J. (2010). Prognostic significance of AEG-1 expression in colorectal carcinoma. *International Journal of Colorectal Disease, 25*, 1201–1209.
Song, H., Li, C., Lu, R., Zhang, Y., & Geng, J. (2010). Expression of astrocyte elevated gene-1: A novel marker of the pathogenesis, progression, and poor prognosis for endometrial cancer. *International Journal of Gynecological Cancer, 20*, 1188–1196.
Song, L., Li, W., Zhang, H., Liao, W., Dai, T., Yu, C., et al. (2009). Over-expression of AEG-1 significantly associates with tumour aggressiveness and poor prognosis in human non-small cell lung cancer. *The Journal of Pathology, 219*, 317–326.
Srivastava, J., Siddiq, A., Emdad, L., Santhekadur, P. K., Chen, D., Gredler, R., et al. (2012). Astrocyte elevated gene-1 promotes hepatocarcinogenesis: Novel insights from a mouse model. *Hepatology, 56*, 1782–1791.
Stratmann, A., Risau, W., & Plate, K. H. (1998). Cell type-specific expression of angiopoietin-1 and angiopoietin-2 suggests a role in glioblastoma angiogenesis. *The American Journal of Pathology, 153*, 1459–1466.
Su, Z. Z., Chen, Y., Kang, D. C., Chao, W., Simm, M., Volsky, D. J., et al. (2003). Customized rapid subtraction hybridization (RaSH) gene microarrays identify overlapping expression changes in human fetal astrocytes resulting from human immunodeficiency virus-1 infection or tumor necrosis factor-alpha treatment. *Gene, 306*, 67–78.
Su, Z. Z., Kang, D. C., Chen, Y., Pekarskaya, O., Chao, W., Volsky, D. J., et al. (2002). Identification and cloning of human astrocyte genes displaying elevated expression after infection with HIV-1 or exposure to HIV-1 envelope glycoprotein by rapid subtraction hybridization, RaSH. *Oncogene, 21*, 3592–3602.
Sun, S., Ke, Z., Wang, F., Li, S., Chen, W., Han, A., et al. (2012). Overexpression of astrocyte-elevated gene-1 is closely correlated with poor prognosis in human non-small cell lung cancer and mediates its metastasis through up-regulation of matrix metalloproteinase-9 expression. *Human Pathology, 43*, 1051–1060.
Sutherland, H. G., Lam, Y. W., Briers, S., Lamond, A. I., & Bickmore, W. A. (2004). 3D3/lyric: A novel transmembrane protein of the endoplasmic reticulum and nuclear envelope, which is also present in the nucleolus. *Experimental Cell Research, 294*, 94–105.

Tan, X., Apte, U., Micsenyi, A., Kotsagrelos, E., Luo, J. H., Ranganathan, S., et al. (2005). Epidermal growth factor receptor: A novel target of the Wnt/beta-catenin pathway in liver. *Gastroenterology, 129*, 285–302.

Tanaka, S., Sugimachi, K., Kameyama, T., Maehara, S., Shirabe, K., Shimada, M., et al. (2003). Human WISP1v, a member of the CCN family, is associated with invasive cholangiocarcinoma. *Hepatology, 37*, 1122–1129.

Testa, J. R., & Bellacosa, A. (2001). AKT plays a central role in tumorigenesis. *Proceedings of the National Academy of Sciences of the United States of America, 98*, 10983–10985.

Thirkettle, H. J., Mills, I. G., Whitaker, H. C., & Neal, D. E. (2009). Nuclear LYRIC/AEG-1 interacts with PLZF and relieves PLZF-mediated repression. *Oncogene, 28*, 3663–3670.

Tokunaga, E., Nakashima, Y., Yamashita, N., Hisamatsu, Y., Okada, S., Akiyoshi, S., et al. (2012). Overexpression of metadherin/MTDH is associated with an aggressive phenotype and a poor prognosis in invasive breast cancer. *Breast Cancer*, http://dx.doi.org/10.1007/s12282-012-0398-2.

Vanden Berghe, W., De Bosscher, K., Boone, E., Plaisance, S., & Haegeman, G. (1999). The nuclear factor-kappaB engages CBP/p300 and histone acetyltransferase activity for transcriptional activation of the interleukin-6 gene promoter. *The Journal of Biological Chemistry, 274*, 32091–32098.

van der Saag, P. T., Caldenhoven, E., & van de Stolpe, A. (1996). Molecular mechanisms of steroid action: A novel type of cross-talk between glucocorticoids and NF-kappa B transcription factors. *The European Respiratory Journal. Supplement, 22*, 146s–153s.

van Landeghem, F. K., Weiss, T., Oehmichen, M., & von Deimling, A. (2006). Decreased expression of glutamate transporters in astrocytes after human traumatic brain injury. *Journal of Neurotrauma, 23*, 1518–1528.

Vartak-Sharma, N., & Ghorpade, A. (2012). Astrocyte elevated gene-1 regulates astrocyte responses to neural injury: Implications for reactive astrogliosis and neurodegeneration. *Journal of Neuroinflammation, 9*, 195.

Vivanco, I., & Sawyers, C. L. (2002). The phosphatidylinositol 3-kinase AKT pathway in human cancer. *Nature Reviews. Cancer, 2*, 489–501.

Wang, N., Du, X., Zang, L., Song, N., Yang, T., Dong, R., et al. (2012). Prognostic impact of Metadherin-SND1 interaction in colon cancer. *Molecular Biology Reports, 39*, 10497–10504.

Wang, F., Ke, Z. F., Sun, S. J., Chen, W. F., Yang, S. C., Li, S. H., et al. (2011). Oncogenic roles of astrocyte elevated gene-1 (AEG-1) in osteosarcoma progression and prognosis. *Cancer Biology & Therapy, 12*, 539–548.

Warr, T., Ward, S., Burrows, J., Harding, B., Wilkins, P., Harkness, W., et al. (2001). Identification of extensive genomic loss and gain by comparative genomic hybridisation in malignant astrocytoma in children and young adults. *Genes, Chromosomes & Cancer, 31*, 15–22.

Wei, J., Pan, X., Pei, Z., Wang, W., Qiu, W., Shi, Z., et al. (2012). The beta-lactam antibiotic, ceftriaxone, provides neuroprotective potential via anti-excitotoxicity and anti-inflammation response in a rat model of traumatic brain injury. *The Journal of Trauma and Acute Care Surgery, 73*, 654–660.

Weis, S. M., & Cheresh, D. A. (2011). Tumor angiogenesis: Molecular pathways and therapeutic targets. *Nature Medicine, 17*, 1359–1370.

Xia, Z., Zhang, N., Jin, H., Yu, Z., Xu, G., & Huang, Z. (2010). Clinical significance of astrocyte elevated gene-1 expression in human oligodendrogliomas. *Clinical Neurology and Neurosurgery, 112*, 413–419.

Yan, J., Zhang, M., Chen, Q., & Zhang, X. (2012). Expression of AEG-1 in human T-cell lymphoma enhances the risk of progression. *Oncology Reports, 28*, 2107–2114.

Yang, Z. Z., Tschopp, O., Baudry, A., Dummler, B., Hynx, D., & Hemmings, B. A. (2004). Physiological functions of protein kinase B/Akt. *Biochemical Society Transactions*, *32*, 350–354.

Yang, J., Valineva, T., Hong, J., Bu, T., Yao, Z., Jensen, O. N., et al. (2007). Transcriptional co-activator protein p100 interacts with snRNP proteins and facilitates the assembly of the spliceosome. *Nucleic Acids Research*, *35*, 4485–4494.

Yap, T. A., Garrett, M. D., Walton, M. I., Raynaud, F., de Bono, J. S., & Workman, P. (2008). Targeting the PI3K-AKT-mTOR pathway: Progress, pitfalls, and promises. *Current Opinion in Pharmacology*, *8*, 393–412.

Yoo, B. K., Chen, D., Su, Z. Z., Gredler, R., Yoo, J., Shah, K., et al. (2010). Molecular mechanism of chemoresistance by astrocyte elevated gene-1. *Cancer Research*, *70*, 3249–3258.

Yoo, B. K., Emdad, L., Lee, S. G., Su, Z. Z., Santhekadur, P., Chen, D., et al. (2011). Astrocyte elevated gene-1 (AEG-1): A multifunctional regulator of normal and abnormal physiology. *Pharmacology & Therapeutics*, *130*, 1–8.

Yoo, B. K., Emdad, L., Su, Z. Z., Villanueva, A., Chiang, D. Y., Mukhopadhyay, N. D., et al. (2009). Astrocyte elevated gene-1 regulates hepatocellular carcinoma development and progression. *The Journal of Clinical Investigation*, *119*, 465–477.

Yoo, B. K., Gredler, R., Vozhilla, N., Su, Z. Z., Chen, D., Forcier, T., et al. (2009). Identification of genes conferring resistance to 5-fluorouracil. *Proceedings of the National Academy of Sciences of the United States of America*, *106*, 12938–12943.

Yoo, B. K., Santhekadur, P. K., Gredler, R., Chen, D., Emdad, L., Bhutia, S., et al. (2011). Increased RNA-induced silencing complex (RISC) activity contributes to hepatocellular carcinoma. *Hepatology*, *53*, 1538–1548.

Yu, C., Chen, K., Zheng, H., Guo, X., Jia, W., Li, M., et al. (2009). Overexpression of astrocyte elevated gene-1 (AEG-1) is associated with esophageal squamous cell carcinoma (ESCC) progression and pathogenesis. *Carcinogenesis*, *30*, 894–901.

Yuan, C., Li, X., Yan, S., Yang, Q., Liu, X., & Kong, B. (2012). The MTDH (-470G>A) polymorphism is associated with ovarian cancer susceptibility. *PLoS One*, *7*, e51561.

Zhang, F., Yang, Q., Meng, F., Shi, H., Li, H., Liang, Y., et al. (2012). Astrocyte elevated gene-1 interacts with beta-catenin and increases migration and invasion of colorectal carcinoma. *Molecular Carcinogenesis*, http://dx.doi.org/10.1002/mc.21894.

Zhao, Y., Kong, X., Li, X., Yan, S., Yuan, C., Hu, W., et al. (2011). Metadherin mediates lipopolysaccharide-induced migration and invasion of breast cancer cells. *PLoS One*, *6*, e29363.

Zhong, H., May, M. J., Jimi, E., & Ghosh, S. (2002). The phosphorylation status of nuclear NF-kappa B determines its association with CBP/p300 or HDAC-1. *Molecular Cell*, *9*, 625–636.

Zhou, J., Li, J., Wang, Z., Yin, C., & Zhang, W. (2012). Metadherin is a novel prognostic marker for bladder cancer progression and overall patient survival. *Asia-Pacific Journal of Clinical Oncology*, *8*, e42–e48.

Zhou, B. P., Liao, Y., Xia, W., Spohn, B., Lee, M. H., & Hung, M. C. (2001). Cytoplasmic localization of p21Cip1/WAF1 by Akt-induced phosphorylation in HER-2/neu-overexpressing cells. *Nature Cell Biology*, *3*, 245–252.

Zhu, K., Dai, Z., Pan, Q., Wang, Z., Yang, G. H., Yu, L., et al. (2011). Metadherin promotes hepatocellular carcinoma metastasis through induction of epithelial-mesenchymal transition. *Clinical Cancer Research*, *17*, 7294–7302.

CHAPTER FOUR

Pleiotropic Roles of AEG-1/MTDH/LYRIC in Breast Cancer

Liling Wan*, **Yibin Kang***,†,1
*Department of Molecular Biology, Princeton University, Princeton, New Jersey, USA
†Cancer Institute of New Jersey, New Brunswick, New Jersey, USA
1Corresponding author: e-mail address: ykang@princeton.edu

Contents

Abstract

Since the initial discovery of AEG-1/MTDH/LYRIC, our appreciation for this novel protein's involvement in cancer has increased dramatically over the past few years. *AEG-1/MTDH/LYRIC* is a key functional target of the 8q22 genomic gain that is frequently observed in poor-prognosis breast cancer, where it plays a dual role in promoting chemoresistance and metastasis. Beyond this, growing evidence from clinical research indicates a strong correlation between AEG-1/MTDH/LYRIC expression and the pathogenesis of a large spectrum of cancer types, and multiple studies employing *in vitro* cell culture systems and *in vivo* xenograft models have revealed multifaceted roles of AEG-1/MTDH/LYRIC in cancer biology, including tumor cell proliferation, apoptosis, angiogenesis, and autophagy. With increasing mechanistic understanding of AEG-1/MTDH/LYRIC, discovery of agents that can block AEG-1/MTDH/LYRIC and its regulated pathways will be beneficial to cancer patients with aberrant expression of AEG-1/MTDH/LYRIC.

Advances in Cancer Research, Volume 120
ISSN 0065-230X
http://dx.doi.org/10.1016/B978-0-12-401676-7.00004-8

1. INTRODUCTION

AEG-1/MTDH/LYRIC was initially reported as a novel gene induced by human immunodeficiency virus-1 in primary human fetal astrocytes (Su et al., 2002). Subsequently, four independent groups cloned *AEG-1/MTDH/LYRIC* (Britt et al., 2004; Brown & Ruoslahti, 2004; Kang et al., 2005; Sutherland, Lam, Briers, Lamond, & Bickmore, 2004). Employing *in vivo* phage display screening, Brown and colleagues identified mouse AEG-1/MTDH/LYRIC as a protein mediating specific adhesion of mouse 4T1 mammary tumor cells to lung vascular endothelium, thus giving it the name "Metadherin" (Brown & Ruoslahti, 2004). The mouse/rat orthologs of AEG-1/MTDH/LYRIC were also found as lysine-rich CEACAM1 coisolated (LYRIC) protein that is associated with tight junctions in polarized prostate epithelial cells (Britt et al., 2004), and as a novel transmembrane protein that is present in the cytoplasm, endoplasmic reticulum, perinuclear regions, and nucleolus by gene-trapping techniques (Sutherland et al., 2004).

The initial identification of AEG-1/MTDH/LYRIC raised broad controversies on the understanding of its biological functions and molecular characteristics, most of which still remain elusive. Nevertheless, some consensus features of AEG-1/MTDH/LYRIC have been recognized. Evolutionally, AEG-1/MTDH/LYRIC orthologs have been identified in most vertebrates with a high degree of evolutionary conservation but not detected in lower invertebrates, indicating that AEG-1/MTDH/LYRIC may have specialized functions that evolve only in higher organisms. At molecular level, the human *AEG-1/MTDH/LYRIC* encodes a 582-amino acid protein with no recognizable domains that could indicate its function, except for three putative lysine-rich nuclear localization signals (Thirkettle, Girling, et al., 2009). Distinct isoforms or modifications of AEG-1/MTDH/LYRIC have long been speculated based on multiple RNA/protein species detected (Britt et al., 2004; Brown & Ruoslahti, 2004; Kang et al., 2005; Sutherland et al., 2004), although the identity and function of these isoforms and/or modifications remain mysterious. The subcellular localization of AEG-1/MTDH/LYRIC is variable and dependent on the cell types examined and detection methods employed. In most cases, endogenous and ectopic expression of AEG-1/MTDH/LYRIC is predominantly cytoplasmic (including endoplasmic reticulum and perinuclear regions) (Blanco et al., 2011; Kang et al., 2005; Li, Zhang, et al., 2008; Meng et al., 2012; Yoo et al., 2011), but it is also found in the nucleus, especially nucleolus (Emdad et al., 2006; Sutherland et al., 2004; Thirkettle, Girling, et al., 2009; Thirkettle, Mills, Whitaker, & Neal, 2009) and plasma membrane (Britt et al., 2004; Brown & Ruoslahti, 2004). How the subcelluar localization of AEG-1/MTDH/LYRIC is regulated and what effects this has on its function are largely unknown.

AEG-1/MTDH/LYRIC is expressed ubiquitously in almost all human and murine tissues at variable levels (Jeon et al., 2010; Kang et al., 2005). In cancer, the expression level of AEG-1/MTDH/LYRIC is dramatically elevated. The first piece of evidence suggesting a functional involvement of AEG-1/MTDH/LYRIC in cancer was found in mouse mammary tumor metastasis (Brown & Ruoslahti, 2004). In this study, an unbiased screen for cell surface proteins that mediate the metastasis of the 4T1 murine mammary tumor cells to the lung identified a "lung-homing domain (LHD)" belonging to AEG-1/MTDH/LYRIC. Later in 2009, the 8q22 genomic region, where *AEG-1/MTDH/LYRIC* resides, was reported to have recurrent amplification in more than 30% of breast cancer, and this genomic alteration was associated with poor clinical outcome, underscoring the potentially crucial role of this gene in breast cancer progression (Hu, Chong, et al., 2009). In addition, numerous studies over the past decade have demonstrated a multifaceted role of AEG-1/MTDH/LYRIC in regulating phenotype characteristics of malignant features, such as aberrant proliferation (Lee et al., 2009; Li et al., 2009; Yu et al., 2009), evasion of apoptosis (Kikuno et al., 2007; Lee et al., 2008), invasion (Emdad et al., 2006; Sarkar et al., 2008), angiogenesis (Emdad et al., 2009), and chemoresistance (Hu, Chong, et al., 2009; Liu et al., 2009; Yoo, Gredler, et al., 2009; Yoo et al., 2010), in multiple cancer types. With a focus on the emerging roles of AEG-1/MTDH/LYRIC in breast cancer in this chapter, we present evidence that the expression level of AEG-1/MTDH/LYRIC correlates with patient outcome, review evidence that AEG-1/MTDH/LYRIC promotes a large spectrum of tumor-related properties of cancer cells, discuss our limited understanding of its mechanisms, and finally, evaluate the potential of therapeutic targeting of AEG-1/MTDH/LYRIC.

2. ABERRATIONS OF AEG-1/MTDH/LYRIC IN BREAST CANCER

The high frequency of AEG-1/MTDH/LYRIC overexpression in many different cancer types underscores the importance of this protein in cancer biology. Compared to normal breast tissues, in which AEG-1/MTDH/LYRIC is almost undetectable by IHC staining, the level of AEG-1/MTDH/LYRIC is drastically increased in breast cancer cell lines and tumors (Brown & Ruoslahti, 2004; Hu, Chong, et al., 2009; Kang et al., 2005; Li, Zhang, et al., 2008; Su, Zhang, & Yang, 2010). In two large cohorts of archived breast cancer samples in China (Li, Zhang, et al., 2008) and the United States (Hu, Chong, et al., 2009), more than 40% of breast cancers have a higher level of AEG-1/MTDH/LYRIC compared to normal counterparts. In most cases, AEG-1/MTDH/LYRIC is observed in cytoplasm, although a minority of primary cancer cells are also stained positive for AEG-1/MTDH/LYRIC in the nucleus (Li, Zhang, et al., 2008).

Breast cancer is a heterogeneous disease that can be classified into different subtypes based on gene expression profiles and molecular markers (Perou et al., 2000; Sorlie et al., 2001). AEG-1/MTDH/LYRIC neither seems to correlate with a specific subtype of breast cancers nor is it significantly associated with other common clinicopathological parameters including age, estrogen receptor (ER), progesterone receptor, HER2, and p53 status (Hu, Chong, et al., 2009; Li, Zhang, et al., 2008). Instead, the abundance of AEG-1/MTDH/LYRIC is positively correlated with advanced clinical stages and clinicopathological features, as well as distant metastasis and poor patient survival (Dalgin et al., 2007; Hu, Chong, et al., 2009; Tokunaga et al., 2012). Multivariate Cox analysis showed that the risk of metastasis was significantly higher with AEG-1/MTDH/LYRIC expression, even when all of the other factors, including ER, PR, HER2, p53, and tumor size were considered (Hu, Chong, et al., 2009). Taken together, the clinical association studies suggest that the expression of AEG-1/MTDH/LYRIC can be used as an independent prognostic indicator for metastasis and survival of breast cancer patients.

3. VARIANTS OF *AEG-1/MTDH/LYRIC* IN BREAST CANCER

Single-nucleotide polymorphisms have emerged as useful tools to help evaluate the susceptibility, prognosis, and treatment response of cancer

(Easton et al., 2007; Orr & Chanock, 2008). Aiming to identify novel variants of *AEG-1/MTDH/LYRIC* in breast cancer, Liu et al. direct sequenced 108 breast cancer samples and 100 normal controls (Liu et al., 2011). This study led to the identification of 13 variants in the control group and 11 in the breast cancer patient group, among which 2 variants were found to be associated with increased susceptibility for breast cancer development. While larger patient populations are required to confirm the correlation of these variants to breast cancer susceptibility and further examine their prognostic value, this study provides new possibilities of how AEG-1/MTDH/LYRIC may contribute to cancer development. In addition to this finding in breast cancer, one recent study assessing *AEG-1/MTDH/LYRIC* gene polymorphisms and their potential relationship to ovarian cancer susceptibility has also been reported (Yuan et al., 2012). By comparing 145 ovarian cancer patients and 254 matched control subjects, it was found that the *AEG-1/MTDH/LYRIC* (−470G>A) polymorphism was statistically correlated with ovarian cancer risk and clinical stage. These data suggest that *AEG-1/MTDH/LYRIC* (−470G>A) could be a useful molecular marker for assessing ovarian cancer risk and for predicting ovarian cancer patient prognosis. The implication of this variant in breast cancer and other cancer types remains to be investigated in future study.

4. REGULATION OF *AEG-1/MTDH/LYRIC* IN BREAST CANCER

Cancer is a genetic disease characterized by rampant genetic instability and massive genetic/epigenetic alterations (Chin & Gray, 2008; Hanahan & Weinberg, 2011). Recurrent DNA copy number alteration is one type of genetic change that has been observed in a wide range of human cancers, and such genetic events often indicate the presence of key mediators of malignancy in the affected genomic loci (Chin & Gray, 2008). Using a computational algorithm to map minimal recurrent genomic alterations that are associated with poor-prognosis breast cancer, Hu et al. identified the poor-prognosis genomic gain of chromosome 8q22, where human *AEG-1/MTDH/LYRIC* is located (Hu, Chong, et al., 2009). Regional gain of 8q22 was further validated in an extensive collection of breast tumor samples and cell lines. As expected, AEG-1/MTDH/LYRIC was found to be overexpressed in breast tumors with 8q22 amplification (Hu, Chong, et al., 2009; Li et al., 2011). Although the elevated level of AEG-1/MTDH/LYRIC can be predominantly attributed to genomic amplification, a substantial fraction of breast tumors with normal copies of *AEG-1/MTDH/LYRIC* also overexpresses the protein, suggesting alternative mechanisms of *AEG-1/MTDH/LYRIC* upregulation in cancer (discussed later). Nevertheless, survival analysis of breast cancer patients showed that AEG-1/MTDH/LYRIC activated by genomic gain or other means led to similar clinical outcome, highlighting the prognostic value of *AEG-1/MTDH/LYRIC* itself, rather than its association with other genes on the 8q22 genetic locus (Hu, Chong, et al., 2009; Li et al., 2011).

Studies accumulated so far have suggested alternative mechanisms for the regulation of *AEG-1/MTDH/LYRIC* in cancer (Lee, Su, Emdad, Sarkar, & Fisher, 2006; Ward et al., 2013; Zhang et al., 2011). One possibility is transcriptional regulation by oncogenic regulatory signals. For instance, in human adult astrocytes, activation of the RAS oncogene and subsequent induction of transcription factor c-Myc lead to increased binding of c-Myc to the *AEG-1/MTDH/LYRIC* promoter region, which augments the transcription level of *AEG-1/MTDH/LYRIC* (Lee et al., 2006). Notably, *c-Myc* is often amplified in breast cancer (Bergamaschi et al., 2006). Yet it remains to be tested whether or not *AEG-1/MTDH/LYRIC* is subjected to transcriptional regulation by c-Myc in human breast tumor tissues.

The second potential way to regulate *AEG-1/MTDH/LYRIC* is through microRNAs (miRNAs). miRNAs have been increasingly recognized as crucial regulators for tumorigenesis (Chen, 2005). In fact, downregulation of miRNAs that target oncogenes is frequently observed in cancer, and this remains one of the key mechanisms of oncogene overexpression. Analysis of the 3′UTR of *AEG-1/MTDH/LYRIC* reveals multiple miRNA-binding sites, and two miRNAs have recently been reported to suppress oncogenic phenotypes by targeting *AEG-1/MTDH/LYRIC* (Nohata et al., 2011; Ward et al., 2013; Zhang et al., 2011). In breast cancer, the level of miR-26a is significantly decreased compared to adjacent normal tissues (Iorio et al., 2005; Zhang et al., 2011). Expression and mutagenesis assays validated *AEG-1/MTDH/LYRIC* and *EZH2* as two of the targets of miR-26a that mediated the proapoptotic effect of miR-26a *in vitro*. Furthermore, a negative correlation between the expression level of miR-26a and MTDH/EZH2 is also observed in a few paired clinical breast cancer samples (Zhang et al., 2011). While large-scale clinical association studies have yet to be conducted, this study provides the first piece of evidence that *AEG-1/MTDH/LYRIC* may be regulated by miRNAs in breast cancer. Besides miR-26a, tumor suppressive miRNA miR-375 was also found to have inversed expression level to that of *AEG-1/MTDH/LYRIC* in breast cancer, hepatocellular carcinoma, head and neck squamous cell carcinoma, and esophageal squamous cell carcinoma (He et al., 2012; Hui et al., 2011; Nohata et al., 2011; Ward et al., 2013). Intriguingly, miR-375 is downregulated in a tamoxifen-resistance clone of human breast cancer cell line MCF7, and this event is associated with epithelial to mesenchymal transition (EMT)-like properties. *AEG-1/MTDH/LYRIC* was shown to be one of the targets of miR-375 to mediate this effect (Ward et al., 2013). Of note, while the level of miR-375 is much higher in ER-positive breast cancer, AEG-1/MTDH/LYRIC expression does not significantly correlate with ER status. This further supports the notion that there are multiple alternative mechanisms underlying the regulation of AEG-1/MTDH/LYRIC in breast cancer, which require further investigation.

5. AEG-1/MTDH/LYRIC IN BREAST CANCER GROWTH CONTROL

AEG-1/MTDH/LYRIC has been demonstrated in different contexts to affect cancer cell growth. In some clinical breast cancer samples, AEG-1/MTDH/LYRIC expression is highly associated with proliferative marker Ki-67 (Li, Zhang, et al., 2008). Functionally, ectopic expression and RNAi silencing of *AEG-1/MTDH/LYRIC* promotes and decreases, respectively, proliferation of breast cancer cell lines (Li et al., 2009). Additionally, key cell cycle inhibitors $p27^{Kip1}$ and $p21^{Cip1}$ were found to be downregulated by AEG-1/MTDH/LYRIC via an increase of phosphorylation and subsequent cytoplasm retention of FOXO1 transcription factor (Li et al., 2009). In this study, the effect of AEG-1/MTDH/LYRIC on FOXO1 phosphorylation is likely to be mediated by PI3K/AKT signaling, as inhibitors of this pathway abolished the growth-promoting function of AEG-1/MTDH/LYRIC. Intriguingly, AEG-1/MTDH/LYRIC has been reported in prostate cancer cells to physically interact with and promote the proteasome degradation of BCCIPα, a CDKN1A and BRCA2 binding protein involved in DNA repair and cell cycle control (Ash, Yang, & Britt, 2008). It is known that BCCIPα binds to p21 and enhances its inhibitory activity toward CDK2, leading to impairment in G1/S cell cycle progression (McShea, Samuel, Eppel, Galloway, & Funk, 2000; Meng, Liu, & Shen, 2004; Meng, Lu, & Shen, 2004). While the interaction between AEG-1/MTDH/LYRIC and BCCIPα has yet to be confirmed in breast cancer, it can be speculated that AEG-1/MTDH/LYRIC may enhance proliferation in breast cancer through its negative regulation of BCCIPα.

It is of note that inconsistencies concerning the involvement of AEG-1/MTDH/LYRIC in breast cancer proliferation do exist. For example, Hu et al. revealed no difference in proliferation or cell growth caused by manipulation of AEG-1/MTDH/LYRIC in multiple breast cancer cell lines (Hu, Chong, et al., 2009). Whether these discrepancies are caused by differences in culture conditions, cell line heterogeneity, or assays employed remains to be addressed.

6. AEG-1/MTDH/LYRIC AND EMT IN BREAST CANCER

EMT, an essential embryonic program, is often aberrantly activated in cancer. This process can endow epithelial cancer cells with a highly motile mesenchymal phenotype associated with increased metastatic capability (Nieto, 2011). In carcinomas, the cancer cells at the invasive front are often stained positive for mesenchymal protein markers such as N-cadherin and Vimentin, suggesting that these cells may have undergone EMT and invaded into surrounding tissues. Two recent studies reported a possible link between AEG-1/MTDH/LYRIC and EMT in breast cancer (Li et al., 2011; Ward et al., 2013). Li and colleagues showed that transient ectopic expression of AEG-1/MTDH/LYRIC in MCF7 breast cancer cells enhanced their migratory and invasive capabilities by inducing EMT (Li et al., 2011). This was evidenced by a switch from epithelial-to-mesenchymal-like cell shape, downregulation of E-cadherin expression, and upregulation of mesenchymal markers Fibronectin and Vimentin. Moreover, transcriptional factors that are important for EMT such as Snail and Slug are also upregulated by AEG-1/MTDH/LYRIC in MCF7 cells. Mechanistically, it was showed that inhibitors of NF-κb pathway or RNAi silencing of NF-κB subunit p65 abolished the effect of AEG-1/MTDH/LYRIC on EMT. Of note, AEG-1/MTDH/LYRIC has been linked to NF-κB signaling pathway in the context of cell growth, survival, and invasion in other cancer types (Emdad et al., 2006; Kikuno et al., 2007; Liu et al., 2010; Sarkar et al., 2008). In those studies, AEG-1/MTDH/LYRIC was hypothesized to act as a transcriptional coactivator in the nucleus through its binding with the NF-κB subunit p65. Given the observation that AEG-1/MTDH/LYRIC is predominantly localized in the cytoplasm of breast cancer cells, it is important to vigorously test whether the interaction with NF-κB pathway is the main mechanism through which AEG-1/MTDH/LYRIC may enhance EMT and invasion of breast cancer cells.

Another study also using MCF7 cells revealed an unexpected connection between AEG-1/MTDH/LYRIC, EMT, and tamoxifen resistance. MCF7 cells are ER positive and thus are sensitive to tamoxifen treatment. However, it has been demonstrated that long-term treatment of MCF7 cells with tamoxifen will result in the appearance of resistant clones, recapitulating clinical observation that ER+ breast cancers often acquire endocrine resistance with prolonged treatment (Musgrove & Sutherland, 2009). Multiple studies demonstrated that tamoxifen-resistant clones of MCF7 have undergone EMT, as indicated by morphological and transcriptional changes (Hiscox et al., 2006; Kim, Choi, Cho, Kim, & Kang, 2009; Ward et al., 2013). An unbiased miRNA screening identified mir-375 as a suppressor of this EMT-associated resistance by targeting *AEG-1/MTDH/LYRIC*. Silencing of *AEG-1/MTDH/LYRIC* in the tamoxifen-resistant clones of MCF7 partially reversed EMT in this context. Endocrine therapies have become the most important treatment options for women with ER-positive breast cancer (which account for 70% of diagnosed cases); however, their efficacy is limited by intrinsic and acquired therapeutic resistance. While more thorough mechanistic studies are needed to further prove the link between AEG-1/MTDH/LYRIC and endocrine resistance, this study suggests a potential use of AEG-1/MTDH/LYRIC as both a diagnosis marker for treatment resistance and a target for combination therapy to enhance drug efficacy.

7. AEG-1/MTDH/LYRIC AND BREAST CANCER METASTASIS

Poor prognosis of breast cancer at the time of diagnosis or surgery indicates a higher probability of death, mainly as the result of metastasis to vital organs. The strong clinical correlation between AEG-1/MTDH/LYRIC expression and shorter metastasis-free survival suggests that AEG-1/MTDH/LYRIC may function as a metastasis gene with great prognostic potential and therapeutic value. Two studies so far have independently identified and demonstrated the functional involvement of AEG-1/MTDH/LYRIC in breast cancer metastasis (Brown & Ruoslahti, 2004; Hu, Chong, et al., 2009). Brown et al. used *in vivo* phage screening of cDNAs library from metastatic breast carcinoma to identify protein domains that preferentially bind to lung vasculatures. An extracellular LHD (a.a. 378–440 in mouse or 381–443 in human) in AEG-1/MTDH/LYRIC has been uncovered to mediate the adhesion of 4T1 murine mammary tumor cells to lung endothelium and thus promote lung metastasis. Neutralizing antibodies against this LHD- or siRNA-mediated silencing of *AEG-1/MTDH/LYRIC* reduce experimental metastasis of 4T1 cells. On the contrary, ectopic expression of *AEG-1/MTDH/LYRIC* in human embryonic kidney cells HEK293 results in increased localization of these cells to the lung vasculatures. The effect of AEG-1/MTDH/LYRIC on metastasis has also been validated using the MDA-MB-231 xenograft model of breast cancer metastasis (Hu, Chong, et al., 2009). In this model system, AEG-1/MTDH/LYRIC promotes experimental metastasis not only to the lung but also to the bone, albeit to a lesser extent.

Mechanistically, it is conceivable that the LHD of AEG-1/MTDH/LYRIC may function as an adhesion molecule to direct the association of cancer cells to lung endothelia cells, as it was originally identified (Brown & Ruoslahti, 2004). In fact, it has been showed that antibodies against LHD bound to nonpermeabilized cells, confirming the presence of this LHD of AEG-1/MTDH/LYRIC at the cell surface of tumor cells where it would be able to bind to vascular targets during metastasis. However, a significant amount of AEG-1/MTDH/LYRIC was also detected in the cytoplasm when cells were permeabilized (Brown & Ruoslahti, 2004). This is consistent with subsequent studies in which AEG-1/MTDH/LYRIC has been shown to predominantly localize in the cytoplasm in breast tumor tissues and cancer cell lines (Blanco et al., 2011; Li, Zhang, et al., 2008). Therefore, a better understanding of the function of AEG-1/MTDH/LYRIC requires a higher resolution of its possible dynamic changes in subcellular localization under various pathological conditions and during cancer progression. It also remains to be tested whether AEG-1/MTDH/LYRIC may affect the lung-homing properties of cancer cells by inducing adhesion molecules through its interaction with other pathways such as NF-κB, a known pathway that has been shown to activate a cohort of adhesion genes. Alternatively, AEG-1/MTDH/LYRIC may affect different steps during metastatic cascades, such as cancer cell dissemination through EMT, rather than homing cancer cells to the lung.

8. AEG-1/MTDH/LYRIC AND CHEMORESISTANCE

Chemoresistance remains one of the biggest challenges for clinical management of breast cancer. Current standard treatment for breast cancer uses a combination of surgical removal of primary tumors and adjuvant chemotherapy to prevent systematic spreading. However, recurrent cancers inevitably acquire resistance to treatment and often spread to distant organs, which accounts for over 90% of cancer-related deaths. This suggests that metastatic cancers not only have to overcome numerous obstacles during the multistep process of metastasis but also need to acquire the ability of survival under antineoplastic stresses such as those imposed by standard adjuvant therapy. Pharmacogenomic analysis of the NCI60 panel of cancer cell lines revealed that the genomic gains of 8q22 strongly correlate with a higher mean GI50 (the drug concentration for 50% growth inhibition) for nearly half of the drugs used in this database (Hu, Chong, et al., 2009). Moreover, 8q22 amplification, and the resulting overexpression of genes located in this region, significantly associates with cancer recurrence despite of adjuvant chemotherapy (Li et al., 2010). When individual genes on the 8q22 region were examined, only the mRNA level of *AEG-1/MTDH/LYRIC* was found to significantly correlate with higher drug resistance of NCI60 cell lines (Hu, Chong, et al., 2009), although another two genes, *LAMPTM4B* and *YWHAZ*, were demonstrated in a separate study to correlate with and functionally contribute to chemoresistance of breast cancer cell lines (Li et al., 2010). *In vitro* and *in vivo* silencing of *AEG-1/MTDH/LYRIC* sensitize human breast cancer cells to a broad spectrum of chemotherapy drugs, including paclitacxel, doxorubicin, cisplatin, and hydrogen peroxide (Hu, Chong, et al., 2009; Li et al., 2010). Importantly, in a therapeutic model, AEG-1/MTDH/LYRIC-based DNA vaccine was proved to increase chemosensitivity to doxorubicin and inhibit breast cancer lung metastasis (Qian et al., 2011). Mechanistically, AEG-1/MTDH/LYRIC does not seem to affect the uptake and/or retention of drugs in breast cancer cells but rather confers a survival advantage under antineoplastic stress (Hu, Chong, et al., 2009). This could be mediated by survival pathways such as PI3K and NF-κB, which have been connected to AEG-1/MTDH/LYRIC in other cancer types (Hu, Wei, & Kang, 2009), or by direct downstream targets of AEG-1/MTDH/LYRIC. In fact, microarray profiling in breast cancers revealed that AEG-1/MTDH/LYRIC increases drug resistance genes such as *ALDH3A1* and *MET* and downregulates proapoptotic genes such as *TRAIL*.

The chemoresistance function of AEG-1/MTDH/LYRIC has also expanded to other cancer types, including neuroblastoma (Liu et al., 2009), hepatocellular carcinoma (Yoo et al., 2010), ovarian (Li et al., 2012; Meng et al., 2011), and endometrial cancers (Meng et al., 2011). Despite accumulating evidence for the broad chemoresistance effect of AEG-1/MTDH/LYRIC, the underlying mechanisms seem to vary among different cancer types or depend on chemotherapeutic drugs tested. For example, in contrast to what has been shown in breast cancer (Hu, Chong, et al., 2009), AEG-1/MTDH/LYRIC decreases doxorubicin uptake and retention in the hepatocellular carcinoma cells through regulation of multidrug resistance gene MDR1 at posttranscriptional level (Yoo et al., 2010). Alternatively, AEG-1/MTDH/LYRIC may confer resistance to chemotherapeutic drugs by regulating genes that are critical for drug action and catabolism, as in the case of 5-FU. It has been reported that AEG-1/MTDH/LYRIC confers resistance to 5-FU in hepatocellular carcinoma cells by augmenting the expression of the transcription factor LSF that induces thymidylate synthase, the substrate for 5-FU, and by increasing the 5-FU-catabolizing enzyme dihydropyrimidine dehydrogenase, DPYD (Yoo, Gredler, et al., 2009). Future studies are necessary to gain a better understanding of how AEG-1/MTDH/LYRIC could confer resistance to a diverse range of chemotherapeutic drugs through distinct mechanisms.

9. MOLECULAR UNDERSTANDING OF AEG-1/MTDH/LYRIC

9.1. Integration of oncogenic pathways

In sharp contrast to the rapidly growing clinical and phenotypic studies of AEG-1/MTDH/LYRIC, our molecular understanding of this novel protein remains poor. Depending on the cancer cell types tested, AEG-1/MTDH/LYRIC can activate multiple major oncogenic pathways, including PI3K/AKT, Wnt/β-catenin, and NF-κB signaling pathways, to promote different aspects of tumor malignancy (Fig. 4.1). Overexpression of AEG-1/MTDH/LYRIC inhibits serum starvation-induced cell death of human fetal astrocytes through the activation of PI3K–AKT signaling pathway and its downstream substrates, such as GSK3β/c-Myc, MDM2/p53, and Bad (Lee et al., 2008). On the other hand, silencing of *AEG-1/MTDH/LYRIC* results in apoptosis of prostate cancer cell lines through upregulation of forkhead box 3a activity, which is dependent on the reduction of AKT signaling (Kikuno et al., 2007). How AEG-1/MTDH/LYRIC regulates PI3K/AKT pathway is currently unknown. In addition, AEG-1/MTDH/LYRIC promotes anchorage-independent growth and invasion of Hela cells by enhancing the nuclear accumulation, DNA binding, and transcriptional activities of the NF-κB subunit p65 (Emdad et al., 2006). This may be achieved by directly binding of AEG-1/MTDH/LYRIC to p65 (Sarkar et al., 2008), although this model requires further elucidation, as

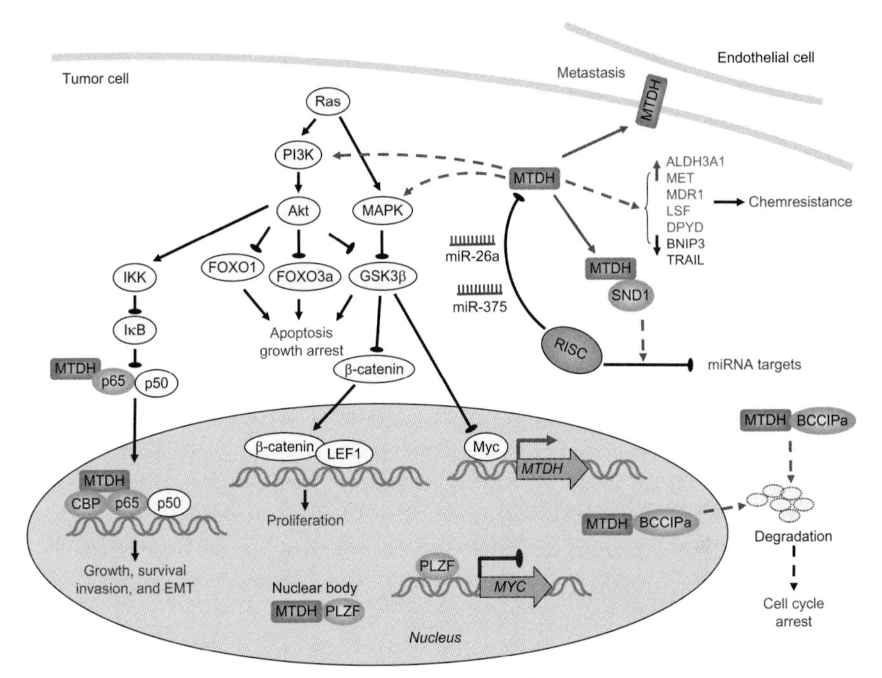

Figure 4.1 A schematic depiction of AEG-1/MTDH/LYRIC functions in cancer. AEG-1/MTDH/LYRIC can be regulated at transcriptional level by oncogenic Ha-Ras through the activation of PI3K/AKT signaling, or alternatively, at posttranscriptional level by miRNAs such as miR-375 and miR-26a. AEG-1/MTDH/LYRIC activates PI3K/AKT, Wnt/β-catenin, and NFκB pathways to promote growth, survival, EMT, and invasion under different conditions. The activation of NFκB pathway by AEG-1/MTDH/LYRIC is in part through the direct interaction of AEG-1/MTDH/LYRIC with p65 and CBP, a general transcriptional coactivator. AEG-1/MTDH/LYRIC confers resistance to a broad spectrum of chemotherapeutic agents by regulating a number of downstream genes. The prometastasis function of AEG-1/MTDH/LYRIC could be mediated by the interaction of the lung-homing domain of AEG-1/MTDH/LYRIC with an unknown receptor in endothelial cells. In the cytoplasm, AEG-1/MTDH/LYRIC binds to SND1 and thus associates with RNA-induced silencing complex (RISC). AEG-1/MTDH/LYRIC also interacts with and targets BCCIPα for proteasome-mediated degradation. Nuclear AEG-1/MTDH/LYRIC binds to PLZF and relieves PLZF-mediated repression on genes such as *c-Myc*. Proteins with direct interactions with AEG-1/MTDH/LYRIC are shown in green, while proteins with possible interactions with AEG-1/MTDH/LYRIC are shown in blue. Red-dotted line indicates pathways/mechanisms yet to be fully validated and/or characterized. Additional oncogenic phenotypes and related mechanisms of AEG-1/MTDH/LYRIC not discussed in this review are not highlighted here. (See Page 5 in Color Section at the back of the book.)

the interacting domain of AEG-1/MTDH/LYRIC to p65 failed to mediate its effect on NF-κB activation. Alternatively, AEG-1/MTDH/LYRIC may function to bridge the interaction between p65 and CBP, a transcriptional coactivator of NF-κB in glioma (Sarkar et al., 2008). In hepatocellular carcinoma, AEG-1/MTDH/LYRIC has been connected to the Wnt/β-catenin pathway through the activation of the Raf/MEK/MAPK branch of the Ras signaling pathway (Yoo, Emdad, et al., 2009).

Despite the evidence presented in these above cancer types, little is known about whether or not AEG-1/MTDH/LYRIC exerts its roles in breast cancer through cross talk with these well-known oncogenic pathways. The observation that AEG-1/MTDH/LYRIC displays diverse oncogenic phenotypes in a highly context-dependent manner may well suggest that AEG-1/MTDH/LYRIC is a multifunctional protein that works through different mechanisms.

9.2. Interacting partners of AEG-1/MTDH/LYRIC

Despite the high level of interest in AEG-1/MTDH/LYRIC due to its diverse oncogenic phenotypes, the lack of any recognizable functional domains in this protein remains as a major obstacle for a better understanding of its molecular mechanisms. Multiple groups have employed unbiased screening methods to uncover binding partners of AEG-1/MTDH/LYRIC in a variety of cell types and physiological conditions (Fig. 4.1) (Ash et al., 2008; Blanco et al., 2011; Meng et al., 2012; Thirkettle, Mills, et al., 2009; Yoo et al., 2011). AEG-1/MTDH/LYRIC is a dynamic protein that localizes in multiple subcellular compartments, and in different locations, it binds to specific proteins. In the nucleus, AEG-1/MTDH/LYRIC was first shown to bind to the p65 subunit of NF-κB, and this interaction augments the transcriptional activity of p65 (Sarkar et al., 2008). Two additional AEG-1/MTDH/LYRIC-interacting proteins, PLZF and BCCIPα, were later identified in prostate cancer cells using yeast two-hybrid screening (Ash et al., 2008; Thirkettle, Mills, et al., 2009). AEG-1/MTDH/LYRIC binds to PLZF via the NH_2-terminus (a.a. 1–285) and the COOH-terminus (a.a. 487–582) and localizes with PLZF transcriptional repressor machinery within nuclear bodies. Overexpression of AEG-1/MTDH/LYRIC reduces PLZF-mediated repression by abrogating the ability of PLZF to bind DNA, perhaps by sequestering PLZF to nuclear bodies (Thirkettle, Mills, et al., 2009). The AEG-1/MTDH/LYRIC–BCCIPα interaction was mapped to the region amino acid 72–192 of AEG-1/MTDH/LYRIC. Instead of affecting transcriptional activities of its binding partners, AEG-1/MTDH/LYRIC targets BCCIPα for proteasomal degradation (Ash et al., 2008).

In human metastatic breast cancer cell lines, protein coimmunoprecipitation followed by mass spectrometry sequencing led to the identification of *staphylococcal* nuclease domain-containing 1 (SND1) as a major AEG-1/MTDH/LYRIC-interacting protein in the cytoplasm (Blanco et al., 2011). SND1 is a multifunctional protein that contains four N-terminal SN domain repeats, which harbor nuclease activity (Li, Yang, Chen, & Yuan, 2008), and a C-terminal Tudor-SN hybrid domain, which has been shown to interact with methylated protein substrates (Shaw et al., 2007). The multifaceted roles of SND1 depend on its subcellular localizations. When localized in the nucleus, SND1 couples transcription and splicing via interactions with key components of both processes. On the one hand, it can act as a transcriptional coactivator to enhance transcriptional activity of EBNA-2 (Tong, Drapkin, Yalamanchili, Mosialos, & Kieff, 1995), c-Myb (Leverson et al., 1998), STAT5 (Paukku, Yang, & Silvennoinen, 2003), and STAT6 (Yang et al., 2002). On the other hand, it interacts with the spliceosome machinery (Shaw et al., 2007) and thus facilitates mRNA splicing (Yang et al., 2007). In the cytoplasm, SND1 functions as a nuclease in the RNA-induced silencing complex (RISC), where it promotes cleavage of double-stranded RNA and hyperedited double-stranded RNA substrates (Caudy et al., 2003). Furthermore, SND1 has been implicated in other cytoplasmic processes such as programmed cell death (Sundstrom et al., 2009) and the formation of stress granules upon induction of various cellular stresses (Gao et al., 2010). Though various intracellular localizations have been reported for AEG-1/MTDH/LYRIC and SND1, endogenous AEG-1/MTDH/LYRIC and SND1 were each observed to predominantly localize to regions of the ER with a diffuse signal observed in the cytoplasm of human breast cancer cells (Blanco et al., 2011). Immunofluorescence analysis revealed colocalization of AEG-1/MTDH/LYRIC and SND1 in punctate patterns to the ER/cytoplasm regions. Domain-mapping experiments with a series of AEG-1/MTDH/LYRIC deletion mutant constructs demonstrated that amino acid 364–470 region of AEG-1/MTDH/LYRIC is required for its interaction with SND1, whereas both SN domains and Tudor-SN domain of SND1 are sufficient to bind to AEG-1/MTDH/LYRIC (Blanco et al., 2011). To explore the potential functionality of SND1 in the context as a novel AEG-1/MTDH/LYRIC-interacting protein in breast cancers, Blanco et al. conducted *in vivo* functional studies and demonstrated that SND1 is itself a strong promoter for metastasis. However, similar to other AEG-1/MTDH/LYRIC-binding partners, the functional significance of the interaction remains to be elucidated.

Two other groups have also independently uncovered the AEG-1/MTDH/LYRIC–SND1 interaction in different cancer types, underscoring the potential importance of this binding (Meng et al., 2012; Yoo et al., 2011). Intriguingly, Yoo et al. mapped the domain of AEG-1/MTDH/LYRIC–SND1 interaction to amino acid 101–205 of AEG-1/MTDH/LYRIC, a region that does not overlap with what has been reported. This discrepancy remains to be further clarified using biochemical and structural studies. Given the observation that SND1 could interact with AEG-1/MTDH/LYRIC with both SN domain and Tudor-SN domains (Blanco et al., 2011), it can be speculated that more than one domain of AEG-1/MTDH/LYRIC is sufficient to interact with SND1. Functionally, AEG-1/MTDH/LYRIC was documented to associate with RISC through its physical interaction with SND1 in hepatocellular carcinoma, and AEG-1/MTDH/LYRIC-mediated activation of RISC increased degradation of tumor suppressor mRNAs that are targets of oncomiRs (Yoo et al., 2011). Still, it is unclear how AEG-1/MTDH/LYRIC affects RISC activities and how this action results in net oncogenic effects of miRNAs in hepatocellular cancer cells. Besides SND1, AEG-1/MTDH/LYRIC interacts with multiple other cytoplasmic proteins including NPM1 and RPL4 (Meng et al., 2012). Interestingly, treatment with Benzonase nuclease abolished the binding of AEG-1/MTDH/LYRIC with SND1 and RPL4, prompting the hypothesis that AEG-1/MTDH/LYRIC functions as an mRNA-binding protein.

Taken together, identification of AEG-1/MTDH/LYRIC-interacting proteins has provided interesting insights into how AEG-1/MTDH/LYRIC may orchestrate gene regulation and exert its function in a highly context-dependent manner. Future studies are urgently needed to further elucidate the functional significance of these interactions to AEG-1/MTDH/LYRIC-mediated oncogenic effects.

10. THERAPEUTIC TARGETING OF AEG-1/MTDH/LYRIC

Breast cancer patients harboring an elevated level of AEG-1/MTDH/LYRIC are more likely to suffer from metastatic recurrence and acquire resistance to chemotherapeutic treatment. Therefore, AEG-1/MTDH/LYRIC can be used as a biomarker to identify subgroups of patients that require closer monitoring for signs of relapse for early clinical intervention.

For these high-risk patients, more aggressive adjuvant treatment, coupled with molecular targeting of *AEG-1/MTDH/LYRIC*, may be required to enhance therapeutic efficacy and achieve optimal clinical outcome.

There are several possible avenues to develop novel cancer therapies through molecular targeting of AEG-1/MTDH/LYRIC. Based on the initial observation that AEG-1/MTDH/LYRIC mediates the adhesion of cancer cells to lung endothelium, one can envision that neutralizing antibodies against AEG-1/MTDH/LYRIC can be used to block the early establishment of lung metastasis. In fact, antibodies reacting to the LHD of AEG-1/MTDH/LYRIC have been reported to reduce lung metastasis when coinjected with 4T1 murine mammary tumor cells (Brown & Ruoslahti, 2004). This needs to be expanded and tested in preclinical models of human breast cancer to ensure that the feasibility and effectiveness before human monoclonal antibodies against AEG-1/MTDH/LYRIC are developed. In addition, studies are required to examine the effect of AEG-1/MTDH/LYRIC on established metastasis in order to fully explore the potential of targeting AEG-1/MTDH/LYRIC to treat metastatic diseases. Another alternative is silencing *AEG-1/MTDH/LYRIC* by RNA interference or miRNAs if high-efficiency *in vivo* delivery can be achieved and nonspecific immune response can be overcome. Indeed, previous studies have shown that RNAi silencing of *AEG-1/MTDH/LYRIC* sensitizes breast cancer cells to chemotherapeutic drugs and reduces lung metastasis (Hu, Chong, et al., 2009). More recently, an artificial miRNA engineered to target *AEG-1/MTDH/LYRIC* has been established and proved to inhibit AEG-1/MTDH/LYRIC expression and its oncogenic effect *in vitro* (Wang, Shu, Cai, Bao, & Liang, 2012). Different from directly targeting the gene of interest by neutralizing antibodies or RNA interference, immunotherapy harnesses the host immune system to attack and destroy tumor cells, and it has emerged as an attractive approach for clinical management of different cancer types. Recently gene-based vaccines have become a favored immunotherapeutic strategy to activate effective immunity against cancer (Rice, Ottensmeier, & Stevenson, 2008). An AEG-1/MTDH/LYRIC-based DNA vaccine was recently developed and tested in preclinical models of human breast cancer (Qian et al., 2011). Delivered by attenuated salmonella typhimurium, this vaccine effectively suppressed tumor growth in a prophylactic model and metastasis of 4T1 cells in both prophylactic and therapeutic models, without obvious side effects observed in the mice. Mechanistically, this vaccine evoked strong CD8+ cytotoxic T cell-mediated immune response both *in vitro* and *in vivo*. Although our knowledge about the normal physiological function of AEG-1/MTDH/LYRIC is currently lacking, and uncertainties exist regarding to the safety of AEG-1/MTDH/LYRIC-based DNA vaccines, this study offers new possibilities to target AEG-1/MTDH/LYRIC. Finally, identification and functional characterization of interactions between AEG-1/MTDH/LYRIC and its partners may lead to the discovery of small molecules that can effectively block the functionality of AEG-1/MTDH/LYRIC. A better understanding of the molecular mechanisms of AEG-1/MTDH/LYRIC will also help reveal more alternative targets that are mediators of AEG-1/MTDH/LYRIC's oncogenic effects.

11. CONCLUDING REMARKS AND FUTURE PERSPECTIVES

Accumulated clinical and functional studies have demonstrated AEG-1/MTDH/LYRIC as a strong indicator and promoter of tumor malignancy in multiple organs, including the breast. Notably, AEG-1/MTDH/LYRIC exerts its multifaceted roles in a highly context-dependent manner, and the underlying mechanisms remain poorly understood. Future investigations are urgently needed to clarify the following questions. Although a great abundance of evidence points to important roles of AEG-1/MTDH/LYRIC in cancer, little is known about its biological significance in normal physiological conditions. Studies employing relevant animal models are essential to provide insights into the role of this conserved molecule in development. At the molecular level, while recent progress has been made to elucidate the subcellular localization of AEG-1/MTDH/LYRIC, it remains a mystery what signaling events lead to the distribution of AEG-1/MTDH/LYRIC and whether subcellular localizations contribute to the function of AEG-1/MTDH/LYRIC. In addition, it has long been speculated that AEG-1/MTDH/LYRIC harbors many different isoforms/modifications based on sequence prediction and multiple RNA/Protein detected; however, none of them have been adequately examined. Another layer of complexity within our understanding of AEG-1/MTDH/LYRIC is the board range of proteins it interacts with in different cellular components. Evidence is urgently needed to illustrate whether AEG-1/MTDH/LYRIC exerts its diverse functions through interacting and modifying its binding partners, or whether it functions as a scaffold protein in a signaling complex. Only by the understanding of these basic cellular and biochemical properties of AEG-1/MTDH/LYRIC can we gain better understanding of its actions in cancer and develop effective inhibitors.

REFERENCES

Ash, S. C., Yang, D. Q., & Britt, D. E. (2008). LYRIC/AEG-1 overexpression modulates BCCIPalpha protein levels in prostate tumor cells. *Biochemical and Biophysical Research Communications*, *371*, 333–338.

Bergamaschi, A., Kim, Y. H., Wang, P., Sorlie, T., Hernandez-Boussard, T., Lonning, P. E., et al. (2006). Distinct patterns of DNA copy number alteration are associated with different clinicopathological features and gene-expression subtypes of breast cancer. *Genes, Chromosomes & Cancer*, *45*, 1033–1040.

Blanco, M. A., Aleckovic, M., Hua, Y., Li, T., Wei, Y., Xu, Z., et al. (2011). Identification of staphylococcal nuclease domain-containing 1 (SND1) as a Metadherin-interacting protein with metastasis-promoting functions. *The Journal of Biological Chemistry*, *286*, 19982–19992.

Britt, D. E., Yang, D. F., Yang, D. Q., Flanagan, D., Callanan, H., Lim, Y. P., et al. (2004). Identification of a novel protein, LYRIC, localized to tight junctions of polarized epithelial cells. *Experimental Cell Research*, *300*, 134–148.

Brown, D. M., & Ruoslahti, E. (2004). Metadherin, a cell surface protein in breast tumors that mediates lung metastasis. *Cancer Cell*, *5*, 365–374.

Caudy, A. A., Ketting, R. F., Hammond, S. M., Denli, A. M., Bathoorn, A. M., Tops, B. B., et al. (2003). A micrococcal nuclease homologue in RNAi effector complexes. *Nature*, *425*, 411–414.

Chen, C. Z. (2005). MicroRNAs as oncogenes and tumor suppressors. *The New England Journal of Medicine*, *353*, 1768–1771.

Chin, L., & Gray, J. W. (2008). Translating insights from the cancer genome into clinical practice. *Nature*, *452*, 553–563.

Dalgin, G. S., Alexe, G., Scanfeld, D., Tamayo, P., Mesirov, J. P., Ganesan, S., et al. (2007). Portraits of breast cancer progression. *BMC Bioinformatics*, *8*, 291.

Easton, D. F., Pooley, K. A., Dunning, A. M., Pharoah, P. D., Thompson, D., Ballinger, D. G., et al. (2007). Genome-wide association study identifies novel breast cancer susceptibility loci. *Nature*, *447*, 1087–1093.

Emdad, L., Lee, S. G., Su, Z. Z., Jeon, H. Y., Boukerche, H., Sarkar, D., et al. (2009). Astrocyte elevated gene-1 (AEG-1) functions as an oncogene and regulates angiogenesis. *Proceedings of the National Academy of Sciences of the United States of America*, *106*, 21300–21305.

Emdad, L., Sarkar, D., Su, Z. Z., Randolph, A., Boukerche, H., Valerie, K., et al. (2006). Activation of the nuclear factor kappaB pathway by astrocyte elevated gene-1: Implications for tumor progression and metastasis. *Cancer Research*, *66*, 1509–1516.

Gao, X., Ge, L., Shao, J., Su, C., Zhao, H., Saarikettu, J., et al. (2010). Tudor-SN interacts with and co-localizes with G3BP in stress granules under stress conditions. *FEBS Letters*, *584*, 3525–3532.

Hanahan, D., & Weinberg, R. A. (2011). Hallmarks of cancer: The next generation. *Cell*, *144*, 646–674.

He, X. X., Chang, Y., Meng, F. Y., Wang, M. Y., Xie, Q. H., Tang, F., et al. (2012). MicroRNA-375 targets AEG-1 in hepatocellular carcinoma and suppresses liver cancer cell growth in vitro and in vivo. *Oncogene*, *31*, 3357–3369.

Hiscox, S., Jiang, W. G., Obermeier, K., Taylor, K., Morgan, L., Burmi, R., et al. (2006). Tamoxifen resistance in MCF7 cells promotes EMT-like behaviour and involves modulation of beta-catenin phosphorylation. *International Journal of Cancer*, *118*, 290–301.

Hu, G., Chong, R. A., Yang, Q., Wei, Y., Blanco, M. A., Li, F., et al. (2009). MTDH activation by 8q22 genomic gain promotes chemoresistance and metastasis of poor-prognosis breast cancer. *Cancer Cell*, *15*, 9–20.

Hu, G., Wei, Y., & Kang, Y. (2009). The multifaceted role of MTDH/AEG-1 in cancer progression. *Clinical Cancer Research*, *15*, 5615–5620.

Hui, A. B., Bruce, J. P., Alajez, N. M., Shi, W., Yue, S., Perez-Ordonez, B., et al. (2011). Significance of dysregulated metadherin and microRNA-375 in head and neck cancer. *Clinical Cancer Research, 17*, 7539–7550.

Iorio, M. V., Ferracin, M., Liu, C. G., Veronese, A., Spizzo, R., Sabbioni, S., et al. (2005). MicroRNA gene expression deregulation in human breast cancer. *Cancer Research, 65*, 7065–7070.

Jeon, H. Y., Choi, M., Howlett, E. L., Vozhilla, N., Yoo, B. K., Lloyd, J. A., et al. (2010). Expression patterns of astrocyte elevated gene-1 (AEG-1) during development of the mouse embryo. *Gene Expression Patterns, 10*, 361–367.

Kang, D. C., Su, Z. Z., Sarkar, D., Emdad, L., Volsky, D. J., & Fisher, P. B. (2005). Cloning and characterization of HIV-1-inducible astrocyte elevated gene-1, AEG-1. *Gene, 353*, 8–15.

Kikuno, N., Shiina, H., Urakami, S., Kawamoto, K., Hirata, H., Tanaka, Y., et al. (2007). Knockdown of astrocyte-elevated gene-1 inhibits prostate cancer progression through upregulation of FOXO3a activity. *Oncogene, 26*, 7647–7655.

Kim, M. R., Choi, H. K., Cho, K. B., Kim, H. S., & Kang, K. W. (2009). Involvement of Pin1 induction in epithelial-mesenchymal transition of tamoxifen-resistant breast cancer cells. *Cancer Science, 100*, 1834–1841.

Lee, S. G., Jeon, H. Y., Su, Z. Z., Richards, J. E., Vozhilla, N., Sarkar, D., et al. (2009). Astrocyte elevated gene-1 contributes to the pathogenesis of neuroblastoma. *Oncogene, 28*, 2476–2484.

Lee, S. G., Su, Z. Z., Emdad, L., Sarkar, D., & Fisher, P. B. (2006). Astrocyte elevated gene-1 (AEG-1) is a target gene of oncogenic Ha-ras requiring phosphatidylinositol 3-kinase and c-Myc. *Proceedings of the National Academy of Sciences of the United States of America, 103*, 17390–17395.

Lee, S. G., Su, Z. Z., Emdad, L., Sarkar, D., Franke, T. F., & Fisher, P. B. (2008). Astrocyte elevated gene-1 activates cell survival pathways through PI3K-Akt signaling. *Oncogene, 27*, 1114–1121.

Leverson, J. D., Koskinen, P. J., Orrico, F. C., Rainio, E. M., Jalkanen, K. J., Dash, A. B., et al. (1998). Pim-1 kinase and p100 cooperate to enhance c-Myb activity. *Molecular Cell, 2*, 417–425.

Li, X., Kong, X., Huo, Q., Guo, H., Yan, S., Yuan, C., et al. (2011). Metadherin enhances the invasiveness of breast cancer cells by inducing epithelial to mesenchymal transition. *Cancer Science, 102*, 1151–1157.

Li, C., Li, Y., Wang, X., Wang, Z., Cai, J., Wang, L., et al. (2012). Elevated expression of astrocyte elevated gene-1 (AEG-1) is correlated with cisplatin-based chemoresistance and shortened outcome in patients with stages III-IV serous ovarian carcinoma. *Histopathology, 60*, 953–963.

Li, C. L., Yang, W. Z., Chen, Y. P., & Yuan, H. S. (2008). Structural and functional insights into human Tudor-SN, a key component linking RNA interference and editing. *Nucleic Acids Research, 36*, 3579–3589.

Li, J., Yang, L., Song, L., Xiong, H., Wang, L., Yan, X., et al. (2009). Astrocyte elevated gene-1 is a proliferation promoter in breast cancer via suppressing transcriptional factor FOXO1. *Oncogene, 28*, 3188–3196.

Li, J., Zhang, N., Song, L. B., Liao, W. T., Jiang, L. L., Gong, L. Y., et al. (2008). Astrocyte elevated gene-1 is a novel prognostic marker for breast cancer progression and overall patient survival. *Clinical Cancer Research, 14*, 3319–3326.

Li, Y., Zou, L., Li, Q., Haibe-Kains, B., Tian, R., Desmedt, C., et al. (2010). Amplification of LAPTM4B and YWHAZ contributes to chemotherapy resistance and recurrence of breast cancer. *Nature Medicine, 16*, 214–218.

Liu, H., Song, X., Liu, C., Xie, L., Wei, L., & Sun, R. (2009). Knockdown of astrocyte elevated gene-1 inhibits proliferation and enhancing chemo-sensitivity to cisplatin or doxorubicin in neuroblastoma cells. *Journal of Experimental & Clinical Cancer Research, 28*, 19.

Liu, L., Wu, J., Ying, Z., Chen, B., Han, A., Liang, Y., et al. (2010). Astrocyte elevated gene-1 upregulates matrix metalloproteinase-9 and induces human glioma invasion. *Cancer Research, 70*, 3750–3759.

Liu, X., Zhang, N., Li, X., Moran, M. S., Yuan, C., Yan, S., et al. (2011). Identification of novel variants of metadherin in breast cancer. *PLoS One, 6*, e17582.

McShea, A., Samuel, T., Eppel, J. T., Galloway, D. A., & Funk, J. O. (2000). Identification of CIP-1-associated regulator of cyclin B (CARB), a novel p21-binding protein acting in the G2 phase of the cell cycle. *The Journal of Biological Chemistry, 275*, 23181–23186.

Meng, X., Brachova, P., Yang, S., Xiong, Z., Zhang, Y., Thiel, K. W., et al. (2011). Knockdown of MTDH sensitizes endometrial cancer cells to cell death induction by death receptor ligand TRAIL and HDAC inhibitor LBH589 co-treatment. *PLoS One, 6*, e20920.

Meng, X., Liu, J., & Shen, Z. (2004). Inhibition of G1 to S cell cycle progression by BCCIP beta. *Cell Cycle, 3*, 343–348.

Meng, X., Lu, H., & Shen, Z. (2004). BCCIP functions through p53 to regulate the expression of p21Waf1/Cip1. *Cell Cycle, 3*, 1457–1462.

Meng, X., Zhu, D., Yang, S., Wang, X., Xiong, Z., Zhang, Y., et al. (2012). Cytoplasmic Metadherin (MTDH) provides survival advantage under conditions of stress by acting as RNA-binding protein. *The Journal of Biological Chemistry, 287*, 4485–4491.

Musgrove, E. A., & Sutherland, R. L. (2009). Biological determinants of endocrine resistance in breast cancer. *Nature Reviews. Cancer, 9*, 631–643.

Nieto, M. A. (2011). The ins and outs of the epithelial to mesenchymal transition in health and disease. *Annual Review of Cell and Developmental Biology, 27*, 347–376.

Nohata, N., Hanazawa, T., Kikkawa, N., Mutallip, M., Sakurai, D., Fujimura, L., et al. (2011). Tumor suppressive microRNA-375 regulates oncogene AEG-1/MTDH in head and neck squamous cell carcinoma (HNSCC). *Journal of Human Genetics, 56*, 595–601.

Orr, N., & Chanock, S. (2008). Common genetic variation and human disease. *Advances in Genetics, 62*, 1–32.

Paukku, K., Yang, J., & Silvennoinen, O. (2003). Tudor and nuclease-like domains containing protein p100 function as coactivators for signal transducer and activator of transcription 5. *Molecular Endocrinology, 17*, 1805–1814.

Perou, C. M., Sorlie, T., Eisen, M. B., van de Rijn, M., Jeffrey, S. S., Rees, C. A., et al. (2000). Molecular portraits of human breast tumours. *Nature, 406*, 747–752.

Qian, B. J., Yan, F., Li, N., Liu, Q. L., Lin, Y. H., Liu, C. M., et al. (2011). MTDH/AEG-1-based DNA vaccine suppresses lung metastasis and enhances chemosensitivity to doxorubicin in breast cancer. *Cancer Immunology, Immunotherapy, 60*, 883–893.

Rice, J., Ottensmeier, C. H., & Stevenson, F. K. (2008). DNA vaccines: Precision tools for activating effective immunity against cancer. *Nature Reviews. Cancer, 8*, 108–120.

Sarkar, D., Park, E. S., Emdad, L., Lee, S. G., Su, Z. Z., & Fisher, P. B. (2008). Molecular basis of nuclear factor-kappaB activation by astrocyte elevated gene-1. *Cancer Research, 68*, 1478–1484.

Shaw, N., Zhao, M., Cheng, C., Xu, H., Saarikettu, J., Li, Y., et al. (2007). The multifunctional human p100 protein 'hooks' methylated ligands. *Nature Structural & Molecular Biology, 14*, 779–784.

Sorlie, T., Perou, C. M., Tibshirani, R., Aas, T., Geisler, S., Johnsen, H., et al. (2001). Gene expression patterns of breast carcinomas distinguish tumor subclasses with clinical implications. *Proceedings of the National Academy of Sciences of the United States of America, 98*, 10869–10874.

Su, Z. Z., Kang, D. C., Chen, Y., Pekarskaya, O., Chao, W., Volsky, D. J., et al. (2002). Identification and cloning of human astrocyte genes displaying elevated expression after infection with HIV-1 or exposure to HIV-1 envelope glycoprotein by rapid subtraction hybridization, RaSH. *Oncogene, 21*, 3592–3602.

Su, P., Zhang, Q., & Yang, Q. (2010). Immunohistochemical analysis of Metadherin in proliferative and cancerous breast tissue. *Diagnostic Pathology*, *5*, 38.

Sundstrom, J. F., Vaculova, A., Smertenko, A. P., Savenkov, E. I., Golovko, A., Minina, E., et al. (2009). Tudor staphylococcal nuclease is an evolutionarily conserved component of the programmed cell death degradome. *Nature Cell Biology*, *11*, 1347–1354.

Sutherland, H. G., Lam, Y. W., Briers, S., Lamond, A. I., & Bickmore, W. A. (2004). 3D3/lyric: A novel transmembrane protein of the endoplasmic reticulum and nuclear envelope, which is also present in the nucleolus. *Experimental Cell Research*, *294*, 94–105.

Thirkettle, H. J., Girling, J., Warren, A. Y., Mills, I. G., Sahadevan, K., Leung, H., et al. (2009). LYRIC/AEG-1 is targeted to different subcellular compartments by ubiquitinylation and intrinsic nuclear localization signals. *Clinical Cancer Research*, *15*, 3003–3013.

Thirkettle, H. J., Mills, I. G., Whitaker, H. C., & Neal, D. E. (2009). Nuclear LYRIC/AEG-1 interacts with PLZF and relieves PLZF-mediated repression. *Oncogene*, *28*, 3663–3670.

Tokunaga, E., Nakashima, Y., Yamashita, N., Hisamatsu, Y., Okada, S., Akiyoshi, S., et al. (2012). Overexpression of metadherin/MTDH is associated with an aggressive phenotype and a poor prognosis in invasive breast cancer. *Breast Cancer*, http://dx.doi.org/10.1007/s12282-012-0398-2.

Tong, X., Drapkin, R., Yalamanchili, R., Mosialos, G., & Kieff, E. (1995). The Epstein-Barr virus nuclear protein 2 acidic domain forms a complex with a novel cellular coactivator that can interact with TFIIE. *Molecular and Cellular Biology*, *15*, 4735–4744.

Wang, S., Shu, J. Z., Cai, Y., Bao, Z., & Liang, Q. M. (2012). Establishment and characterization of MTDH knockdown by artificial MicroRNA interference—Functions as a potential tumor suppressor in breast cancer. *Asian Pacific Journal of Cancer Prevention*, *13*, 2813–2818.

Ward, A., Balwierz, A., Zhang, J. D., Kublbeck, M., Pawitan, Y., Hielscher, T., et al. (2013). Re-expression of microRNA-375 reverses both tamoxifen resistance and accompanying EMT-like properties in breast cancer. *Oncogene*, *32*, 1173–1182.

Yang, J., Aittomaki, S., Pesu, M., Carter, K., Saarinen, J., Kalkkinen, N., et al. (2002). Identification of p100 as a coactivator for STAT6 that bridges STAT6 with RNA polymerase II. *The EMBO Journal*, *21*, 4950–4958.

Yang, J., Valineva, T., Hong, J., Bu, T., Yao, Z., Jensen, O. N., et al. (2007). Transcriptional co-activator protein p100 interacts with snRNP proteins and facilitates the assembly of the spliceosome. *Nucleic Acids Research*, *35*, 4485–4494.

Yoo, B. K., Chen, D., Su, Z. Z., Gredler, R., Yoo, J., Shah, K., et al. (2010). Molecular mechanism of chemoresistance by astrocyte elevated gene-1. *Cancer Research*, *70*, 3249–3258.

Yoo, B. K., Emdad, L., Su, Z. Z., Villanueva, A., Chiang, D. Y., Mukhopadhyay, N. D., et al. (2009). Astrocyte elevated gene-1 regulates hepatocellular carcinoma development and progression. *The Journal of Clinical Investigation*, *119*, 465–477.

Yoo, B. K., Gredler, R., Vozhilla, N., Su, Z. Z., Chen, D., Forcier, T., et al. (2009). Identification of genes conferring resistance to 5-fluorouracil. *Proceedings of the National Academy of Sciences of the United States of America*, *106*, 12938–12943.

Yoo, B. K., Santhekadur, P. K., Gredler, R., Chen, D., Emdad, L., Bhutia, S., et al. (2011). Increased RNA-induced silencing complex (RISC) activity contributes to hepatocellular carcinoma. *Hepatology*, *53*, 1538–1548.

Yu, C., Chen, K., Zheng, H., Guo, X., Jia, W., Li, M., et al. (2009). Overexpression of astrocyte elevated gene-1 (AEG-1) is associated with esophageal squamous cell carcinoma (ESCC) progression and pathogenesis. *Carcinogenesis*, *30*, 894–901.

Yuan, C., Li, X., Yan, S., Yang, Q., Liu, X., & Kong, B. (2012). The MTDH (-470G>A) polymorphism is associated with ovarian cancer susceptibility. *PLoS One*, *7*, e51561.

Zhang, B., Liu, X. X., He, J. R., Zhou, C. X., Guo, M., He, M., et al. (2011). Pathologically decreased miR-26a antagonizes apoptosis and facilitates carcinogenesis by targeting MTDH and EZH2 in breast cancer. *Carcinogenesis*, *32*, 2–9.

CHAPTER FIVE

Drug Resistance Mediated by AEG-1/MTDH/LYRIC

Xiangbing Meng[*,†,1], **Kristina W. Thiel**[*], **Kimberly K. Leslie**[*,†]
[*]Department of Obstetrics and Gynecology, The University of Iowa, Iowa City, Iowa, USA
[†]Holden Comprehensive Cancer Center, The University of Iowa, Iowa City, Iowa, USA
[1]Corresponding author: e-mail address: xiangbing-meng@uiowa.edu

Contents

Abstract

AEG-1/MTDH/LYRIC has been shown to promote cancer progression and development. Overexpression of AEG-1/MTDH/LYRIC correlates with angiogenesis, metastasis, and chemoresistance to various chemotherapy agents in cancer cells originating from a variety of tissues. In this chapter, we focus on the role of AEG-1/MTDH/LYRIC in drug resistance. Mechanistic studies have shown that AEG-1/MTDH/LYRIC is involved in classical oncogenic pathways including Ha-Ras, myc, NFκB, and PI3K/Akt. AEG-1/MTDH/LYRIC also promotes protective autophagy by activating AMP kinase and autophagy-related gene 5. Another reported mechanism by which AEG-1/MTDH/LYRIC regulates drug resistance is by increasing loading of multidrug resistance gene (MDR) 1 mRNA to the polysome, thereby facilitating MDR1 protein translation. More recently, a novel

Advances in Cancer Research, Volume 120
ISSN 0065-230X
http://dx.doi.org/10.1016/B978-0-12-401676-7.00005-X

function for AEG-1/MTDH/LYRIC as an RNA-binding protein was elucidated, which has the potential to impact expression of drug sensitivity or resistance genes. Finally, AEG-1/MTDH/LYRIC acts in microRNA-directed gene silencing via an interaction with staphylococcal nuclease and tudor domain containing 1, a component of the RNA-induced silencing complex. Altered microRNA expression and activity induced by AEG-1/MTDH/LYRIC represent an additional way that AEG-1/MTDH/LYRIC may cause drug resistance in cancer. The multiple functions of AEG-1/MTDH/LYRIC in drug resistance highlight that it is a viable target as an anticancer agent for a wide variety of cancers.

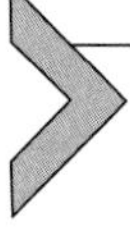

1. INTRODUCTION: AEG-1/MTDH/LYRIC, A GENE INVOLVED IN CANCER METASTASIS AND DRUG RESISTANCE

Novel treatment strategies are necessary to improve outcomes for cancer patients with drug resistance and metastasis, which combined are responsible for greater than 90% of cancer-related deaths (Ahmad, Sakr, & Rahman, 2012). Drug resistance can be classified into two categories: intrinsic resistance and acquired resistance. Intrinsic resistance refers to resistance mechanism(s) that are present in the tumor prior to treatment, whereas acquired resistance develops during the course of treatment. Mechanisms of drug resistance include genetic mutations, epigenetic alterations, induced expression of drug targets, overexpression of drug resistance genes (i.e., genes involved in drug efflux), increased repair of DNA damage, reduced apoptosis, drug-induced mutations, altered metabolism of the drug, and expression of microRNAs. Given this wide range of causes of drug resistance, future treatment strategies must be tailored to the specific mediator of drug resistance. In this review, we discuss one such specific target, AEG-1/MTDH/LYRIC, and the current knowledge regarding its contribution to drug resistance.

AEG-1/MTDH/LYRIC was recently identified as an oncogene that functions in both drug resistance and metastasis (Nestal de Moraes et al., 2012). Originally discovered as an HIV-1-inducible gene in fetal astrocytes (Kang et al., 2005), AEG-1/MTDH/LYRIC overexpression has now been documented in many cancers, including breast, prostate, esophageal, gastric, renal, colorectal, lung, hepatocellular, ovarian and endometrial cancers, neuroblastoma, glioma, and sarcoma (Bhutia et al., 2010; Hu, Wei, & Kang, 2009; Li, Li, et al., 2012; Meng, Luo, Ma, Hu, & Lou, 2011; Qian et al., 2011; Srivastava et al., 2012; Thirkettle, Mills, Whitaker, & Neal, 2009; Wang & Yang, 2011; Ying, Li, & Li, 2011; Yoo, Emdad, et al., 2011;

Yoo, Gredler, et al., 2009; Yoo et al., 2010). AEG-1/MTDH/LYRIC is located on chromosome 8q22, a region that is frequently amplified in various cancers, including breast and hepatocellular carcinoma (HCC) (Bergamaschi et al., 2006; Hu, Chong, et al., 2009; Poon et al., 2006). Overexpression of AEG-1/MTDH/LYRIC has been observed in >90% of neuroblastomas and HCC (Lee et al., 2009). AEG-1/MTDH/LYRIC was identified as a metastasis gene in a phage expression library screen from metastatic breast carcinoma, which was designed to identify proteins that bind to the vasculature of the lung (Brown & Ruoslahti, 2004). Further study revealed AEG-1/MTDH/LYRIC overexpression to be independently associated with a poor disease-free survival rate and a poor distant metastasis-free survival rate in breast cancer (Tokunaga et al., 2012).

The first evidence of a role for AEG-1/MTDH/LYRIC in chemoresistance came from pharmacogenomic analysis of the NCI-60 panel of cancer cells. In this study, the AEG-1/MTDH/LYRIC DNA copy number was found to inversely correlate with sensitivity to chemotherapeutic agents (Hu, Chong, et al., 2009). We now understand that overexpression of AEG-1/MTDH/LYRIC confers broad drug resistance to chemotherapeutic agents, including 5-fluorouracil (5-FU), doxorubicin, paclitaxel, and cisplatin, as well as to targeted therapies (Table 5.1). In this review, we will

Table 5.1 Various cancers in which AEG-1/MTDH/LYRIC overexpression has been associated with drug resistance

Cancer sites	Drug	References
Breast cancer	Doxorubicin, paclitaxel, cisplatin	Hu, Chong, et al. (2009)
	Tamoxifen	Ward et al. (2012)
	AZD6244	Kong, Moran, Zhao, and Yang (2012)
HCC	5-Fluorouracil	Yoo, Emdad, et al. (2009) and Yoo, Gredler, et al., 2009
	Doxorubicin	Yoo et al. (2010)
Ovarian cancer	Cisplatin	Li, Li, et al. (2012)
Endometrial cancer	TRAIL, HDAC inhibitor, mitomycin C, BIBF1120	Meng, Brachova, et al. (2011) and Meng et al. (2012)
Neuroblastoma	Cisplatin, doxorubicin	Liu et al. (2009)

discuss the known mechanisms of AEG-1/MTDH/LYRIC-mediated drug resistance to date as well as the contribution of other genes near AEG-1/MTDH/LYRIC on chromosome 8q22 that are also amplified in tumors.

2. MECHANISMS OF AEG-1/MTDH/LYRIC-MEDIATED DRUG RESISTANCE

2.1. Inhibition of apoptosis

Given the high rate of AEG-1/MTDH/LYRIC overexpression in many types of cancer, along with its association with resistance to multiple types of therapy, many studies have explored how AEG-1/MTDH/LYRIC mediates these detrimental effects. One potential mechanism is through activation of prosurvival pathways. For example, overexpression of AEG-1/MTDH/LYRIC increases cell survival in response to serum deprivation (Kikuno et al., 2007; Lee et al., 2008). AEG-1/MTDH/LYRIC-mediated activation of PI3K/Akt prosurvival signaling and downregulation of Bad, p21, p27, and FOXO3a may be involved in this process (Kikuno et al., 2007; Lee et al., 2008). In addition, knockdown of AEG-1/MTDH/LYRIC increases expression and activation of FOXO3a by promoting its translocation to nucleus via an AEG-1/MTDH/LYRIC/ERK1/2 pathway rather than an AEG-1/MTDH/LYRIC/Akt pathway (Wilson, Brosens, Schwenen, & Lam, 2011). Thus, AEG-1/MTDH/LYRIC accomplishes escape from apoptosis through multiple mechanisms, all of which may contribute to its role in chemoresistance.

In addition to activating prosurvival pathways, AEG-1/MTDH/LYRIC overexpression also mediates resistance to therapies that mediate apoptosis, whether it is chemotherapy or targeted therapy such as death receptor ligand TRAIL (tumor necrosis factor-related apoptosis-inducing ligand), angiogenesis inhibitor BIBF1120, or MAPK/ERK kinase (MEK) inhibitor AZD6244 (Kong et al., 2012). TRAIL is the ligand that activates the death receptor-mediated extrinsic apoptosis pathway (Holoch & Griffith, 2009). TRAIL is important as an anticancer therapy because it induces cancer-specific apoptosis without impairing normal cells (Walczak et al., 1999). TRAIL-mediated apoptosis can be enhanced by inhibition of histone deacetylase (HDAC) activity in a mechanism that includes increased expression of death receptors (Meng, Brachova, et al., 2011). In a recent study, we found that knockdown of AEG-1/MTDH/LYRIC using a specific shRNA can further increase endometrial cancer cell death induced by TRAIL and HDAC inhibitor combination treatment via inhibiting antiapoptotic gene X-linked inhibitor of apoptosis protein and increasing activation of caspases

3 and 8 (Meng, Brachova, et al., 2011). AEG-1/MTDH/LYRIC also contributes to resistance to BIBF1120, an angiogenesis inhibitor that targets multiple proangiogenic receptors including platelet-derived growth factor receptor, vascular endothelial growth factor receptor, and fibroblast growth factor receptor. In endometrial cancer cells that express these angiogenic receptors, silencing of AEG-1/MTDH/LYRIC sensitizes cancer cells to BIBF1120 (Meng et al., 2012). Finally, AEG-1/MTDH/LYRIC is associated with resistance to AZD6244, an ATP-noncompetitive inhibitor of MAPK/MEK1/2 that has been used in clinical trials (Kong et al., 2012). Resistance to AZD6244 has been reversed by depleting AEG-1/MTDH/LYRIC in breast cancer cell lines (Kong et al., 2012). These examples highlight the potential to restore sensitivity to multiple distinct types of therapy by inhibiting AEG-1/MTDH/LYRIC function.

2.2. Role in protective autophagy

Autophagy is a lysosomal degradation pathway that participates in the degradation of cytosolic proteins, macromolecules, organelles, and protein aggregates to maintain cellular homeostasis (Lum et al., 2005). As such, activation of autophagy may function in a tumor-suppressive capacity by degrading defective cells. However, the autophagy process may also be co-opted by cancer cells to survive during periods of stress, such as exposure to chemotherapy. In immortalized primary human fetal astrocyte cells, AEG-1/MTDH/LYRIC induces autophagy by decreasing the ATP/AMP ratio, which, in turn, activates AMP kinase (AMPK) and a non-canonical autophagy pathway (Bhutia et al., 2010). Autophagy can also be surveyed by monitoring the accumulation of LC3-II, an autophagy marker. Silencing AEG-1/MTDH/LYRIC in multiple cancer cell lines (TG98, HeLa, MDA-MB-231, HO-1, and MIA PaCa 2 cells) restores chemosensitization by decreasing accumulation of LC3-II, indicative of a decrease in activation of protective autophagy (Bhutia et al., 2010). Collectively, these data support a role for AEG-1/MTDH/LYRIC in the induction of protective autophagy following cellular stress through multiple possible mechanisms.

2.3. Activation of transcription factor NFκB

AEG-1/MTDH/LYRIC is expressed in multiple cellular compartments, including the nucleus. Some of the first studies of AEG-1/MTDH/LYRIC function identified a role as a transcription cofactor based on its interaction with nuclear factor κB (NFκB) p65 subunit (Emdad et al., 2006). Overexpression of AEG-1/MTDH/LYRIC in HeLa cells results in induction of several NFκB downstream genes, including intercellular adhesion molecule (ICAM)-2, ICAM-3, E-, L-, and P-selectin, interleukin (IL)-6, IL-8, Toll-like receptor (TLR)-4, TLR-5, matrix metalloproteinase-9 (MMP9), c-Jun, and c-Fos (Emdad et al., 2006; Kikuno et al., 2007). The region of AEG-1/MTDH/LYRIC that directly binds p65 was mapped to residues 101–201 and is termed the p65-interaction domain. AEG-1/MTDH/LYRIC binding to p65 increases its nuclear translocation and thus transcriptional activity. In addition to modulating NFκB localization, AEG-1/MTDH/LYRIC has also been shown to facilitate the interaction between NFκB and its cofactor cAMP response element-binding protein (CREB)-binding protein (CBP) in glioma cells (Sarkar et al., 2008). In contrast to direct binding to p65, the N-terminal 71 residues of AEG-1/MTDH/LYRIC are required for the interaction with CBP and corresponding NFκB activation (Sarkar et al., 2008). Alternatively, AEG-1/MTDH/LYRIC may indirectly activate NFκB-mediated transcriptional activation in a pathway that includes PI3K/Akt. Specifically, PI3K/Akt promotes the activation of IκB kinase (Nohata et al., 2011), which in turn phosphorylates the NFκB inhibitor IκB. IκB phosphorylation results in its destabilization, thus relieving NFκB inhibition.

By modulating NFκB activation state, AEG-1/MTDH/LYRIC regulates NFκB transcription of target genes. However, AEG-1/MTDH/LYRIC may also play a role in NFκB transcription of microRNAs, thus expanding the milieu of genes that are altered by AEG-1/MTDH/LYRIC. NFκB has been reported to regulate the expression of select microRNAs, including miR-21 and miR-221 (Galardi, Mercatelli, Farace, & Ciafre, 2011). miR-21 targets the tumor suppressor gene PTEN and proapoptotic gene-programmed cell death 4 (PDCD4). Overexpressing staphylococcal nuclease and tudor domain containing 1 (SND1), a AEG-1/MTDH/LYRIC interacting protein, is also associated with an increase in NFκB-mediated miR-221 expression (Santhekadur et al., 2012), providing further evidence for an intimate link between AEG-1/MTDH/LYRIC and NFκB in the mechanism of cell survival.

2.4. Regulation of translation

In addition to nuclear expression, AEG-1/MTDH/LYRIC is highly expressed in the cytoplasm and endoplasmic reticulum, indicating that

AEG-1/MTDH/LYRIC may alter gene expression by regulating translation. Indeed, AEG-1/MTDH/LYRIC has been reported to interact with ribosomal proteins and translation factors to control protein translation (Meng et al., 2012), with potential implications in drug resistance. Other studies have provided evidence that suggest a global role for AEG-1/MTDH/LYRIC in translation. Specifically, AEG-1/MTDH/LYRIC has been shown to increase phosphorylation of eIF4G, which is required for the recruitment of eIF4G to the 5′-cap of mRNA and translation initiation (Yoo et al., 2010).

2.4.1 AEG-1/MTDH/LYRIC promotes mRNA loading in the polysome

Resistance to the anthracycline doxorubicin is a very common event in anticancer treatment. AEG-1/MTDH/LYRIC has been implicated in altering the uptake or retention of chemotherapy drugs in some cancer cells (Yoo et al., 2010). The most frequent mechanism of chemoresistance is increased expression of MDR1, a member of the ATP-binding cassette (ABC) transporter family that mediates drug efflux (Chen & Tiwari, 2011). AEG-1/MTDH/LYRIC has been found to increase MDR1 protein translation by increasing MDR1 mRNA loading in polysomes without altering MDR1 mRNA levels (Yoo et al., 2010). This increase in MDR1 protein expression promotes drug resistance in cancer cells by facilitating export of chemotherapy agents. Treatment of cells with a PI3K inhibitor inhibits AEG-1/MTDH/LYRIC-mediated loading of MDR1 to polysomes (Yoo et al., 2010), implicating PI3K/Akt signaling in the mechanism. Perhaps, more importantly, these data suggest that PI3K inhibitors, which are currently in the clinical pipeline, may represent a strategy to block AEG-1/MTDH/LYRIC function in cancer. Other mRNAs that encode ABC transporters interact with AEG-1/MTDH/LYRIC (Meng et al., 2012), though no studies have provided a link between this association and drug resistance. Interestingly, AEG-1/MTDH/LYRIC also inhibits ubiquitination of MDR1 and subsequent proteasome-mediated degradation by an unknown mechanism (Yoo et al., 2010), suggestive of alternate mechanisms by which AEG-1/MTDH/LYRIC regulates MDR1 to achieve drug resistance.

In addition to regulating MDR1 mRNA association with the polysome, AEG-1/MTDH/LYRIC increases factor XII (FXII) protein translation by increasing its loading to the polysome (Srivastava et al., 2012). FXII is a serum glycoprotein that participates in the initiation of blood coagulation and angiogenesis, indicating that AEG-1/MTDH/LYRIC, through regulation of FXII, may affect drug bioavailability.

2.4.2 AEG-1/MTDH/LYRIC acts as a scaffold protein to form multiprotein and RNA complexes

To date, several intracellular AEG-1/MTDH/LYRIC binding partners have been identified, and some of these interactions explain in part the aggressive phenotype of AEG-1/MTDH/LYRIC-overexpressing cancers. Relevant to the role of AEG-1/MTDH/LYRIC in protein translation, mass spectrometry revealed an association of AEG-1/MTDH/LYRIC with multiple RNA-binding proteins, ribosomal proteins, and translation regulatory proteins (Meng et al., 2012; Yoo, Santhekadur, et al., 2011). Some of these interactions are disrupted by nuclease (i.e., SND1, NPM, and B23) (Meng et al., 2012), leading to the hypothesis that AEG-1/MTDH/LYRIC binds RNA. Consistent with this notion, putative RNA-binding regions have been identified in AEG-1/MTDH/LYRIC. Moreover, RNA-binding protein immunoprecipitation followed by microarray analysis (RIP-chip) revealed that AEG-1/MTDH/LYRIC associates with various mRNA sequences (Meng et al., 2012). The proteins encoded by these mRNAs are implicated in a wide range of cellular processes (Table 5.2), including DNA repair (FANCC, FANCA, FANCD2, FANCI), epigenetic regulation (KDM6A, WHSC1L1), signal transduction (PDCD11, OSMR), and RNA metabolism (DDX6, DDX60, DDX21). While inhibition of PI3K signaling alters the association of AEG-1/MTDH/LYRIC with mRNAs (Meng et al., 2012), the effect of AEG-1/MTDH/LYRIC on translation and stability of these AEG-1/MTDH/LYRIC-associated mRNAs requires further investigation, as well as determining whether AEG-1/MTDH/LYRIC mRNA binding plays a role in drug resistance.

2.5. Inhibition of stress granule formation

In response to cellular stress such as heat shock, the stalled translation complex forms stress granules, which can be detected by staining with Ras-GAP SH3 domain-binding protein (G3BP) (Unsworth, Raguz, Edwards, Higgins, & Yague, 2010). While cytosolic mRNAs aggregate into stress granules, ER-bound transcripts escape sequestration in stress granules. SND1, the AEG-1/MTDH/LYRIC interacting protein, was found in stress granules (Weissbach & Scadden, 2012), though AEG-1/MTDH/LYRIC does not accumulate there. Interestingly, MDR1 is one of the genes that escape from stress granules, thus allowing drug

Table 5.2 Representative mRNAs that associate with AEG-1/MTDH/LYRIC

Gene	**Function**
WHSC1L1	Histone methyltransferase
KDM6A	Histone demethylase
PDCD11	Programmed cell death
FANCI	DNA repair
FANCA	DNA repair
FANCD2	DNA repair
FANCC	DNA repair
OSMR	Oncostatin M receptor
POLQ	DNA polymerase
MMP16	Matrix metalloproteinase
DDX6	RNA helicase
DDX21	RNA helicase
DDX60	RNA helicase
WWP1	E3 ubiquitin ligase
PPP2R1B	Protein phosphatase 2 regulatory subunit A beta
ZNF217	Transcription factor

The full list of RNAs associated with AEG-1/MTDH/LYRIC is provided in Supplementary Tables 2–5 of Meng et al. (2012).
WHSC1L1, Wolf–Hirschhorn syndrome candidate 1-like 1; KDM6A, lysine (K)-specific demethylase 6A; FANC, Fanconi anemia, complementation group; DDX, DEAD (Asp-Glu-Ala-Asp) box helicase; WWP1, WW domain-containing E3 ubiquitin protein ligase 1; ZNF217, zinc finger protein 217.

resistance (Unsworth et al., 2010). As mentioned above, AEG-1/MTDH/LYRIC can increase MDR1 translation and inhibit MDR1 degradation, which is consistent with the absence of AEG-1/MTDH/LYRIC in stress granules. We recently reported that AEG-1/MTDH/LYRIC depletion increases stress granule formation in response to heat shock stimulation (Meng et al., 2012), indicating negative regulation of stress granules by AEG-1/MTDH/LYRIC through an unknown mechanism. These data suggest that prevention of stress granule formation is one mechanism by which AEG-1/MTDH/LYRIC mediates drug resistance.

2.6. Regulation of gene silencing

Several lines of evidence have demonstrated that AEG-1/MTDH/LYRIC regulates microRNA expression. Examples mentioned earlier include regulation of NFκB transcription of select microRNAs. Below, we describe other reported roles for AEG-1/MTDH/LYRIC in posttranscriptional gene silencing.

2.6.1 AEG-1/MTDH/LYRIC regulates RNA interference mediated by siRNAs and microRNAs

A growing body of the literature demonstrates that a single microRNA can impact hundreds of targets and also that a single target can be affected by multiple microRNAs (Finnegan & Pasquinelli, 2013). Both SND1 and AEG-1/MTDH/LYRIC have been implicated in siRNA- and microRNA-directed gene silencing (Yoo, Santhekadur, et al., 2011). In addition to roles in transcription, splicing, and degradation, SND1 is a component of the RNA-induced silencing complex (RISC) (Cieply et al., 2012) that interacts with Ago-2 to regulate RNA interference. *In vitro* experiments have demonstrated the critical roles of AEG-1/MTDH/LYRIC and SND1 in optimal activity of RISC to downregulate tumor suppressor genes by microRNAs. Modulation of AEG-1/MTDH/LYRIC or SND1 expression in HCC cells produces the same effects: overexpression of AEG-1/MTDH/LYRIC or SND1 results in decreased levels of PTEN due to silencing by miR-221 and miR-21, decreased cyclin-dependent kinase inhibitor 1A (CDKN1A, also p21) by miR-106b, decreased sprouty homolog 2 (SPRY2) by miR-21, and decreased transforming growth factor-β receptor II (TGFBR2) by miR-93 (Yoo, Santhekadur, et al., 2011). AEG-1/MTDH/LYRIC or SND1 knockdown produces the anticipated opposing effects. The interaction of AEG-1/MTDH/LYRIC with SND1 has been observed in several types of cancer (Blanco et al., 2011; Yoo, Santhekadur, et al., 2011). Like AEG-1/MTDH/LYRIC, SND1 is recognized as a prometastatic gene in breast cancer (Blanco et al., 2011). SND1 also acts as a transcription cofactor via interactions with signal transducer and activator of transcription 6 (Valineva, Yang, Palovuori, & Silvennoinen, 2005), myb, and the serine/threonine kinase PIM1 (Leverson et al., 1998). Thus, while AEG-1/MTDH/LYRIC and SND1 play clear roles in gene silencing, it is unknown whether AEG-1/MTDH/LYRIC also contributes to the transcription cofactor function of SND1.

2.6.2 AEG-1/MTDH/LYRIC causes drug resistance by regulating microRNAs via cross talk with other pathways

In addition to the above-described regulation of NFκB transcription of miR-21 and miR-221, other mechanisms have been described by which AEG-1/MTDH/LYRIC alters transcription of particular microRNAs. For example, AEG-1/MTDH/LYRIC is intricately linked with c-myc, a transcription factor that has been associated with altered expression of multiple microRNAs (Chang et al., 2008). Not only does c-myc regulate AEG-1/MTDH/LYRIC transcription (Thirkettle, Mills, et al., 2009), but it also induces expression of the miR-17-92 cluster and represses expression of Let-7, and both of these alterations are associated with drug resistance (Lu et al., 2011; Totary-Jain et al., 2013). Expression of c-myc is controlled by β-catenin via the M2 isoform of pyruvate kinase (PKM2) (Tamada, Suematsu, & Saya, 2012). PKM2 is a splice isoform of pyruvate kinase that is overexpressed in cancer and plays a critical role in aerobic glycolysis (also known as the Warburg effect) and nonglycolytic transcriptional regulation of multiple genes, including c-myc. In a positive feedback loop, c-myc increases transcription of heterogeneous nuclear ribonucleoprotein, which, in turn, plays a role in alternative splicing and enhanced PKM2 isoform expression in cancer (Chen, Zhang, & Manley, 2010). Providing further complexity to this feedback loop, AEG-1/MTDH/LYRIC increases c-myc transcription by either impairing PLZF-mediated transcriptional repression or activating β-catenin-mediated c-myc transcription (Thirkettle, Mills, et al., 2009). Thus, in the setting of AEG-1/MTDH/LYRIC overexpression, the end result of these positive feedback loops may be amplification of the c-myc-mediated effects on microRNA expression and corresponding drug resistance.

In addition to c-myc, AEG-1/MTDH/LYRIC expression has been associated with repression of transcription factors FOXO1 and FOXO3a via activating the PI3K/Akt pathway (Kikuno et al., 2007). Repression of FOXO1 and FOXO3a leads to the activation of FoxM1 expression, another transcription factor involved in transcriptional regulation of microRNAs (Wilson et al., 2011). Specifically, FoxM1 transcribes miR-135a, which targets metastasis suppressor 1 (Liu et al., 2012) and is correlated with drug resistance to paclitaxel (Holleman et al., 2011). Thus, AEG-1/MTDH/LYRIC-mediated activation of downstream FoxM1 results in an increase in miR-135 and the emergence of a resistant phenotype.

2.6.3 AEG-1/MTDH/LYRIC is involved in resistance to endocrine therapies for breast cancer by reducing expression of miR-375

Almost 70% of breast cancer cases are estrogen receptor-α positive. For these cancers, endocrine therapies of selective estrogen receptor modulators, such as tamoxifen, are widely applied (Rodriguez Lajusticia, Martin Jimenez, & Lopez-Tarruella Cobo, 2008). Although these treatments reduce breast cancer mortality, and many estrogen receptor-α-positive tumors initially respond well, resistance to treatment usually develops. Studies in an *in vitro* model of tamoxifen resistance in breast cancer cells demonstrate that alterations in microRNA expression are associated with the acquisition of tamoxifen resistance (Ward et al., 2012). miR-375 was one of the most significantly downregulated microRNAs in that study. Dysregulated expression of miR-375 has also been detected in lung cancer, squamous cervical cancer, colon cancer, and glioma (Bierkens et al., 2013; Chang et al., 2012; Dai et al., 2012; He et al., 2012; Isozaki et al., 2012; Kinoshita, Hanazawa, Nohata, Okamoto, & Seki, 2012; Li, Jiang, Xia, Yang, & Hu, 2012; Nohata et al., 2011; Ward et al., 2012). Interestingly, AEG-1/MTDH/LYRIC is a validated target of miR-375 (Ward et al., 2012). Elevated AEG-1/MTDH/LYRIC levels inversely correlate with miR-375 expression and positively correlate with poorer disease-free survival in tamoxifen-treated breast cancer patients (Ward et al., 2012). Reexpression of miR-375 in tamoxifen-resistant breast cancer cells sensitizes cells to tamoxifen and inhibits their invasive capacity (Ward et al., 2012). Moreover, depletion of AEG-1/MTDH/LYRIC also partly reverses resistance to tamoxifen (Ward et al., 2012).

2.7. Regulation of tumor microenvironment

In addition to the tumor, cells in the surrounding stroma (i.e., fibroblasts, endothelial or mesothelial cells, adipose tissue-derived stromal cells, and immune cells) may also contribute to chemoresistance. Overexpression of AEG-1/MTDH/LYRIC increases the expression of molecular markers of angiogenesis, including angiopoietin-1, Tie 2, MMP2, and hypoxia-inducible factor 1-alpha (Emdad et al., 2009). To understand the role of AEG-1/MTDH/LYRIC in HCC etiology, a transgenic mouse model with hepatocyte-specific AEG-1/MTDH/LYRIC overexpression was recently generated (Srivastava et al., 2012). Not only do these mice develop HCC tumors following challenge with *N*-nitrosodiethylamine, but conditioned media obtained after culture of isolated AEG-1/MTDH/LYRIC-overexpressing hepatocytes induce angiogenesis as well as an increase in coagulation factor FXII. Increased FXII loading to the polysome, as described earlier, is responsible for changes in FXII expression. A correlation between increased expression of AEG-1/MTDH/LYRIC and VEGF has also been observed in patients with triple-negative breast cancer (Li, Li, et al., 2011), though no mechanistic studies have been reported.

3. AEG-1/MTDH/LYRIC DOWNSTREAM GENES IDENTIFIED BY MICROARRAY

In order to achieve a better understanding of how AEG-1/MTDH/LYRIC promotes drug resistance, microarray analysis has been used to explore the downstream genes that are altered by AEG-1/MTDH/LYRIC expression in several cancer types, with a focus on genes that have been previously implicated in resistance. In this section, we describe AEG-1/MTDH/LYRIC-mediated gene regulation in breast, hepatocellular, and endometrial cancers, which all have extremely high expression of AEG-1/MTDH/LYRIC in a majority of tumors.

3.1. Breast cancer

In breast cancer cells, AEG-1/MTDH/LYRIC knockdown leads to decreased expression of chemoresistance genes aldehyde dehydrogenase 3 family, member A1 (ALDH3A1), MET (the receptor for hepatocyte growth factor), heat shock protein 90, and heme oxygenase (decycling) 1 (HMOX1) and increased expression of proapoptotic genes BCL2/adenovirus E1B 19-kDa interacting protein 3 (BNIP3) and TRAIL (Hu, Chong, et al., 2009). Simultaneous knockdown of MET and ALDH3A1, an antioxidant enzyme that scavenges free radicals, results in chemosensitization to levels comparable to those observed with AEG-1/MTDH/LYRIC knockdown (Hu, Chong, et al., 2009). Constitutively overexpressing ALDH3A1 or MET partially rescues resistance to paclitaxel, doxorubicin, and 4-hydroxycyclophosphamide in AEG-1/MTDH/LYRIC knockdown LM2 cells (Hu, Chong, et al., 2009) (Fig. 5.1).

3.2. Hepatocellular carcinoma

AEG-1/MTDH/LYRIC is overexpressed in the vast majority (>90%) of HCC tumors. Microarray analysis of AEG-1/MTDH/LYRIC overexpression in HepG3 cells revealed a panel of genes that may also contribute to chemoresistance that were distinct from those identified in breast cancer cells (Yoo, Gredler, et al., 2009). Some of the most significantly altered genes include those that encode enzymes that metabolize chemotherapeutic agents: dihydropyrimidine dehydrogenase (DPYD), cytochrome P450 2B6 (CYP2B6), and dihydrodiol dehydrogenase (AKR1C2). These genes

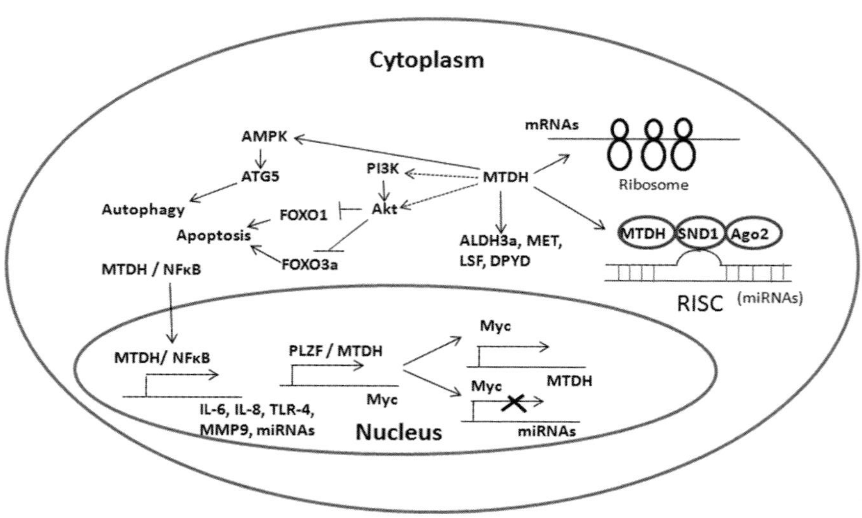

Figure 5.1 Molecular mechanisms of chemoresistance mediated by metadherin (AEG-1/MTDH/LYRIC) in cancer. In the cytoplasm, AEG-1/MTDH/LYRIC has been proposed to regulate mRNA stability, loading in the polysome, and protein translation, putatively by interacting with distinct mRNA sequences (Yoo et al., 2010; Meng et al., 2012). Cytoplasmic AEG-1/MTDH/LYRIC also serves as a component of RISC by interacting with SND1 (Blanco et al., 2011; Yoo, Santhekadur, et al., 2011). AEG-1/MTDH/LYRIC augments the PI3K/Akt signaling pathway to inhibit FOXO1 and FOXO3a-mediated apoptosis (Kikuno et al., 2007; Lee et al., 2008). Protective autophagy is induced via increased activation of AMPK and ATG-5 (Bhutia et al., 2010). In the nucleus, AEG-1/MTDH/LYRIC interacts with PLZF and releases suppression of target genes, including c-myc (Thirkettle, Mills, et al., 2009). In a feedback loop, c-myc increases transcription of AEG-1/MTDH/LYRIC and suppress expression of microRNAs (Wilson et al., 2011). Increased AEG-1/MTDH/LYRIC expression stabilizes the p50/p65 complex of NFκB (Emdad et al., 2006), inducing complex translocation to the nucleus. Downstream genes, including ALDH3A1, MET, LSF, DPYD and MDR1, are up-regulated by AEG-1/MTDH/LYRIC and thus mediate chemoresistance. Dotted lines indicate pathways yet to be fully characterized.

may contribute to the broad-spectrum role of AEG-1/MTDH/LYRIC in chemoresistance in different cancer types. For example, DPYD expression and activity modulate the efficacy of 5-FU, a common chemotherapeutic agent used to treat HCC (Yoo, Gredler, et al., 2009). Consistent with changes in DPYD expression in AEG-1/MTDH/LYRIC-overexpressing cells, it has been shown that overexpression of AEG-1/MTDH/LYRIC increases resistance of HCC cells to 5-FU by induction of DPYD as well as upregulation of thymidylate synthase (TS) by the oncogenic transcription factor late SV40 factor (LSF) (Yoo, Gredler, et al., 2011). In addition to TS, LSF transcribes several genes associated with aggressive cancers, including osteopontin, MMP9, MET, and complement factor H (Yoo, Gredler, et al., 2011). A high-throughput screen identified factor quinolinone inhibitor 1 as an LSF DNA-binding inhibitor that induces apoptosis and inhibits proliferation of HCC cells (Grant et al., 2012). Because specific inhibitors to target AEG-1/MTDH/LYRIC have not been identified, inhibiting AEG-1/MTDH/LYRIC downstream effector LSF may provide a new strategy to treat HCC. Indeed, LSF may very well represent an "Achilles heel" for HCC given the addiction to LSF for survival (Grant et al., 2012).

The microarray analysis of HCC cells also revealed that mRNA levels of the ABC transporter ABCC11, which mediates drug efflux, are increased in cells with AEG-1/MTDH/LYRIC overexpression (Yoo, Emdad, et al., 2009). The mechanism underlying the AEG-1/MTDH/LYRIC-mediated increase in ABCC11 is likely different from that of MDR1 (i.e., polysome loading) given that MDR1 mRNA levels were unchanged in response to alterations in AEG-1/MTDH/LYRIC expression (Yoo et al., 2010).

3.3. Endometrial cancer

Our group has also used microarray to survey the downstream genes under the control of AEG-1/MTDH/LYRIC in endometrial cancer cells. Our data identified calbindin 1 (CALB1), a calcium-binding protein involved in phosphatidylinositol metabolism, and the lectin, galectin-1, among the downstream genes regulated by AEG-1/MTDH/LYRIC (Meng, Brachova, et al., 2011). Both CALB1 (Sun et al., 2011) and galectin-1 (Camby, Le Mercier, Lefranc, & Kiss, 2006) have previously been linked to prosurvival signaling pathways. Taken with the gene array data in breast cancer and HCC, these results indicate that AEG-1/MTDH/LYRIC regulates different downstream genes depending on the cellular context. Thus, a mechanistic understanding of AEG-1/MTDH/LYRIC in each distinct cancer type may be necessary in order to identify downstream pathways that can be targeted to overcome the cellular effects of AEG-1/MTDH/LYRIC.

4. POTENTIAL CONTRIBUTION OF GENES LOCATED NEAR AEG-1/MTDH/LYRIC ON CHROMOSOME 8Q22 LOCUS TO CANCER DRUG RESISTANCE

AEG-1/MTDH/LYRIC overexpression in 40% of breast cancer patients is attributed to a genomic gain of chromosome 8q22 (Hu, Chong, et al., 2009). In addition to AEG-1/MTDH/LYRIC, other genes in this region have been associated with a resistant phenotype, including 14-3-3ζ, lysosomal protein transmembrane 4β (LAPTM4B), E3 identified by differential display (EDD1), and grainyhead-like 2 (GRHL2) (Fig. 5.2).

Increased expression of 14-3-3ζ (also YWHAZ) correlates with resistance to tamoxifen and a shorter time to recurrence (Bergamaschi & Katzenellenbogen, 2012; Bergamaschi et al., 2013). As a member of the 14-3-3 family, 14-3-3ζ regulates multiple cellular proteins by altering protein conformation, stability, activity, localization, and complex formation. Some 14-3-3 isoforms act as tumor suppressors, whereas others, including 14-3-3ζ, act as oncogenes. In addition, some 14-3-3 isoforms have been implicated in chemoresistance (Sinha et al., 2000; Vazquez et al., 2010), with 14-3-3ζ being one such isoform (Murata et al., 2012). Exposure to tamoxifen promotes increased expression of 14-3-3ζ due to downregulation of

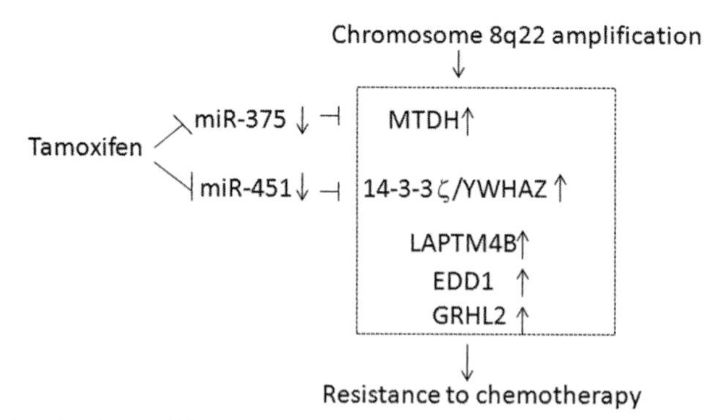

Figure 5.2 Mechanisms of drug resistance induced by chromosome 8q22 amplification. Chromosome 8q22 amplification results in increased expression of AEG-1/MTDH/LYRIC, 14-3-3ζ, LAPTM4B, EDD1, and GRHL2; each of these genes has been implicated in drug resistance. Tamoxifen resistance correlates with increased expression of AEG-1/MTDH/LYRIC and 14-3-3ζ and decreased expression of miR-375 and miR-451.

miRNA-451, which specifically targets 14-3-3ζ (Bergamaschi & Katzenellenbogen, 2012). The increase in 14-3-3ζ results in resistance to tamoxifen (Fig. 5.2) and, not surprisingly, levels of 14-3-3ζ and miR-451 are inversely correlated in tamoxifen-resistant breast cancer cells (Bergamaschi & Katzenellenbogen, 2012).

In addition to early disease recurrence after tamoxifen treatment, amplification of 14-3-3ζ as well as LAPTM4B is associated with early disease recurrence after anthracycline-based chemotherapy (Li, Zhang, et al., 2011). LAPTM4B mediates multidrug resistance by promoting drug efflux for a variety of drugs, including doxorubicin, paclitaxel, and cisplatin, in a mechanism that includes interactions with MDR1 (Li, Wei, et al., 2010). LAPTM4B also interacts with the p85α regulatory subunit of PI3K to activate PI3K/Akt signaling (Li, Wei, et al., 2010). Knockdown of either 14-3-3ζ or LAPTM4B sensitizes cancer cells to anthracyclines, whereas overexpression produces the anticipated resistance (Li, Zou, et al., 2010). In the setting of LAPTM4B overexpression, doxorubicin nuclear localization is delayed. In addition to drug efflux, LAPTM4B plays a key role in lysosomes during autophagic maturation (Li, Zhang, et al., 2011). Consistent with its role in cell survival through PI3K/Akt, overexpression of LAPTM4B promotes protective autophagy and cell survival under stress conditions, whereas LAPTM4B knockdown inhibits maturation of the autophagosome, thereby restoring sensitive to chemotherapy (Li, Zhang, et al., 2011). It is of note that AEG-1/MTDH/LYRIC promotes drug resistance through many of these same mechanisms, that is, activation of PI3K/Akt and induction of protective autophagy.

Deregulation of apoptotic pathways, such as death receptor pathways, is frequently associated with cancer recurrence. An shRNA screen identified several genes including two genes that are adjacent to AEG-1/MTDH/LYRIC, EDD1 (also called UBR5) and GRHL2, as potential targets to restore apoptosis induced by death receptor pathways (Dompe et al., 2011). While it is intriguing to postulate that evasion of apoptosis may be one mechanism for increased drug resistance in tumors with amplified chromosome 8q22, GRHL2 was recently found to suppress oncogenic transformation (Cieply et al., 2012).

5. CONCLUSIONS AND FUTURE PERSPECTIVES

AEG-1/MTDH/LYRIC overexpression contributes to cancer metastasis and drug resistance at multiple levels, including control of transcription, mRNA stability, and translation. A recent study has reported an association between AEG-1/MTDH/LYRIC overexpression and resistance to γ-radiation in cervical cancer (Zhao et al., 2012). It will be interesting to determine the pathways by which AEG-1/MTDH/LYRIC mediates this effect, and whether they are the same or different from those that mediate resistance to DNA-damaging chemotherapy (i.e., anthracyclines, cisplatin). This will be particularly informative for types of cancer in which radiation is the frontline therapy.

While no mutations have been reported in AEG-1/MTDH/LYRIC in cancer specimens, a very recent report identified a polymorphism in AEG-1/MTDH/LYRIC (−470G > A) that predicts the development of ovarian cancer (Yuan et al., 2012). Specifically, patients with the GG allele have an increased risk of developing cancer, though no correlation between the presence of the polymorphism and AEG-1/MTDH/LYRIC overexpression were detected. Future studies should explore whether this polymorphism might serve as a biomarker for drug resistance in addition to tumorigenic potential.

We are only beginning to understand the function of AEG-1/MTDH/LYRIC in normal biological processes. Gene knockout and transgenic mouse models, such as the recently published hepatocyte-specific AEG-1/MTDH/LYRIC transgenic mouse (Srivastava et al., 2012), will provide new insights into AEG-1/MTDH/LYRIC in physiology and pathology. AEG-1/MTDH/LYRIC has no canonical functional domains besides nuclear localization sequences, protein interaction regions, and the putative RNA-binding motifs identified by our group. The lack of obvious functional domains precludes development of small-molecule inhibitors that target AEG-1/MTDH/LYRIC function. Specific inhibitors that disrupt the interactions between AEG-1/MTDH/LYRIC and other proteins or mRNAs have the potential to overcome AEG-1/MTDH/LYRIC-mediated drug resistance. Alternatively, a structural characterization of AEG-1/MTDH/LYRIC may facilitate the development of specific AEG-1/MTDH/LYRIC inhibitors.

It will also be informative to better understand how AEG-1/MTDH/LYRIC activity is regulated. AEG-1/MTDH/LYRIC is present in both the cytoplasm and the nucleus and seems to have independent functions based on its localization. Intracellular distribution of AEG-1/MTDH/LYRIC has been shown to be regulated by ubiquitination (Thirkettle, Girling, et al., 2009). In addition, several phosphoproteomic studies have identified AEG-1/MTDH/LYRIC phosphorylation on seven sites (S84, T143, S298, S308, S415, S426, and S496) (Dephoure et al., 2008; Olsen et al., 2006). These sites are predicted substrates for MAPK, polo-like kinase 1, and ataxia telangiectasia mutated, the latter indicating a potential involvement of AEG-1/MTDH/LYRIC in DNA damage response. However, the effect of these posttranslation modifications on AEG-1/MTDH/LYRIC-mediated drug resistance is unknown. Understanding their functional significance may provide valuable information in reaching the goal of an AEG-1/MTDH/LYRIC-targeted agent.

ACKNOWLEDGMENTS

This work was partially supported by NIH Grant R01CA99908-7 to K. K. L., the Department of Obstetrics and Gynecology Research Development Fund, and Institutional Research Grant Number IRG-77-004-31 from the American Cancer Society to X. M., administered through the Holden Comprehensive Cancer Center at the University of Iowa.

REFERENCES

Ahmad, A., Sakr, W. A., & Rahman, K. M. (2012). Novel targets for detection of cancer and their modulation by chemopreventive natural compounds. *Frontiers in Bioscience (Elite Edition)*, *4*, 410–425.

Bergamaschi, A., Frasor, J., Borgen, K., Stanculescu, A., Johnson, P., Rowland, K., et al. (2013). 14-3-3zeta as a predictor of early time to recurrence and distant metastasis in hormone receptor-positive and -negative breast cancers. *Breast Cancer Research and Treatment*, *137*, 689–696.

Bergamaschi, A., & Katzenellenbogen, B. S. (2012). Tamoxifen downregulation of miR-451 increases 14-3-3zeta and promotes breast cancer cell survival and endocrine resistance. *Oncogene*, *31*, 39–47.

Bergamaschi, A., Kim, Y. H., Wang, P., Sorlie, T., Hernandez-Boussard, T., Lonning, P. E., et al. (2006). Distinct patterns of DNA copy number alteration are associated with different clinicopathological features and gene-expression subtypes of breast cancer. *Genes, Chromosomes & Cancer*, *45*, 1033–1040.

Bhutia, S. K., Kegelman, T. P., Das, S. K., Azab, B., Su, Z. Z., Lee, S. G., et al. (2010). Astrocyte elevated gene-1 induces protective autophagy. *Proceedings of the National Academy of Sciences of the United States of America*, *107*, 22243–22248.

Bierkens, M., Krijgsman, O., Wilting, S. M., Bosch, L., Jaspers, A., Meijer, G. A., et al. (2013). Focal aberrations indicate EYA2 and hsa-miR-375 as oncogene and tumor suppressor in cervical carcinogenesis. *Genes, Chromosomes & Cancer*, *52*, 56–68.

Blanco, M. A., Aleckovic, M., Hua, Y., Li, T., Wei, Y., Xu, Z., et al. (2011). Identification of staphylococcal nuclease domain-containing 1 (SND1) as a Metadherin-interacting protein with metastasis-promoting functions. *The Journal of Biological Chemistry*, *286*, 19982–19992.

Brown, D. M., & Ruoslahti, E. (2004). Metadherin, a cell surface protein in breast tumors that mediates lung metastasis. *Cancer Cell*, *5*, 365–374.

Camby, I., Le Mercier, M., Lefranc, F., & Kiss, R. (2006). Galectin-1: A small protein with major functions. *Glycobiology*, *16*, 137R–157R.

Chang, C., Shi, H., Wang, C., Wang, J., Geng, N., Jiang, X., et al. (2012). Correlation of microRNA-375 downregulation with unfavorable clinical outcome of patients with glioma. *Neuroscience Letters*, *531*, 204–208.

Chang, T. C., Yu, D., Lee, Y. S., Wentzel, E. A., Arking, D. E., West, K. M., et al. (2008). Widespread microRNA repression by Myc contributes to tumorigenesis. *Nature Genetics*, *40*, 43–50.

Chen, Z. S., & Tiwari, A. K. (2011). Multidrug resistance proteins (MRPs/ABCCs) in cancer chemotherapy and genetic diseases. *The FEBS Journal*, *278*, 3226–3245.

Chen, M., Zhang, J., & Manley, J. L. (2010). Turning on a fuel switch of cancer: hnRNP proteins regulate alternative splicing of pyruvate kinase mRNA. *Cancer Research*, *70*, 8977–8980.

Cieply, B., Riley, P., 4th., Pifer, P. M., Widmeyer, J., Addison, J. B., Ivanov, A. V., et al. (2012). Suppression of the epithelial-mesenchymal transition by Grainyhead-like-2. *Cancer Research*, *72*, 2440–2453.

Dai, X., Chiang, Y., Wang, Z., Song, Y., Lu, C., Gao, P., et al. (2012). Expression levels of microRNA-375 in colorectal carcinoma. *Molecular Medicine Reports*, *5*, 1299–1304.

Dephoure, N., Zhou, C., Villen, J., Beausoleil, S. A., Bakalarski, C. E., Elledge, S. J., et al. (2008). A quantitative atlas of mitotic phosphorylation. *Proceedings of the National Academy of Sciences of the United States of America*, *105*, 10762–10767.

Dompe, N., Rivers, C. S., Li, L., Cordes, S., Schwickart, M., Punnoose, E. A., et al. (2011). A whole-genome RNAi screen identifies an 8q22 gene cluster that inhibits death receptor-mediated apoptosis. *Proceedings of the National Academy of Sciences of the United States of America*, *108*, E943–E951.

Emdad, L., Lee, S. G., Su, Z. Z., Jeon, H. Y., Boukerche, H., Sarkar, D., et al. (2009). Astrocyte elevated gene-1 (AEG-1) functions as an oncogene and regulates angiogenesis. *Proceedings of the National Academy of Sciences of the United States of America*, *106*, 21300–21305.

Emdad, L., Sarkar, D., Su, Z. Z., Randolph, A., Boukerche, H., Valerie, K., et al. (2006). Activation of the nuclear factor kappaB pathway by astrocyte elevated gene-1: Implications for tumor progression and metastasis. *Cancer Research*, *66*, 1509–1516.

Finnegan, E. F., & Pasquinelli, A. E. (2013). MicroRNA biogenesis: Regulating the regulators. *Critical Reviews in Biochemistry and Molecular Biology*, *48*, 51–68.

Galardi, S., Mercatelli, N., Farace, M. G., & Ciafre, S. A. (2011). NF-kB and c-Jun induce the expression of the oncogenic miR-221 and miR-222 in prostate carcinoma and glioblastoma cells. *Nucleic Acids Research*, *39*, 3892–3902.

Grant, T. J., Bishop, J. A., Christadore, L. M., Barot, G., Chin, H. G., Woodson, S., et al. (2012). Antiproliferative small-molecule inhibitors of transcription factor LSF reveal oncogene addiction to LSF in hepatocellular carcinoma. *Proceedings of the National Academy of Sciences of the United States of America*, *109*, 4503–4508.

He, X. X., Chang, Y., Meng, F. Y., Wang, M. Y., Xie, Q. H., Tang, F., et al. (2012). MicroRNA-375 targets AEG-1 in hepatocellular carcinoma and suppresses liver cancer cell growth in vitro and in vivo. *Oncogene*, *31*, 3357–3369.

Holleman, A., Chung, I., Olsen, R. R., Kwak, B., Mizokami, A., Saijo, N., et al. (2011). miR-135a contributes to paclitaxel resistance in tumor cells both in vitro and in vivo. *Oncogene*, *30*, 4386–4398.

Holoch, P. A., & Griffith, T. S. (2009). TNF-related apoptosis-inducing ligand (TRAIL): A new path to anti-cancer therapies. *European Journal of Pharmacology*, *625*, 63–72.

Hu, G., Chong, R. A., Yang, Q., Wei, Y., Blanco, M. A., Li, F., et al. (2009). MTDH activation by 8q22 genomic gain promotes chemoresistance and metastasis of poor-prognosis breast cancer. *Cancer Cell*, *15*, 9–20.

Hu, G., Wei, Y., & Kang, Y. (2009). The multifaceted role of MTDH/AEG-1 in cancer progression. *Clinical Cancer Research*, *15*, 5615–5620.

Isozaki, Y., Hoshino, I., Nohata, N., Kinoshita, T., Akutsu, Y., Hanari, N., et al. (2012). Identification of novel molecular targets regulated by tumor suppressive miR-375 induced by histone acetylation in esophageal squamous cell carcinoma. *International Journal of Oncology*, *41*, 985–994.

Kang, D. C., Su, Z. Z., Sarkar, D., Emdad, L., Volsky, D. J., & Fisher, P. B. (2005). Cloning and characterization of HIV-1-inducible astrocyte elevated gene-1, AEG-1. *Gene*, *353*, 8–15.

Kikuno, N., Shiina, H., Urakami, S., Kawamoto, K., Hirata, H., Tanaka, Y., et al. (2007). Knockdown of astrocyte-elevated gene-1 inhibits prostate cancer progression through upregulation of FOXO3a activity. *Oncogene, 26*, 7647–7655.

Kinoshita, T., Hanazawa, T., Nohata, N., Okamoto, Y., & Seki, N. (2012). The functional significance of microRNA-375 in human squamous cell carcinoma: Aberrant expression and effects on cancer pathways. *Journal of Human Genetics, 57*, 556–563.

Kong, X., Moran, M. S., Zhao, Y., & Yang, Q. (2012). Inhibition of metadherin sensitizes breast cancer cells to AZD6244. *Cancer Biology & Therapy, 13*, 43–49.

Lee, S. G., Jeon, H. Y., Su, Z. Z., Richards, J. E., Vozhilla, N., Sarkar, D., et al. (2009). Astrocyte elevated gene-1 contributes to the pathogenesis of neuroblastoma. *Oncogene, 28*, 2476–2484.

Lee, S. G., Su, Z. Z., Emdad, L., Sarkar, D., Franke, T. F., & Fisher, P. B. (2008). Astrocyte elevated gene-1 activates cell survival pathways through PI3K-Akt signaling. *Oncogene, 27*, 1114–1121.

Leverson, J. D., Koskinen, P. J., Orrico, F. C., Rainio, E. M., Jalkanen, K. J., Dash, A. B., et al. (1998). Pim-1 kinase and p100 cooperate to enhance c-Myb activity. *Molecular Cell, 2*, 417–425.

Li, Y., Jiang, Q., Xia, N., Yang, H., & Hu, C. (2012). Decreased expression of MicroRNA-375 in nonsmall cell lung cancer and its clinical significance. *The Journal of International Medical Research, 40*, 1662–1669.

Li, C., Li, R., Song, H., Wang, D., Feng, T., Yu, X., et al. (2011). Significance of AEG-1 expression in correlation with VEGF, microvessel density and clinicopathological characteristics in triple-negative breast cancer. *Journal of Surgical Oncology, 103*, 184–192.

Li, C., Li, Y., Wang, X., Wang, Z., Cai, J., Wang, L., et al. (2012). Elevated expression of astrocyte elevated gene-1 (AEG-1) is correlated with cisplatin-based chemoresistance and shortened outcome in patients with stages III-IV serous ovarian carcinoma. *Histopathology, 60*, 953–963.

Li, L., Wei, X. H., Pan, Y. P., Li, H. C., Yang, H., He, Q. H., et al. (2010). LAPTM4B: A novel cancer-associated gene motivates multidrug resistance through efflux and activating PI3K/AKT signaling. *Oncogene, 29*, 5785–5795.

Li, Y., Zhang, Q., Tian, R., Wang, Q., Zhao, J. J., Iglehart, J. D., et al. (2011). Lysosomal transmembrane protein LAPTM4B promotes autophagy and tolerance to metabolic stress in cancer cells. *Cancer Research, 71*, 7481–7489.

Li, Y., Zou, L., Li, Q., Haibe-Kains, B., Tian, R., Desmedt, C., et al. (2010). Amplification of LAPTM4B and YWHAZ contributes to chemotherapy resistance and recurrence of breast cancer. *Nature Medicine, 16*, 214–218.

Liu, S., Guo, W., Shi, J., Li, N., Yu, X., Xue, J., et al. (2012). MicroRNA-135a contributes to the development of portal vein tumor thrombus by promoting metastasis in hepatocellular carcinoma. *Journal of Hepatology, 56*, 389–396.

Liu, H., Song, X., Liu, C., Xie, L., Wei, L., & Sun, R. (2009). Knockdown of astrocyte elevated gene-1 inhibits proliferation and enhancing chemo-sensitivity to cisplatin or doxorubicin in neuroblastoma cells. *Journal of Experimental & Clinical Cancer Research, 28*, 19.

Lu, L., Schwartz, P., Scarampi, L., Rutherford, T., Canuto, E. M., Yu, H., et al. (2011). MicroRNA let-7a: A potential marker for selection of paclitaxel in ovarian cancer management. *Gynecologic Oncology, 122*, 366–371.

Lum, J. J., Bauer, D. E., Kong, M., Harris, M. H., Li, C., Lindsten, T., et al. (2005). Growth factor regulation of autophagy and cell survival in the absence of apoptosis. *Cell, 120*, 237–248.

Meng, X., Brachova, P., Yang, S., Xiong, Z., Zhang, Y., Thiel, K. W., et al. (2011). Knockdown of MTDH sensitizes endometrial cancer cells to cell death induction by death receptor ligand TRAIL and HDAC inhibitor LBH589 co-treatment. *PLoS one, 6*, e20920.

Meng, F., Luo, C., Ma, L., Hu, Y., & Lou, G. (2011). Clinical significance of astrocyte elevated gene-1 expression in human epithelial ovarian carcinoma. *International Journal of Gynecological Pathology, 30,* 145–150.

Meng, X., Zhu, D., Yang, S., Wang, X., Xiong, Z., Zhang, Y., et al. (2012). Cytoplasmic Metadherin (MTDH) provides survival advantage under conditions of stress by acting as RNA-binding protein. *The Journal of Biological Chemistry, 287,* 4485–4491.

Murata, T., Takayama, K., Urano, T., Fujimura, T., Ashikari, D., Obinata, D., et al. (2012). 14-3-3zeta, a novel androgen-responsive gene, is upregulated in prostate cancer and promotes prostate cancer cell proliferation and survival. *Clinical Cancer Research, 18,* 5617–5627.

Nestal de Moraes, G., Souza, P. S., Costas, F. C., Vasconcelos, F. C., Reis, F. R., & Maia, R. C. (2012). The interface between BCR-ABL-dependent and -independent resistance signaling pathways in chronic myeloid leukemia. *Leukemia Research and Treatment, 2012,* 671702.

Nohata, N., Hanazawa, T., Kikkawa, N., Mutallip, M., Sakurai, D., Fujimura, L., et al. (2011). Tumor suppressive microRNA-375 regulates oncogene AEG-1/MTDH in head and neck squamous cell carcinoma (HNSCC). *Journal of Human Genetics, 56,* 595–601.

Olsen, J. V., Blagoev, B., Gnad, F., Macek, B., Kumar, C., Mortensen, P., et al. (2006). Global, in vivo, and site-specific phosphorylation dynamics in signaling networks. *Cell, 127,* 635–648.

Poon, T. C., Wong, N., Lai, P. B., Rattray, M., Johnson, P. J., & Sung, J. J. (2006). A tumor progression model for hepatocellular carcinoma: Bioinformatic analysis of genomic data. *Gastroenterology, 131,* 1262–1270.

Qian, B. J., Yan, F., Li, N., Liu, Q. L., Lin, Y. H., Liu, C. M., et al. (2011). MTDH/AEG-1-based DNA vaccine suppresses lung metastasis and enhances chemosensitivity to doxorubicin in breast cancer. *Cancer Immunology, Immunotherapy, 60,* 883–893.

Rodriguez Lajusticia, L., Martin Jimenez, M., & Lopez-Tarruella Cobo, S. (2008). Endocrine therapy of metastatic breast cancer. *Clinical & Translational Oncology, 10,* 462–467.

Santhekadur, P. K., Das, S. K., Gredler, R., Chen, D., Srivastava, J., Robertson, C., et al. (2012). Multifunction protein staphylococcal nuclease domain containing 1 (SND1) promotes tumor angiogenesis in human hepatocellular carcinoma through novel pathway that involves nuclear factor kappaB and miR-221. *The Journal of Biological Chemistry, 287,* 13952–13958.

Sarkar, D., Park, E. S., Emdad, L., Lee, S. G., Su, Z. Z., & Fisher, P. B. (2008). Molecular basis of nuclear factor-kappaB activation by astrocyte elevated gene-1. *Cancer Research, 68,* 1478–1484.

Sinha, P., Kohl, S., Fischer, J., Hutter, G., Kern, M., Kottgen, E., et al. (2000). Identification of novel proteins associated with the development of chemoresistance in malignant melanoma using two-dimensional electrophoresis. *Electrophoresis, 21,* 3048–3057.

Srivastava, J., Siddiq, A., Emdad, L., Santhekadur, P. K., Chen, D., Gredler, R., et al. (2012). Astrocyte elevated gene-1 promotes hepatocarcinogenesis: Novel insights from a mouse model. *Hepatology, 56,* 1782–1791.

Sun, S., Li, F., Gao, X., Zhu, Y., Chen, J., Zhu, X., et al. (2011). Calbindin-D28K inhibits apoptosis in dopaminergic neurons by activation of the PI3-kinase-Akt signaling pathway. *Neuroscience, 199,* 359–367.

Tamada, M., Suematsu, M., & Saya, H. (2012). Pyruvate kinase m2: Multiple faces for conferring benefits on cancer cells. *Clinical Cancer Research, 18,* 5554–5561.

Thirkettle, H. J., Girling, J., Warren, A. Y., Mills, I. G., Sahadevan, K., Leung, H., et al. (2009). LYRIC/AEG-1 is targeted to different subcellular compartments by ubiquitinylation and intrinsic nuclear localization signals. *Clinical Cancer Research, 15,* 3003–3013.

Thirkettle, H. J., Mills, I. G., Whitaker, H. C., & Neal, D. E. (2009). Nuclear LYRIC/AEG-1 interacts with PLZF and relieves PLZF-mediated repression. *Oncogene, 28,* 3663–3670.

Tokunaga, E., Nakashima, Y., Yamashita, N., Hisamatsu, Y., Okada, S., Akiyoshi, S., et al. (2012). Overexpression of metadherin/MTDH is associated with an aggressive phenotype and a poor prognosis in invasive breast cancer. *Breast Cancer*, http://dx.doi.org/10.1007/s12282-012-0398-2.

Totary-Jain, H., Sanoudou, D., Ben-Dov, I. Z., Dautriche, C. N., Guarnieri, P., Marx, S. O., et al. (2013). Reprogramming of the microRNA transcriptome mediates resistance to rapamycin. *The Journal of Biological Chemistry*, *288*, 6034–6044.

Unsworth, H., Raguz, S., Edwards, H. J., Higgins, C. F., & Yague, E. (2010). mRNA escape from stress granule sequestration is dictated by localization to the endoplasmic reticulum. *The FASEB Journal*, *24*, 3370–3380.

Valineva, T., Yang, J., Palovuori, R., & Silvennoinen, O. (2005). The transcriptional co-activator protein p100 recruits histone acetyltransferase activity to STAT6 and mediates interaction between the CREB-binding protein and STAT6. *The Journal of Biological Chemistry*, *280*, 14989–14996.

Vazquez, A., Grochola, L. F., Bond, E. E., Levine, A. J., Taubert, H., Muller, T. H., et al. (2010). Chemosensitivity profiles identify polymorphisms in the p53 network genes 14-3-3tau and CD44 that affect sarcoma incidence and survival. *Cancer Research*, *70*, 172–180.

Walczak, H., Miller, R. E., Ariail, K., Gliniak, B., Griffith, T. S., Kubin, M., et al. (1999). Tumoricidal activity of tumor necrosis factor-related apoptosis-inducing ligand in vivo. *Nature Medicine*, *5*, 157–163.

Wang, C., & Yang, Q. (2011). Astrocyte elevated gene-1 and breast cancer (Review). *Oncology Letters*, *2*, 399–405.

Ward, A., Balwierz, A., Zhang, J. D., Kublbeck, M., Pawitan, Y., Hielscher, T., et al. (2012). Re-expression of microRNA-375 reverses both tamoxifen resistance and accompanying EMT-like properties in breast cancer. *Oncogene*, *32*, 1173–1182.

Weissbach, R., & Scadden, A. D. (2012). Tudor-SN and ADAR1 are components of cytoplasmic stress granules. *RNA*, *18*, 462–471.

Wilson, M. S., Brosens, J. J., Schwenen, H. D., & Lam, E. W. (2011). FOXO and FOXM1 in cancer: The FOXO-FOXM1 axis shapes the outcome of cancer chemotherapy. *Current Drug Targets*, *12*, 1256–1266.

Ying, Z., Li, J., & Li, M. (2011). Astrocyte elevated gene 1: Biological functions and molecular mechanism in cancer and beyond. *Cell & Bioscience*, *1*, 36.

Yoo, B. K., Chen, D., Su, Z. Z., Gredler, R., Yoo, J., Shah, K., et al. (2010). Molecular mechanism of chemoresistance by astrocyte elevated gene-1. *Cancer Research*, *70*, 3249–3258.

Yoo, B. K., Emdad, L., Lee, S. G., Su, Z. Z., Santhekadur, P., Chen, D., et al. (2011). Astrocyte elevated gene-1 (AEG-1): A multifunctional regulator of normal and abnormal physiology. *Pharmacology & Therapeutics*, *130*, 1–8.

Yoo, B. K., Emdad, L., Su, Z. Z., Villanueva, A., Chiang, D. Y., Mukhopadhyay, N. D., et al. (2009). Astrocyte elevated gene-1 regulates hepatocellular carcinoma development and progression. *The Journal of Clinical Investigation*, *119*, 465–477.

Yoo, B. K., Gredler, R., Chen, D., Santhekadur, P. K., Fisher, P. B., & Sarkar, D. (2011). c-Met activation through a novel pathway involving osteopontin mediates oncogenesis by the transcription factor LSF. *Journal of Hepatology*, *55*, 1317–1324.

Yoo, B. K., Gredler, R., Vozhilla, N., Su, Z. Z., Chen, D., Forcier, T., et al. (2009). Identification of genes conferring resistance to 5-fluorouracil. *Proceedings of the National Academy of Sciences of the United States of America*, *106*, 12938–12943.

Yoo, B. K., Santhekadur, P. K., Gredler, R., Chen, D., Emdad, L., Bhutia, S., et al. (2011). Increased RNA-induced silencing complex (RISC) activity contributes to hepatocellular carcinoma. *Hepatology*, *53*, 1538–1548.

Yuan, C., Li, X., Yan, S., Yang, Q., Liu, X., & Kong, B. (2012). The MTDH (-470G>A) polymorphism is associated with ovarian cancer susceptibility. *PLoS one*, *7*, e51561.

Zhao, Y., Moran, M. S., Yang, Q., Liu, Q., Yuan, C., Hong, S., et al. (2012). Metadherin regulates radioresistance in cervical cancer cells. *Oncology Reports*, *27*, 1520–1526.

CHAPTER SIX

The Role of AEG-1/MTDH/LYRIC in the Pathogenesis of Central Nervous System Disease

Evan K. Noch[1], Kamel Khalili

Department of Neurology and Neuroscience, Weill-Cornell Medical Center-New York Presbyterian Hospital, New York, NY USA

[1]Corresponding author: e-mail address: noch@temple.edu

Contents

Abstract

Astrocyte-elevated gene-1 (AEG-1/MTDH/LYRIC) is a potent oncogene that regulates key cellular processes underlying disease of the central nervous system (CNS). From its involvement in human immunodeficiency virus (HIV)-1 infection to its role in neurodegenerative disease and malignant brain tumors, AEG-1/MTDH/LYRIC facilitates

Advances in Cancer Research, Volume 120
ISSN 0065-230X
http://dx.doi.org/10.1016/B978-0-12-401676-7.00006-1

cellular survival and proliferation through the control of a multitude of molecular signaling cascades. AEG-1/MTDH/LYRIC induction by HIV-1 and TNF highlights its importance in viral infection, and its incorporation into viral vesicles supports its potential role in active viral replication. Overexpression of AEG-1/MTDH/LYRIC in the brains of Huntington's disease patients suggests its function in neurodegenerative disease, and its association with genetic polymorphisms in large genome-wide association studies of migraine patients suggests a possible role in the pathogenesis of migraine headaches. In the field of cancer, AEG-1/MTDH/LYRIC promotes angiogenesis, migration, invasion, and enhanced tumor metabolism through key oncogenic signaling cascades. In response to external stress cues and cellular mechanisms to inhibit further growth, AEG-1/MTDH/LYRIC activates pathways that bypass cell checkpoints and potentiates signals to enhance survival and tumorigenesis. As an oncogene that promotes aberrant cellular processes within the CNS, AEG-1/MTDH/LYRIC represents an important therapeutic target for the treatment of neurological disease.

1. INTRODUCTION

The central nervous system (CNS) represents a significant focus for biomedical research due to the relative lack of understanding of the variety of disease pathology that lies within this compartment. A thorough understanding of neurological disease and the discovery and validation of treatment targets have lagged behind many other organ systems within the human body. This relative shortcoming results from the difficulty in studying the CNS, in creating *in vitro* and *in vivo* models for CNS disease, and in producing therapeutics that are able to achieve sufficient concentration in targeted areas of the brain. For example, the treatment of neurodegenerative disease remains a challenge due to the lack of validated biomarkers, targets for treatment, and the ability to reverse disease pathogenesis with therapeutic compounds. Similarly, migraines remain an area of particular therapeutic challenge due to a relative lack of understanding of their etiology and molecular pathogenesis. Soon after infection with the human immunodeficiency virus (HIV), the CNS becomes a reservoir of latent virus that remains resistant to systemic treatment using traditional antiretroviral drugs. As a result, a more complete understanding of the way in which HIV enters the CNS and continues to replicate at low levels and the proteins that are responsible for this process could pave the way for enhanced therapeutic efficacy in the future.

In the area of CNS neoplasms, there has been limited progress to improve clinical prognoses. Glioblastoma represents one of the most malignant human cancers, with 5-year survival rates of just 6.9% Central Brain

Tumor Registry of the United States (CBTRUS) (Dolecek et al., 2012) and median survival rates that remain around 12–15 months from the time of diagnosis. Despite optimal treatment with surgical debulking, chemotherapeutic regimens, and radiation protocols, glioblastoma is highly aggressive and typically recurs following treatment. These tumors exhibit a dramatic propensity for mutation, chemo- and radioresistance, and microscopic spread throughout the CNS that provides a difficult substrate for therapeutic intervention. Therefore, this tumor entity represents a particularly attractive target for research in order to achieve progress in overall patient survival.

Astrocyte-elevated gene-1 (AEG-1/MTDH/LYRIC) is an HIV- and tumor necrosis factor (TNF)-inducible oncogene that is upregulated in all human cancers thus far profiled. Initially characterized as a responsive gene to HIV infection, the role of this gene has greatly expanded in the realm of CNS disease and has been shown to regulate neurodegenerative disease, migraines, glutamate signaling, and tumorigenesis. Through a variety of cellular signaling pathways, AEG-1/MTDH/LYRIC promotes cellular survival, the oncogenic state, and coordinates a host of cellular processes that encourage angiogenesis, migration, invasion, and metastasis. Also known as lysine-rich CEACAM-1-associated protein (LYRIC) in rats (Sutherland, Lam, Briers, Lamond, & Bickmore, 2004) and metadherin (MTDH) in humans (Brown & Ruoslahti, 2004), AEG-1/MTDH/LYRIC expression is ubiquitous in vertebrates and may contribute to normal cellular processes that govern growth. However, little is known about the function of AEG-1/MTDH/LYRIC under normal physiological circumstances in nontransformed cells within the CNS.

As a potent regulator of key cellular pathways mediating glioma proliferation, angiogenesis, and chemoresistance, AEG-1/MTDH/LYRIC serves a primary role in well-characterized signaling cascades known to promote tumorigenesis. AEG-1/MTDH/LYRIC enhances the aggressive phenotype in glioblastoma and facilitates its expansion within the skull case through the dissolution of extracellular matrix and death of surrounding cells. By hijacking cellular machinery that interferes with normal checkpoint regulation, AEG-1/MTDH/LYRIC promotes dysregulated tumor growth and associates with poor prognosis in glioma. With evidence suggesting its use as a biomarker in cancer (Chen et al., 2012) and its utility in therapeutic cancer vaccine intervention (Qian et al., 2011), further research into this gene at the center of many signaling cascades may shed light on molecular mechanisms of neurological disease and may provide improved diagnostic and therapeutic tools for patients in the future.

2. THE ROLE OF AEG-1/MTDH/LYRIC IN CNS DISEASE PATHOGENESIS

2.1. The association of AEG-1/MTDH/LYRIC with HIV-1-associated neurocognitive disorder through glutamate excitotoxicity and reactive gliosis

The HIV-1 pandemic continues to represent a significant public health burden and remains a therapeutic challenge. Soon after HIV-1 infection, virus likely enters the CNS through the activation of monocytes in bone marrow and subsequent circulation in the blood. This reservoir of HIV-1 within macrophages is resistant to systemic antiviral treatment and is potentially responsible for the maintenance of low levels of viral replication during treatment. In addition, HIV-1 infection in the CNS eventually leads to the development of HIV-associated neurocognitive disorder and HIV-associated dementia (HAD), which carry significant morbidity for affected patients (Ghafouri, Amini, Khalili, & Sawaya, 2006; McArthur, Steiner, Sacktor, & Nath, 2010). The ability to target this reservoir of viral replication and develop therapeutic options to treat neurological sequelae of HIV-1 infection may prove useful in eliminating latent HIV-1 and low-level replication and improving quality of life in affected patients.

AEG-1/MTDH/LYRIC was initially cloned through a rapid subtraction hybridization approach in primary human fetal astrocytes (PHFA) infected by HIV-1 (Kang et al., 2005). PHFA exhibit an approximately threefold increase in AEG-1/MTDH/LYRIC RNA levels following infection by HIV-1 or treatment with TNF-α (Su et al., 2003), indicating that AEG-1/MTDH/LYRIC may promote HIV-1 infection and its pathogenesis. AEG-1/MTDH/LYRIC/LYRIC physically interacts with HIV-1 Gag and is incorporated into viral particles and cleaved by HIV-1 PR (Engeland et al., 2011). Though the role of AEG-1/MTDH/LYRIC in this context is unknown, it is possible that AEG-1/MTDH/LYRIC thus regulates virion assembly, maturation, or release (Fig. 6.1).

There is compelling evidence that AEG-1/MTDH/LYRIC generally encourages virus entry into the CNS through the promotion of vascular permeability. Through molecular mimicry, AEG-1/MTDH/LYRIC acts as the target of the dengue virus (DENV) and enhances vascular permeability, potentially allowing for the spread of virus (Liu, Chiu, Chen, & Wu, 2011). Neurological manifestations, including viral encephalitis, are present in 1–5% of patients with symptomatic systemic DENV, and therefore,

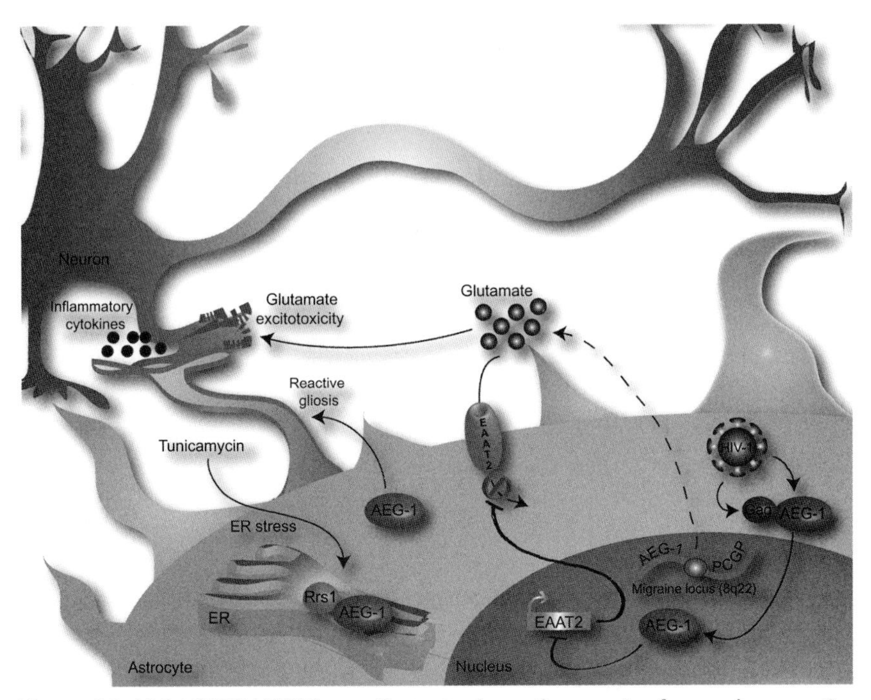

Figure 6.1 AEG-1/MTDH/LYRIC contributes to the pathogenesis of neurodegenerative disease, glutamate excitotoxicity, reactive gliosis, and migraine headaches. AEG-1/MTDH/LYRIC is induced by HIV-1 in primary human fetal astrocytes (PHFA) and interacts with HIV-1 Gag protein, potentially promoting HIV-1 virion assembly, maturation, and release. In the nucleus, AEG-1/MTDH/LYRIC also downregulates the EAAT2 promoter, resulting in decreased transporter expression, reduced glutamate uptake from the extracellular space, and a glutamate excitotoxic environment that is toxic to neighboring neurons. AEG-1/MTDH/LYRIC, as well as Rrs1, a protein involved in ribosomal assembly, is induced by the ER stress compound, tunicamycin, and colocalize in the ER, where these proteins may promote tumorigenic functions. In addition, AEG-1/MTDH/LYRIC promotes reactive gliosis following injury and is involved in astrocyte activation and subsequent neurodegeneration. In the area of migraine, a susceptibility locus on 8q21-22 between the genes encoding *AEG-1/MTDH/LYRIC* and *PGCP*, a glutamate regulatory molecule, may implicate glutamate excitotoxicity in the pathogenesis of migraine. AEG-1/MTDH/LYRIC, astrocyte-elevated gene-1; EAAT2, excitatory amino acid transporter 2; Rrs1, regulatory of ribosomal synthesis 1, PGCP, plasma glutamate carboxypeptidase. (See Page 5 in Color Section at the back of the book.)

AEG-1/MTDH/LYRIC may also promote DENV entry or its spread into the CNS. Though AEG-1/MTDH/LYRIC in this case represents a permissive proponent for vascular dispersion, it is conceivable that AEG-1/MTDH/LYRIC may share sequence homology with other viral antigens to enhance virus entry, survival, and spread after infection.

The death of neuronal cells is a known consequence of HIV infection and leads to the development of HAD. One clear mechanism for this neuronal destruction is the accumulation of toxic levels of glutamate, a process termed glutamate excitotoxicity, that is associated with a number of neurological diseases, such as amyotrophic lateral sclerosis (ALS), Alzheimer's disease, Parkinson's disease, and epilepsy. Glutamate is an essential amino acid for cellular growth and is required for a multitude of normal cellular processes, such as neurotransmission within the CNS. However, dysregulated control of extracellular glutamate concentrations can lead to an excitotoxic environment that is detrimental to surrounding cells. In the brain, the primary mechanism by which glutamate mediates its excitotoxic effects is through the expression of the excitatory amino acid transporter-2 (EAAT2), the principal transmembrane ion channel that regulates the concentration of glutamate through extracellular uptake. AEG-1/MTDH/LYRIC has been shown to downregulate the promoter of the EAAT2, rendering these cells more susceptible to glutamate excitotoxicity (Kang et al., 2005). In addition, this regulation appears to be phosphatidyl inositol 3-kinase (PI3K)-dependent, as PTEN overexpression abrogates this promoter silencing. As EAAT2 promoter activity is dependent on epidermal growth factor (EGF) and cAMP, it is possible that AEG-1/MTDH/LYRIC works in a similar fashion to modify glutamate transport.

In addition to glutamate excitotoxicity, a critical mediator of neuronal injury in HIV-1 infection is the process of reactive gliosis, whereby resting astrocytes are activated and produce inflammatory cytokines that damage surrounding normal brain parenchyma (Maragakis & Rothstein, 2006). This process underlies cell damage pathways in many neurodegenerative diseases and is thought to be responsible for some of the CNS complications of HIV-1 infection. AEG-1/MTDH/LYRIC promotes reactive gliosis *in vivo* and is responsible for astrocyte migration following injury (Vartak-Sharma & Ghorpade, 2012). In addition, the subcellular localization of AEG-1/MTDH/LYRIC is altered following brain injury, with a significant increase in cytoplasmic AEG-1/MTDH/LYRIC staining and no change in overall expression levels. However, during wound healing, AEG-1/MTDH/LYRIC colocalizes with the nucleolar marker, fibrillarin, indicating that AEG-1/MTDH/LYRIC translocates to the nucleolus, where it may cooperate in transcriptional transactivation of target genes necessary for proliferation and migration. Given the importance of restricted HIV-1 infection of astrocytes in the generation of HIV-1-associated neuropathies and dementia (Sabri, Titanji, De Milito, & Chiodi, 2003), this represents an alternative mechanism by which AEG-1/MTDH/LYRIC promotes CNS pathogenesis of HIV-1 infection.

2.2. The role of AEG-1/MTDH/LYRIC in neurodegenerative disease and migraine

Neurodegenerative diseases compose a major subset of neurological disease and are associated with significant morbidity without any currently available disease-modifying therapeutic options. Huntington's disease (HD) is an inherited autosomal-dominant neurodegenerative disorder caused by trinucleotide repeat expansion in the HD gene, huntingtin, and characterized clinically by progressive neurological deterioration and then death in the setting of chorea and psychiatric perturbation. In a model of HD using immortalized striatal neuronal cells, LYRIC was found to colocalize with the regulator of ribosome synthesis (Rrs1), a protein important for the maturation, nuclear export, and assembly of ribosomes, in the nucleolus and endoplasmic reticulum (ER) membrane. Moreover, in response to ER stress *in vitro*, both genes are induced (Carnemolla et al., 2009). Immunohistochemical analysis of HD brains demonstrates that AEG-1/MTDH/LYRIC is upregulated 1.7-fold over control brains, suggesting that its overexpression may be associated with the pathogenesis of HD. It was hypothesized that given its localization to both ER and nucleolar compartments and the proximity of ER and nucleolar membranes (Fricker, Hollinshead, White, & Vaux, 1997), LYRIC may act as a sensor of ER stress in this context and transmit these signals to the nucleolus. Though AEG-1/MTDH/LYRIC has not been studied in the context of other neurodegenerative diseases, the fact that many of these diseases, such as Alzheimer's disease, Parkinson's disease, and ALS, are marked by ER stress (Roussel et al., 2013), AEG-1/MTDH/LYRIC may serve a pathogenic function in a host of chronic progressive CNS diseases.

Like neurodegenerative disease, the pathogenesis of migraine headaches is poorly understood, and few effective therapies are available. As a result, genome-wide association studies have been performed to better investigate the genetic predisposition to migraine and stratify patients based on heritability markers. In a recent genome-wide association study (GWAS) study, AEG-1/MTDH/LYRIC was shown to hold a potential role in migraine pathogenesis, with a susceptibility allele for migraine being identified in the region of AEG-1/MTDH/LYRIC on chromosome 8q22 (Anttila et al., 2010). This study also showed that AEG-1/MTDH/LYRIC transcript levels correlated with the rs1835740 genotype and that this genotype may be a *cis* regulator of AEG-1/MTDH/LYRIC. The migraine susceptibility allele identified in this study is flanked by AEG-1/MTDH/LYRIC and the glutamate regulatory gene, plasma glutamate carboxypeptidase (PGCP), which increases glutamate concentrations in the extracellular space through the cleavage of *N*-acetyl-L-aspartyl-L-glutamate into N-acetyl aspartate (NAA) and glutamate. Therefore, it is possible that glutamate homeostasis is critical for migraine pathogenesis and that AEG-1/MTDH/LYRIC may regulate these pathways through its downregulation of the glutamate transporter, EAAT2 (Kang et al., 2005; Noch & Khalili, 2009). In another GWAS meta-analysis by the Dutch Icelandic migraine genetics consortium, the first population-based GWAS for common migraine, gene-based analysis also identified a possible association of MTDH with migraine (Ligthart et al., 2011). Like the prior study, the common SNPs were located between MTDH and PGCP on 8q21, further supporting dysregulated glutamate signaling in the pathogenesis of migraine. As AEG-1/MTDH/LYRIC regulates glutamate uptake in normal brain, this strengthens the argument that a relative glutamate excitotoxic environment, either through increased release or reduced uptake, stimulates the occurrence of migraine attacks.

3. STRUCTURE AND LOCALIZATION OF ASTROCYTE-ELEVATED GENE-1 IN THE CNS

The full-length AEG-1/MTDH/LYRIC cDNA consists of 3611 bp, excluding the poly-A tail (Kang et al., 2005). The open reading frame from 220 to 1968 nts encodes a putative 582-amino acid protein with a calculated molecular mass of 64 kDa with a p*I* of 9.33. The similar homologue identified in mouse breast cancer cells, named metadherin (MTDH) (metastasis adhesion protein), contains 579 amino acids (Brown & Ruoslahti, 2004). Genomic blast search demonstrated that the AEG-1/MTDH/LYRIC gene consists of 12 exons/11 introns and is located at 8q22 where cytogenetic analysis of human gliomas indicated recurrent amplifications. Protein motif analysis, such as simple modular architecture research tool (SMART), predicted a single-transmembrane domain at the N-terminus of the protein (amino acids 50–70) that includes putative dileucine repeats that are thought to be important for protein trafficking (Kirchhausen, 2000) and was supported by three independent transmembrane protein prediction methods (PSORT II, TMPred, and Hidden Markov Model for TOpology Prediction, HMMTOP). However, PSORT II and TMpred predicted AEG-1/MTDH/LYRIC as a type Ib protein (C-terminal inside) (like 3D3/Lyric), whereas TMHMM and TopPred 2 predicted a type II protein (C-terminal outside), which was verified with metadherin protein (Brown & Ruoslahti, 2004). Analysis of the amino acid sequence of AEG-1/MTDH/LYRIC further revealed the presence of putative, either monopartite or bipartite, nuclear localization signals (NLS) between amino acids 79–91, 432–451, and 561–580, suggesting import into the nucleus. The existence of a cleavable signal peptide was not evident based on the motif analyses. AEG-1/MTDH/LYRIC is a highly basic protein with many lysine (12.3%) and serine (11.6%) residues, indicating that posttranslational modification of AEG-1/MTDH/LYRIC may be critical for its nuclear import.

Initial characterization of AEG-1/MTDH/LYRIC demonstrated its expression in the perinuclear region and in ER-like structures (Kang et al., 2005). As the ER membranes are contiguous with the nuclear envelope, this expression pattern favors type Ib membrane morphology. In addition to its localization in these areas, AEG-1/MTDH/LYRIC is also located in the nucleolus (Sutherland et al., 2004) and at tight junctions (Britt et al., 2004), possibly due to tight junction assembly mediated by protein kinase C (Andreeva, Krause, Müller, Blasig, & Utepbergenov, 2001; Sakakibara, Furuse, Saitou, Ando-Akatsuka, & Tsukita, 1997). However, despite its three NLS, deletion or mutation of these signals does not prevent the functional activity of AEG-1/MTDH/LYRIC, indicating that there are other molecules critical for AEG-1/MTDH/LYRIC trafficking to the nucleus, such as TNF, that induces its nuclear translocation. Interestingly, fully nuclear AEG-1/MTDH/LYRIC is rarely seen possibly because of the leucine-rich putative nuclear export signal from amino acids 61 to 68. Moreover, predominantly monoubiquitination restricts expression of AEG-1/MTDH/LYRIC to the cytoplasmic compartment after posttranslational modification (Thirkettle, Girling, et al., 2009), an expression pattern associated with poor prognosis in cancer (Thirkettle, Girling, et al., 2009). LYRIC is phosphorylated at Ser297 with 2,3,7,8-tetrachlorodibenzo-*p*-dioxin treatment (Schulz et al., 2013), and AEG-1/MTDH/LYRIC is phosphorylated at Thr143 (Dephoure et al., 2008), Ser298 (Villen, Beausoleil, Gerber, & Gygi, 2007), Ser84, Ser415, Ser426, Ser308 (Olsen et al., 2006), and Ser496 (Olsen et al., 2010), which may govern several of its downstream interaction with oncogenic cascades.

Expression of AEG-1/MTDH/LYRIC is evident in many normal and transformed cell types within the CNS. AEG-1/MTDH/LYRIC is expressed in astrocytes at low levels (Kang et al., 2005) and is also present in transformed striatal neurons as well (Carnemolla et al., 2009). Positive expression of AEG-1/MTDH/LYRIC has been demonstrated in normal blood vessels within the CNS (unpublished results) and in human umbilical vein endothelial cells (Liu et al., 2011). In addition, AEG-1/MTDH/LYRIC is developmentally regulated, with expression being noted in the mid-to-hindbrain, frontonasal processes, and pharyngeal arches early in development (Jeon et al., 2010). Subsequently, AEG-1/MTDH/LYRIC is expressed in the brain and olfactory systems, which suggests a potential role in neurogenesis and early brain modeling.

4. THE ROLE OF AEG-1/MTDH/LYRIC IN BRAIN TUMOR PATHOGENESIS

4.1. The molecular interaction of AEG-1/MTDH/LYRIC with tumorigenic cell signaling cascades

Like other systemic malignancies, tumors within the CNS are marked by genetic and acquired mutations that support dysregulated cellular growth. Many of these mutations, such as those involving p53, PTEN, and isocitrate dehydrogenase 1 (IDH1), have been examined extensively in glioblastoma (Purow & Schiff, 2009). In addition, studies of overactive oncogenic signaling pathways, such as PI3K, EGFR, and MAPK, and defects in DNA repair machinery have led to the development of novel therapeutics that have entered clinical trials (Johnson & Chang, 2012). Aside from temozolomide, which has demonstrated clear efficacy in phase III clinical trial for the treatment of newly diagnosed glioblastoma (Stupp et al., 2005), no other chemotherapeutic agent alone has prolonged overall survival in patients with this aggressive neoplasm. Improved therapeutic efficacy requires targeting of related yet synergistic pathways that promote glioblastoma, such as angiogenesis and tumor metabolism. Therefore, it is probable that as genomic and metabolic analyses improve our understanding of pathways that support glioblastoma, future efforts will focus on combination therapy to control aberrant tumor signaling and may achieve greater success.

As a TNF-inducible gene with elevated expression in astrocytes transformed by SV40 T-antigen, telomerase (hTERT), and T24 Ha-*Ras* (Kang et al., 2005; Rich et al., 2001), AEG-1/MTDH/LYRIC was originally thought to play a significant role in the process of tumorigenesis. Since that time, AEG-1/MTDH/LYRIC has been shown to be upregulated in a variety of CNS malignancies, including neuroblastoma (Lee et al., 2009), oligodendroglioma (Xia et al., 2010), and glioblastoma (Emdad et al., 2010; Liu et al., 2010) and correlates with patient prognosis as well (Liu, Liu, Han, Zhang, & Sun, 2012). In some malignancies, such as ovarian and breast cancer, certain AEG-1/MTDH/LYRIC alleles are associated with increased risk of cancer, though protein expression levels do not differ between allele types (Yuan et al., 2012). The initial characterization of AEG-1/MTDH/LYRIC showed that this protein promotes anchorage-independent growth (Kang et al., 2005). Subsequently, AEG-1/MTDH/LYRIC was determined to act as an oncogene and promotes tumor formation when overexpressed in fibroblast cells alone (Emdad et al., 2009).

On a molecular level, AEG-1/MTDH/LYRIC interacts with a variety of cell signaling pathways that promote tumorigenesis and interrupts normal checkpoints involved in the control of cellular growth (Fig. 6.2). Within the nuclear factor kappa B (NFκB) pathway, AEG-1/MTDH/LYRIC activates binding of the two components of NFκB, p50 and p65, to NFκB consensus sequences in the promoters of target genes. As a result, AEG-1/MTDH/LYRIC upregulates several downstream genes, such as intercellular adhesion molecule (ICAM)-3, ICAM-2, selectin E, selectin P ligand, selectin L, toll-like receptor (TLR)-4, TLR-5, FOS, JUN, and IL-8 (Emdad et al., 2006), all of which have known functions in cancer. It appears that AEG-1/MTDH/LYRIC promotes NFκB activation by inducing degradation of the NFκB repressor, IκB. Additionally, AEG-1/MTDH/LYRIC promotes nuclear translocation of NFκB and binds to p65 through amino acids 101–205 (Sarkar et al., 2008). The mechanism by which AEG-1/MTDH/LYRIC promotes NFκB-mediated oncogenic activity involves cooperative binding with cyclic AMP-responsive element-binding protein (CREB)-binding protein (CBP), a known activator of NFκB signaling. In this context, AEG-1/MTDH/LYRIC may act as a bridging molecule between CBP and the NFκB promoter to support transactivation of downstream genes.

The PI3K/Akt pathway and a variety of antiapoptotic pathways are critical to initiate and maintain the tumorigenic phenotype in malignant brain tumors (Lino & Merlo, 2011), and AEG-1/MTDH/LYRIC serves an important positive regulatory role in these pathways. AEG-1/MTDH/LYRIC overexpression induces phosphorylation of Akt (Lee et al., 2008), a signaling event that is required for AEG-1/MTDH/LYRIC-mediated tumor induction. Further, AEG-1/MTDH/LYRIC induces phosphorylation and inactivation

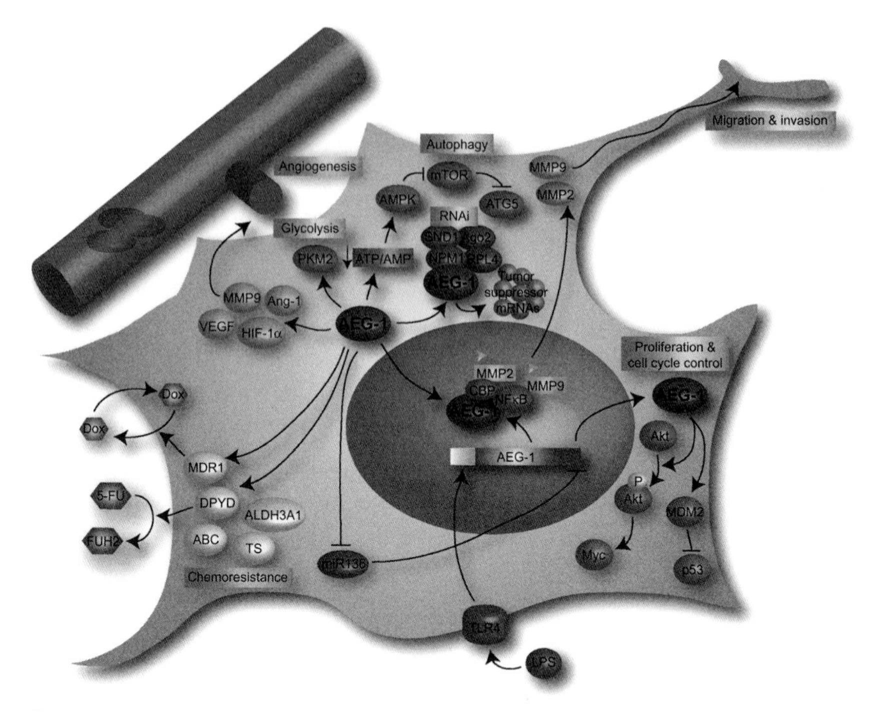

Figure 6.2 AEG-1/MTDH/LYRIC promotes proliferation, angiogenesis, migration and invasion, tumor-associated cellular metabolism, and chemoresistance in glioblastoma. AEG-1/MTDH/LYRIC promotes angiogenesis through the induction of MMP9, Ang-1, HIF-1α, and VEGF. The final rate-limiting enzyme in glycolysis, PKM2, is induced by AEG-1/MTDH/LYRIC, which promotes anaerobic glycolysis in glioblastoma. AEG-1/MTDH/LYRIC accelerates ATP consumption and reduces the ATP/AMP ratio, thereby activating AMPK and promoting autophagy through mTOR inhibition and ATG5 activation. Migration and invasion are governed by NFκB-dependent AEG-1/MTDH/LYRIC activation of the MMP2 and MMP9 promoters, a process which is supported by TLR4 activity. In the cytoplasm, AEG-1/MTDH/LYRIC associates with several members of the RNA-induced silencing complex (RISC), including Ago 2, SND1, RPL4, and NPM1 and promotes sequestration of tumor suppressor mRNAs into stress granules. AEG-1/MTDH/LYRIC supports the chemoresistant phenotype through activation of the ABC transporter superfamily, thymidylate synthase, and ALDH3A1. In addition, AEG-1/MTDH/LYRIC promotes doxorubicin efflux through activation of MDR1 and 5-FU metabolism through DPYD activation. AEG-1/MTDH/LYRIC, astrocyte-elevated gene-1; ATP, adenosine trisphosphate; AMPK, 5′-adenosine monophosphate (AMP)-activated protein kinase; mTOR, mammalian target of rapamycin; ATG5, autophagy protein 5; NFκB, nuclear factor kappa B; MMP2, matrix metalloproteinase 2; TLR4, toll-like receptor 4; SND1, staphylococcal nuclease domain-containing 1; RPL4, ribosomal protein L4; NPM1, nucleophosmin 1; TS, thymidylate synthase; ALDH3A1, aldehyde dehydrogenase 3A1; MDR1, multidrug resistance 1; 5-FU, 5-fluorouracil; DPYD, dihydropyridine dehydrogenase. (See Page 6 in Color Section at the back of the book.)

of glycogen synthase kinase 3β (GSK3β) and concomitant inhibition of p21/mda-6 and the proapoptotic protein, bad. Similarly, expression levels of p53 are reduced in AEG-1/MTDH/LYRIC-overexpressing PHFA along with phosphorylation and activation of the negative regulator of p53, murine double minute 2. AEG-1/MTDH/LYRIC negatively regulates the proapoptotic tumor suppressor, forkhead box class O (FOXO) 3a transcription factor in prostate cancer (Kikuno et al., 2007), which is itself phosphorylated and inhibited by Akt (Brunet et al., 1999), and may thus prevent transcriptional activation of various oncogenic pathways. The Akt target, c-myc, is also induced by AEG-1/MTDH/LYRIC in PHFA (Lee, Su, Emdad, Sarkar, & Fisher, 2006), and AEG-1/MTDH/LYRIC induces n-myc expression in neuroblastoma (Lee et al., 2009). Functionally, AEG-1/MTDH/LYRIC prevents serum starvation-induced apoptosis through a PI3K-dependent mechanism in PHFA. This pathway is dependent on Akt but involves phosphatase and tensin homolog (PTEN), implicating a process downstream of PI3K. In the related extracellular signal-regulated kinase (ERK) pathway, AEG-1/MTDH/LYRIC induces phosphorylation of ERK42/44, p38 mitogen-activated protein kinase (MAPK), and Akt in hepatocellular carcinoma cells, pathways that are critical for AEG-1/MTDH/LYRIC-mediated invasion and anchorage-independent growth (Yoo, Emdad, et al., 2009). Further, AEG-1/MTDH/LYRIC binds to the transcriptional repressor involved in cell cycle control, promyelocytic leukemia zinc finger, and suppresses its transcriptional repression of target genes (Thirkettle, Mills, Whitaker, & Neal, 2009), thereby preventing apoptosis (Bernardo, Yelo, Gimeno, Campillo, & Parrado, 2007). Together, these findings implicate AEG-1/MTDH/LYRIC as a prosurvival and antiapoptotic protein that coordinates with PI3K/Akt signaling to promote tumorigenesis.

Molecules within the Wnt signaling pathway, which regulate patterning and development of the CNS and are also important for certain oncogenic signaling cascades, are modulated by AEG-1/MTDH/LYRIC as well. Lymphoid enhancer-binding factor 1 (LEF1), the transcription factor activated by Wnt signaling, is upregulated, whereas the negative regulators of the Wnt pathway, C-terminal-binding protein 2 and adenomatous polyposis coli, are significantly downregulated by AEG-1/MTDH/LYRIC. Within the Wnt pathway, AEG-1/MTDH/LYRIC enhances phosphorylation of GSK3β and causes nuclear translocation of β-catenin, inducing transcription of the LEF1 promoter (Jian-bo et al., 2011; Yoo, Emdad, et al., 2009). In hepatocellular carcinoma, LEF1 inhibition abrogates AEG-1/MTDH/LYRIC-induced invasion, highlighting the importance of the Wnt pathway in mediating the tumorigenic effects of AEG-1/MTDH/LYRIC. Given that the Wnt pathway is also dysregulated in glioblastoma (Kaur et al., 2013), it is likely that similar signaling cascades mediate the activity of AEG-1/MTDH/LYRIC in CNS neoplasia.

On a transcriptional level, *AEG-1/MTDH/LYRIC* contains several E-box elements within its promoter that are targets for oncogenic proteins. The *AEG-1/MTDH/LYRIC* promoter has positive (−459/−302) and negative (−738/−460) regulatory regions, with Sp1, E-box element, CREB, and Ets-2 necessary for basal promoter activity (Lee et al., 2006). Two of the E-box elements appear to be necessary for basal and Ras-induced promoter activation in immortalized PHFA, and the negative regulatory region contains putative RAR-α and YY1 binding sites that may repress transcriptional activity. Similarly, oncogenic signaling through the Ha-ras pathway requires the participation of AEG-1/MTDH/LYRIC through c-myc binding to the AEG-1/MTDH/LYRIC promoter (Lee et al., 2006). Therefore, as a c-myc-dependent target of oncogenic Ras, AEG-1/MTDH/LYRIC may facilitate the coordinated activity of these oncogenic pathways in tumorigenesis. Moreover, the transcriptional regulatory processes that induce AEG-1/MTDH/LYRIC may be suitable substrates for therapeutic investigation.

4.2. The role of AEG-1/MTDH/LYRIC in cell cycle dysregulation in glioblastoma

The dysregulation of cell cycle progression is critical to the foundation of cancer, and gliomas control nearly every phase of the cell cycle to bolster cellular growth (Alexander, Pinnell, Wen, & D'Andrea, 2012). AEG-1/MTDH/LYRIC has been shown to act as a negative regulator of the BRCA2- and CDKN1A ($p21^{Cip1/Waf-1}$)-interacting protein, BCCIPα, that binds to the cell cycle regulatory protein, p21, and supports p21-mediated cyclin-dependent kinase 2 (Cdk2) inhibition (Ash, Yang, & Britt, 2008). Without such BCCIPα activity, G1/S cell cycle checkpoint activation following DNA damage is impaired. Moreover, BCCIPα plays a role in homologous recombination after DNA damage (Lu et al., 2005) and contributes to chromosomal stability (Meng, Fan, & Shen, 2007). AEG-1/MTDH/LYRIC physically interacts with BCCIPα, and the amino-terminal domain of AEG-1/MTDH/LYRIC is critical for this interaction. In addition, AEG-1/MTDH/LYRIC overexpression reduces levels of the cyclin-dependent kinase (CDK) inhibitors, $p21^{Cip1}$ and $p27^{Kip1}$, as well as phosphorylation of Rb, and enhances progression through the cell cycle following serum starvation (Li et al., 2009). This may occur through regulation of forkhead box-containing O subfamily

(FoxO)-1, which acts as a tumor suppressor by transactivating the promoters of CDK inhibitors (Nakamura et al., 2000; Roeb, Boyer, Cavenee, & Arden, 2007; Seoane, Le, Shen, Anderson, & Massagué, 2004). Knockdown of AEG-1/MTDH/LYRIC reduces the overall percentage of S- and G2/M-phase cells (Liu et al., 2009), which promotes tumor proliferation. AEG-1/MTDH/LYRIC-mediated induction of LEF1 may also control expression of key cell cycle regulatory molecules, including c-myc and Cyclin D1 (Yoo, Emdad, et al., 2009) that promote gliomagenesis (Wang et al., 2012). By circumventing checkpoints that control cellular growth, AEG-1/MTDH/LYRIC hijacks cell cycle function and accelerates glioma progression.

4.3. AEG-1/MTDH/LYRIC regulates of angiogenesis in malignant brain tumors

One of the hallmark features of glioblastoma is vascular proliferation, which is marked by an increase in overall vascular density as well as microvascular proliferation in the form of glomeruloid vessels that surround areas of necrosis (Louis, 2006). These vessels support accelerated growth through oxygen and nutrient retrieval in glioblastoma. In addition, they promote several pathological clinical features, including blood–brain barrier breakdown and tumoral edema, through an increase in vascular permeability mediated by vascular endothelial growth factor (VEGF). Cell stress events, such as hypoxia and glucose deprivation, which lead to tumor necrosis, are also known triggers for VEGF activation, which promotes cell survival under metabolic stress (Raza et al., 2002).

AEG-1/MTDH/LYRIC has been shown to regulate a variety of pathways involved in angiogenesis and blood vessel remodeling (Fig. 6.2). AEG-1/MTDH/LYRIC overexpression in fibroblast cells induces expression of angiopoietin-1 (Ang-1), matrix metalloproteinase (MMP)-2, and HIF-1α, and increases microvessel density (Emdad et al., 2009). Similarly, AEG-1/MTDH/LYRIC overexpression in human hepatocellular carcinoma cells upregulates VEGF, placental growth factor, and fibroblast growth factor-α (Yoo, Emdad, et al., 2009). In addition, AEG-1/MTDH/LYRIC promotes tube formation in cultured endothelial cells, while its knockdown prevents neovascularization in a chicken chorioallantoic membrane model of angiogenesis. As with cellular signaling pathways that promote the AEG-1/MTDH/LYRIC-mediated tumorigenic phenotype, the PI3K-Akt pathway is also critical for enhanced angiogenesis, indicating that there are intermediate pathways that regulate these processes in tumor cells. On a histological level, AEG-1/MTDH/LYRIC is positively associated with expression patterns of molecules regulating invasion, such as MMP-2 and MMP-9, as well as the marker of angiogenesis, CD31 (Emdad et al., 2010). The expression of AEG-1/MTDH/LYRIC also correlates with that of VEGF in breast cancer cells (Li et al., 2011).

The targeting of VEGF has been of moderate success in recurrent glioblastoma since the discovery and testing of the monoclonal antibody, bevacizumab, directed against this growth factor (Friedman et al., 2009). Bevacizumab treatment leads to tumor shrinkage and 6-month progression-free survival that exceeds 40%, and this therapy may be useful against radiation necrosis by reducing edema, mass effect, and leakiness of the blood–brain barrier. However, these benefits are only incremental, and monotherapy against angiogenesis is unlikely to modify disease progression. As AEG-1/MTDH/LYRIC targets extracellular matrix regulatory proteins, invasion, and angiogenesis, the use of small-molecular inhibitors of this pathway along with VEGF inhibition may yield greater efficacy against robust proangiogenic signals in glioblastoma. Further, evidence that AEG-1/MTDH/LYRIC promotes both blood–brain barrier leakiness in virus studies and CNS vascular pathology from migraine studies indicates that AEG-1/MTDH/LYRIC may serve as an efficacious target against tumor angiogenesis.

4.4. AEG-1/MTDH/LYRIC promotes tumor-specific glucose metabolism and autophagy in glioblastoma

The metabolic phenotype of cancer is one of the key mechanisms regulating the unrestricted growth of tumor cells. Unlike their normal counterparts, tumor cells preferentially undergo aerobic glycolysis for energy metabolism, a phenomenon first described by Warburg in 1956 known as the Warburg effect (Warburg, 1956). As a result, enhanced glycolytic flux allows tumor cells to produce sufficient adenosine triphosphate (ATP) to fulfill metabolic demands and leads to increased glucose consumption, decreased oxidative phosphorylation, and increased lactate production. Further, this relative surge in glycolytic flux spares tumor cells the production of detrimental reactive oxygen species (ROS) from oxidative phosphorylation and also enhances pentose phosphate pathway flux, which yields reducing equivalents in the form of nicotinamide adenine dinucleotide phosphate. Though glycolysis produces much less energy in the form of ATP than oxidative phosphorylation, several glycolytic enzymes are highly active in tumor cells and support enhanced anaerobic glycolysis and energy production.

Specifically, the rate-limiting enzyme that catalyzes the final step in glycolysis, pyruvate kinase, is upregulated in tumor cells. Interestingly, tumor cells exclusively express the M2 isoform of this enzyme (M2PK) (Mazurek, Boschek, Hugo, & Eigenbrodt, 2005), which is typically expressed only during embryonic development, rather than the M1 isoform, expressed in most normal adult tissues. Microarray analysis in hepatocellular carcinoma cells shows that AEG-1/MTDH/LYRIC upregulates pyruvate kinase (Yoo, Emdad, et al., 2009). Additionally, AEG-1/MTDH/LYRIC is a key downstream molecule activated by cell stress, and specifically, is upregulated by hypoxia and glucose deprivation in glioblastoma cells in a ROS-dependent manner (Noch, Bookland, & Khalili, 2011). Moreover, AEG-1/MTDH/LYRIC overexpression allows glioblastoma cells to survive periods of glucose deprivation, whereas its downregulation hastens cell death.

At the center of a host of metabolic pathways is the 5′-adenosine monophosphate (AMP)-activated protein kinase (AMPK), which controls the usage of macromolecular precursors for cellular metabolism. AMPK promotes anabolic processes and the Warburg effect in cancer through induction of various glycolytic enzymes. In glioblastoma cells, AEG-1/MTDH/LYRIC reduces the ATP/AMP ratio and activates AMPK and its downstream targets (Bhutia et al., 2010), implicating AEG-1/MTDH/LYRIC as a glycolytic regulator (Fig. 6.2). In this way, AEG-1/MTDH/LYRIC may enhance glycolytic flux in glioblastoma cells by sensing the cellular energy state and signaling through AMPK to induce elevated levels of PKM2, thereby maintaining sufficient ATP production for tumor proliferation. Further, AEG-1/MTDH/LYRIC may indirectly regulate cellular metabolism through various signaling cascades, such as PI3K/Akt, Wnt/B-catenin, and MAPK/ERK, which all occupy central positions in the oncometabolome and the regulation of oxidative stress. An example is MAPK, which is activated by AEG-1/MTDH/LYRIC and functions in oncogene-driven and oxidative stress-promoted tumorigenesis as well as the induction of glucose transporter expression (Fujishiro et al., 2001).

AMPK is also critical to the process of autophagy, which allows tumor cells to recycle key metabolic end products for future use. In immortalized primary human fetal astrocytes, AEG-1/MTDH/LYRIC promotes autophagy in an AMPK-dependent manner (Bhutia et al., 2010). This is evident with either genetic knockdown of AMPK or pharmacological inhibition. With AEG-1/MTDH/LYRIC overexpression, cellular ATP levels decrease in a dose- and time-dependent manner. However, when cellular ATP stores are replenished, this AEG-1/MTDH/LYRIC-dependent autophagy is reversed, indicating that AEG-1/MTDH/LYRIC acts in this pathway to salvage metabolic intermediates. In normal cells, AEG-1/MTDH/LYRIC promotes autophagy to protect these cells from apoptosis, lending support to a prosurvival function for AEG-1/MTDH/LYRIC in autophagy. In *atg5*-deficient glioblastoma cells, AEG-1/MTDH/LYRIC overexpression rescues the chemoresistant phenotype, confirming that AEG-1/MTDH/LYRIC-mediated chemoresistance is mediated through autophagy in this context.

On a molecular level, the AEG-1/MTDH/LYRIC-mediated process of autophagy is independent of canonical signaling through beclin 1 and its partner, class III PI3K. Rather, this cascade proceeds through an autophagy protein 5 (ATG5)-dependent pathway through inhibition of mammalian target of rapamycin (mTOR). mTOR is inactivated by AMPK during periods of metabolic stress, which leads to increased ATG5 expression and subsequent autophagy. This complex signaling pathway involving AEG-1/MTDH/LYRIC, AMPK, and mTOR is able to sense the energy status of the cell and the metabolic profile within the tumor microenvironment to coordinately regulate tumor glucose utilization. In this model, AEG-1/MTDH/LYRIC was also shown paradoxically not to increase the production of ROS, which typically induces protective autophagy. Given the role of AEG-1/MTDH/LYRIC in stromal communication, AEG-1/MTDH/LYRIC may also induce the "Reverse Warburg Effect" in cancer, whereby transformed cells utilize oxidative stress to promote autophagy in surrounding stromal cells, thereby providing metabolic precursors that can be shuttled back to cancer cells to promote accelerated growth (Martinez-Outschoorn, Lin, et al., 2011; Martinez-Outschoorn, Pavlides, et al., 2011; Martinez-Outschoorn et al., 2010; Sotgia et al., 2011). As a result, AEG-1/MTDH/LYRIC may activate both intrinsic and extrinsic autophagy pathways that both promote tumor cell survival and proliferation.

4.5. The role of AEG-1/MTDH/LYRIC in tumor invasion and migration

Unlike other soft-tissue masses that exhibit unlimited expansive capacity, brain tumors are restricted by the skull case in which they grow and must either destroy surrounding cells or invade areas of lower cellular density to expand. Gliomas are well known to intercalate through surrounding intercellular crevices through the regulation of chloride flux (Haas & Sontheimer, 2010; Ransom, O'Neal, & Sontheimer, 2001), autocrine and paracrine chemokine signaling (Hoelzinger, Demuth, & Berens, 2007; Wick, Platten, & Weller, 2001), and proangiogenic cues. Growth along white matter tracts, around neurons, along blood vessels, and beneath pial margins is typical of glioma, which permits satellite expansion and distant propagation (Louis, 2006; Scherer, 1938). Invasion through surrounding brain parenchyma is accomplished through the dissolution of extracellular matrix and the activity of various proteases, such as MMPs, a disintegrin and metalloproteinase (ADAM), and ADAMs with thrombospondin motifs (Mentlein, Hattermann, & Held-Feindt, 2012; Rubenstein & Kaufman, 2008). These proteases promote the degradation of microarchitectural scaffolds within normal brain and support oncogenic chemokine and growth factor signaling.

AEG-1/MTDH/LYRIC regulates tumor invasion through a variety of downstream signaling cascades (Fig. 6.2). AEG-1/MTDH/LYRIC overexpression upregulates genes associated with invasion, such as claudin 4 and tetraspanin 8, and downregulates transgrelin, a suppressor of MMP-9, which together promote adhesion. AEG-1/MTDH/LYRIC overexpression supports the invasive phenotype in neuroblastoma cells (Liu et al., 2010), whereas AEG-1/MTDH/LYRIC knockdown inhibits invasion and decreases MMP-2 and MMP-9 activation through gel zymography in glioma cells. AEG-1/MTDH/LYRIC also activates the MMP-9 promoter (Liu et al., 2010), while AEG-1/MTDH/LYRIC knockdown depresses MMP-2 and MMP-9 promoter activity, indicating a transcriptional regulatory role for AEG-1/MTDH/LYRIC in tumor invasion and adhesion. Similar to transactivation of other promoters, this regulation appears to depend on NFκB, indicating that AEG-1/MTDH/LYRIC may serve as a cooperative regulator of NFκB promoter regulation. These changes in MMP transcription and activation translate to decreased growth of mouse intracranial glioma xenografts, highlighting these pathways as critical to the proliferation and expansion of glioma cells in the brain.

Through coordinated control with lipopolysaccharide (LPS), the main component of the outer membrane of Gram-negative bacteria, AEG-1/MTDH/LYRIC also promotes invasion in malignant brain tumors. Despite a well-known role of LPS in mediating septic shock through the TLR4 (Poltorak et al., 1998), LPS has been implicated in tumor invasion through NFκB pathway activation. In addition, TLR4 activation can lead to enhanced migration and invasion in glioma (Thuringer et al., 2011) and enhances LPS-induced cytokine production (Carpenter et al., 2009). AEG-1/MTDH/LYRIC is induced by LPS through TLR-4 signaling and NFκB phosphorylation (Khuda et al., 2009; Zhao et al., 2011) and is required for LPS-mediated tumor migration and invasion. The production of IL-8 and MMP9 is reduced by AEG-1/MTDH/LYRIC knockdown but is rescued by LPS treatment. AEG-1/MTDH/LYRIC also regulates expression of the LPS receptor, TLR4, as well as TIR domain-containing adaptor inducing IFN-β (TRIF), within the TLR4 signaling pathway. By enhancing the production of tissue-destructive cytokines, such as MMP9, LPS induction of AEG-1/MTDH/LYRIC fosters increased tumor growth through synergistic TLR4 activation.

In light of the importance of invasion to the process of tumor metastasis, MTDH/AEG-1/MTDH/LYRIC has been demonstrated to induce greater chemotaxis and migration in hepatocellular carcinoma cells, highlighting the strength of this protein in promoting the aggressively invasive phenotype in cancer (Zhou et al., 2012). Further, AEG-1/MTDH/LYRIC induces adhesion of breast cancer cells to the endothelium (Brown & Ruoslahti, 2004; Hu et al., 2009), which represents a critical component of perivascular spread in glioma (Scherer, 1938). Moreover, AEG-1/MTDH/LYRIC knockdown reduces expression of molecules that support the metastatic phenotype, including N-cadherin and snail, and reduces pulmonary and abdominal metastases (Zhu et al., 2011). N-cadherin is a key molecule mediating cell–cell adhesion and regulates tumor migration and invasion (Gravdal, Halvorsen, Haukaas, & Akslen, 2007; Hazan, Phillips, Qiao, Norton, & Aaronson, 2000), and therefore, AEG-1/MTDH/LYRIC may control tumor progression through an N-cadherin-dependent mechanism. In glioblastoma, cytoplasmic polyadenylation element-binding protein 1 (CPEB1), which controls the fate of mRNA precursors through the 3′-polyadenylation of target mRNAs, binds to MTDH/AEG-1/MTDH/LYRIC and regulates its expression (Kochanek & Wells, 2013). As CPEB1 modulates directed cellular migration, potentially through trafficking of mature mRNA molecules to the leading edge of dividing cells, AEG-1/MTDH/LYRIC mRNA binding may influence the local expression of AEG-1/MTDH/LYRIC and could implicate this pathway in tumor cellular polarity during invasion. Further, given that CPEB1 shuttles between the nucleus and cytoplasm, this protein may be responsible for context-dependent subcellular localization of AEG-1/MTDH/LYRIC, which carries prognostic importance for patients (Thirkettle, Girling, et al., 2009).

4.6. The function of AEG-1/MTDH/LYRIC in RNA interference in malignant glioma

RNA interference serves a pivotal role in regulating the expression of a multitude of human genes and has been utilized for experimental manipulation since its discovery in *Caenorhabditis elegans* (Fire et al., 1998). In tumor biology, microRNAs (miRs) act as tumor suppressors and/or oncogenes and modulate tumor cellular processes. The RNA-induced silencing complex (RISC), which facilitates mRNA silencing through association with mature miRNA molecules and assistor proteins, is itself an important mediator of tumorigenesis through associated miRs that can promote cancer (Moser & Fritzler, 2010). AEG-1/MTDH/LYRIC interacts with several integral proteins that form the RISC complex, including staphylococcal nuclease domain-containing 1 (SND1), ribosomal protein L4 (RPL4), and nucleophosmin 1 (NPM1) (Blanco et al., 2011; Meng et al., 2012; Yoo, Santhekadur, et al., 2011) (Fig. 6.2). In addition, AEG-1/MTDH/LYRIC itself is a component of the RISC complex and colocalizes in the cytoplasm not just with SND1 but also with another major RISC component, Ago2 (Hutvagner & Simard, 2008). AEG-1/MTDH/LYRIC overexpression enhances RISC activity, whereas its knockdown reduces activity in both hepatocellular carcinoma and glioblastoma cells. SND1 inhibition abrogates AEG-1/MTDH/LYRIC function as well, highlighting the tumorigenic capacity of this protein. Interestingly, the region of AEG-1/MTDH/LYRIC that is required for SND1 binding (aa 101–205) is the same region responsible for NFκB p65 (aa 72–169) and BCCIPα interaction, though a specific protein domain has not been identified in this area. As a result, AEG-1/MTDH/LYRIC may utilize a similar binding motif with p65 protein recruitment to traffic other proteins to the RISC complex for cleavage activity.

In this role, AEG-1/MTDH/LYRIC may also regulate the stability and function of miRs involved in tumor activity. AEG-1/MTDH/LYRIC overexpression in glioma cells reduces the tumor-suppressive activity of miR-136, which has been found to be downregulated in human glioma (Yang et al., 2012). Interestingly, miR-136 directly targets the 3′-UTR of AEG-1/MTDH/LYRIC, indicating a potential feedback loop whereby AEG-1/MTDH/LYRIC directs the degradation of a potent suppressor miRNA that would limit its oncogenic function. AEG-1/MTDH/LYRIC suppression is also a result of miR375 activity in head and neck squamous cell carcinoma and esophageal carcinoma (Isozaki et al., 2012; Nohata et al., 2011).

A further regulatory of AEG-1/MTDH/LYRIC in the processing of mature RNA molecules comes from the study of stress granules. Stress granules are cytoplasmic RNA–protein complexes that are formed when the translation initiation process is halted in response to cellular stress (Buchan & Parker, 2009). AEG-1/MTDH/LYRIC knockdown accentuates the formation of stress granules in response to heat shock and may be responsible for the sequestration of tumor suppressor mRNA during cellular stress (Meng et al., 2012). By blocking the cellular response to stress signals through inhibition of tumor suppressor function, AEG-1/MTDH/LYRIC may encourage aberrant cellular response to these signals, such as impaired DNA repair, redirected metabolism, and enhanced migration and invasion. A more thorough investigation of the RNA-binding activity of AEG-1/MTDH/LYRIC along with its function in RNAi and stress granule formation may facilitate the development of treatments that target these pathways.

4.7. The promotion of chemoresistance by AEG-1/MTDH/LYRIC in malignant brain tumors

A critical detriment to the effective treatment of malignant brain tumors is the lack of validated targets, the rapid and inevitable resistance to chemotherapeutic agents, and the difficulty in traversing and achieving therapeutic concentrations within the blood–brain barrier. In malignant gliomas, the chemoresistant phenotype has been studied in relation to a variety of pathways, including antiapoptotic pathways, drug efflux pumps, multidrug resistant genes, and cancer stem cell activity (Lu & Shervington, 2008). The ATP-binding cassette (ABC) superfamily composes one of the largest protein families that is responsible for chemoresistance in glioblastoma (Benyahia et al., 2004; Spiegl-Kreinecker et al., 2002). The ABC family members transport molecules across cell membranes in an ATP-dependent fashion or through the formation of membrane channels and are expressed in both endothelial cells at the blood–brain barrier and in glioma cells (Bronger et al., 2005). The multidrug resistance gene-1 (MDR-1) protein is overexpressed in high-grade versus low-grade astrocytomas (Kirches et al., 1997; Spiegl-Kreinecker et al., 2002), contributes to chemoresistance by cancer stem cells (Nakai et al., 2009), and is responsible for treatment outcomes with standard temozolamide chemotherapy (Schaich et al., 2009). Therefore, a better understanding of these proteins and their mechanism of regulation will restrict cellular mechanisms of resistance, such as drug metabolism and efflux that limit therapeutic efficacy.

In addition to its role in promoting tumorigenesis, AEG-1/MTDH/LYRIC also serves an important role in each of these pathways and induces the chemoresistant phenotype (Fig. 6.2). Most *in vitro* models have utilized methods of AEG-1/MTDH/LYRIC downregulation to study the therapeutic benefit in promoting sensitivity to chemotherapeutic agents. In a model of hepatocellular carcinoma, AEG-1/MTDH/LYRIC was shown to potently activate dihydropyridine dehydrogenase (DPYD), which converts

5-fluorouracil (5-FU) to its inactive metabolite, fluro-5,6-dihydro uracil (FUH_2) (Yoo, Gredler, et al., 2009). In addition, in this study, AEG-1/MTDH/LYRIC induced expression of thymidylate synthase, which encodes the rate-limiting enzyme in the production of dTTP. AEG-1/MTDH/LYRIC also promotes expression of drug-metabolizing enzymes as well as ABC members, which causes efflux of various chemotherapeutic agents (Yoo, Emdad, et al., 2009). In glioma cells, AEG-1/MTDH/LYRIC knockdown promotes chemosensitivity (Emdad et al., 2010) and enhances translation and stability of MDR-1 (Yoo et al., 2010). Functionally, AEG-1/MTDH/LYRIC promotes chemoresistance to the DNA-alkylating agent, doxorubicin, in an MDR-1-dependent manner and promotes doxorubicin efflux. Moreover, AEG-1/MTDH/LYRIC knockdown in conjunction with treatment using doxorubicin significantly decreases growth of tumor xenografts, indicating a potential therapeutic target to slow tumor growth. Similarly, downregulation of AEG-1/MTDH/LYRIC sensitizes neuroblastoma cells to cisplatin and doxorubicin (Liu et al., 2009), indicating that AEG-1/MTDH/LYRIC is responsible for chemoresistance in neural crest-derived neoplasms.

AEG-1/MTDH/LYRIC also interacts with multiple oncogenic proteins that are known to mediate chemoresistance in other cancer cell types. In a model of breast cancer, MTDH knockdown was shown to significantly enhance the cytotoxic effects of the potent oral inhibitor of MEK1/2 kinase, selumetinib (AZD6244), possibly through the release of FOXO3a from cytoplasmic proteasomal degradation (Kong, Moran, Zhao, & Yang, 2012). These results emphasize the regulation of protein stability by AEG-1/MTDH/LYRIC and highlight a potential activity that would be suitable for therapeutic intervention.

Metabolic inhibitors, such as 2-deoxyglucose and lonidamine, are also being examined either alone or in combination with traditional chemotherapeutic agents for the treatment of malignant brain tumors (El Mjiyad, Caro-Maldonado, Ramirez-Peinado, & Munoz-Pinedo, 2011). The prevention of metabolic stress by these tumors is a primary strategy to survive periods of cell stress within the tumor microenvironment and to resist antimetabolic therapies. AEG-1/MTDH/LYRIC knockdown suppresses expression of aldehyde dehydrogenase 3 family, member A1 (ALDH3A1) in breast cancer cells, which protects against oxidative stress and also promotes the chemoresistant phenotype (Hu et al., 2009). Additionally, AEG-1/MTDH/LYRIC downregulation sensitizes breast cancer cells to combination treatment with TNF-α-related apoptosis-inducing ligand and the histone deacetylase inhibitor,

LBH 589 (Meng et al., 2011). Similar to alkylating agent therapy, it is possible that AEG-1/MTDH/LYRIC knockdown may synergize with metabolic inhibitors to elicit greater cytotoxicity in glioblastoma.

4.8. Diagnostic utility of AEG-1/MTDH/LYRIC expression and antibody production in malignant brain tumors

Due to the propensity for the sudden clinical presentation of malignant gliomas through a variety of emergent neurological signs and symptoms that indicate early and extensive invasion, there is a greater need for earlier diagnosis with the use of easily procured patient markers. Much work has been completed in this context for systemic cancers, but progress has been disappointing in the field of malignant brain tumors. This relative shortcoming may result from the considerable difficulty in obtaining useful samples from the CNS compartment and/or the inability to identify serum markers that reflect the status of the tumor and its microenvironment within the CNS. Most traditional biomarkers, such as platelet-derived growth factor, epidermal growth factor receptor (EGFR), O^6-methylguanine-DNA methyltransferase, and IDH1, must be measured from tumor tissue itself, while many identified serum biomarkers do not correlate with those in cerebrospinal fluid (CSF). In glioblastoma, in particular, some potential biomarkers may be highly expressed in normal brain, making it difficult to differentiate tumor aggressiveness from normal brain processes.

A trove of evidence exists that AEG-1/MTDH/LYRIC correlates with tumor grade in many cancers (Yoo, Emdad, et al., 2011), and similar data have been demonstrated in glioblastoma (Emdad et al., 2010; Liu et al., 2010). Therefore, it is evident that AEG-1/MTDH/LYRIC expression alone may serve as a stratifying marker for malignant potential. However, this strategy is only useful once tumor tissue is obtained and is not valuable prior to diagnosis or with inoperable or difficult-to-biopsy tumors. Though AEG-1/MTDH/LYRIC is not predominantly localized to the cell membrane, antibodies to this protein have been discovered in the serum of patients with cancer (Chen et al., 2012). Moreover, levels of this antibody correlate with tumor grade and therefore reflect the aggressive nature of the tumor itself. Despite the lack of evidence of these serum antibodies in glioblastoma patients, it is possible that they do exist, if not in the serum, then in the CSF that bathes glioblastoma cells. Identification of these antibodies in brain tumor patients may allow noninvasive longitudinal follow-up that could supplement imaging studies after standard therapeutic paradigms.

However, given the propensity for late presentation, the role of early detection in glioblastoma is unclear, other than in select patients who suffer from familial disorders that predispose to the development of malignant brain tumors.

4.9. Treatment strategies for malignant brain tumors derived from the AEG-1/MTDH/LYRIC pathway

The development of novel therapies for malignant brain tumors is largely based on the validation of targets that have been previously identified. In the field of glioma, few therapies have led to any significant success over median survivals of 12–15 months established with earlier therapies (Stupp et al., 2005). These shortcomings are due to the inability to achieve sufficient therapeutic concentrations within the CNS, to rapid development of chemoresistance, and to the oncogenic mechanisms to subvert disease control with various biologic and nonbiologic treatment strategies.

The discovery of DNA vaccines as a potential strategy to treat cancer has led to their use in a variety of systemic malignancies and multiple clinical trials testing their efficacy (Rice, Ottensmeier, & Stevenson, 2008). As this therapeutic strategy involves the use of a DNA vaccine that is long-lived *in vivo* and relatively easy to deliver, it may be possible to utilize such therapy for the treatment of malignant brain tumors as well. The appropriate choice for DNA vaccine targets is critical as many tumor antigens are present on normal cells, with a resultant vaccine response yielding either low efficacy or autoimmunity. Given the overexpression of AEG-1/MTDH/LYRIC in human cancer, an AEG-1/MTDH/LYRIC-based, DNA-based vaccine was developed for use in breast cancer. The AEG-1/MTDH/LYRIC-based DNA vaccine inhibits the growth of breast tumors in mice, promotes a robust cytotoxic T cell response, and enhances overall survival (Qian et al., 2011). Such methods could easily be applied to glioblastoma to assess vaccine efficacy within the CNS. Indeed, DNA vaccines have been effective against well-characterized targets in glioblastoma, such as VEGF, that is critical to angiogenesis (Niethammer et al., 2002) and have been useful in models of metastatic melanoma (Pertl et al., 2003). In addition, peptide vaccines specifically targeted against the EGFR have achieved some success in the treatment of glioblastoma in phase I studies (Sampson et al., 2010), with phase II trials currently underway.

It is evident that AEG-1/MTDH/LYRIC knockdown inhibits the growth of glioblastoma and improves overall survival in animal models (Emdad et al., 2010). Knockdown strategies primarily involve the use of

RNAi to achieve gene silencing and may also point to therapeutic possibilities in humans. Small-molecule inhibitors of AEG-1/MTDH/LYRIC may be relevant in targeting this protein given its ubiquitous expression in cells and its central role in oncogenic signaling cascades. Moreover, given the intimate interaction of AEG-1/MTDH/LYRIC with other tumorigenic pathways, small-molecule inhibitors of those pathways may also target AEG-1/MTDH/LYRIC. Such evidence has been demonstrated with the use of MMP (Liu et al., 2010) and CPEB inhibition (Kochanek & Wells, 2013). What is not currently known is the role of AEG-1/MTDH/LYRIC in normal cells, and the effects of such targeted therapies against AEG-1/MTDH/LYRIC may lead to significant adverse reactions when investigated in animal models. Further examination of the role of AEG-1/MTDH/LYRIC under normal conditions and a more expansive understanding of its expression pattern in normal brain, in particular, may aid in the development of AEG-1/MTDH/LYRIC-based therapies for malignant brain tumors.

5. CONCLUSIONS

The burden of neurological disease in the human population continues to increase with a relative lack of efficacious treatments that improve patient morbidity and mortality. From migraines to neurodegenerative disease and from infectious diseases to cancer, the CNS remains at the center of concerted efforts in biomedical research, both to understand the pathogenesis of neurological disease and to develop innovative therapies. As the human population ages, the incidence of neurodegenerative disease will increase, and there will be a greater need for disease-modifying therapies. Similarly, without robust targeted treatment strategies that improve patient survival, malignant tumors of the CNS will continue to carry single-digit 5-year survivals. A greater thrust to identify common pathways underlying CNS disease and points along these cascades that are amenable to therapeutic intervention is critical to reducing the devastating consequences of CNS disease.

As a potent oncogene that is central to diverse cellular signaling pathways, AEG-1/MTDH/LYRIC represents an important gene that contributes to the pathogenesis of diseases within the CNS. By promoting glutamate excitotoxicity and reactive gliosis, AEG-1/MTDH/LYRIC contributes to the neurological sequelae of HIV-1 infection. Its association with a migraine susceptibility locus on chromosome 8q22 suggests its involvement with the pathophysiology of vascular disease within the CNS. Perhaps, the best studied of the effects of AEG-1/MTDH/LYRIC with regard to neurological disease, however, is its involvement in CNS tumorigenesis. AEG-1/MTDH/LYRIC coordinates a network of oncogenic signaling cascades that contribute to aggressive growth, migration and invasion, angiogenesis, and tumor-specific metabolism that facilitates rapid expansion. In addition, AEG-1/MTDH/LYRIC overexpression in glioblastoma supports the chemoresistant phenotype and potentiates RNA interference-directed silencing of tumor suppressor genes. In light of its importance in tumorigenesis, mounting evidence suggests that antibodies are generated against AEG-1/MTDH/LYRIC in the serum of cancer patients, and AEG-1/MTDH/LYRIC downregulation using DNA vaccination may prolong overall survival.

At the time of initial cloning and characterization, AEG-1/MTDH/LYRIC was thought to contribute to several distinct but central processes underlying HIV-1 infection, glutamate regulation, and tumorigenesis. It is now clear that AEG-1/MTDH/LYRIC contributes significantly to disease within the CNS and represents a potentially valuable target for therapeutic investigation. Though much is now known regarding the role of this oncogenic protein within the CNS compartment, further examination into its precise mechanism of regulation will facilitate the development of treatments that target this pathway or interrelated processes that govern AEG-1/MTDH/LYRIC activity. With an increasing number of identified and validated targets within the AEG-1/MTDH/LYRIC signaling pathway, it is conceivable that AEG-1/MTDH/LYRIC-specific molecules could exhibit efficacy in the treatment of and improve outcomes in neurological disease.

REFERENCES

Alexander, B. M., Pinnell, N., Wen, P. Y., & D'Andrea, A. (2012). Targeting DNA repair and the cell cycle in glioblastoma. *Journal of Neuro-Oncology*, *107*, 463–477.

Andreeva, A. Y., Krause, E., Müller, E. C., Blasig, I. E., & Utepbergenov, D. I. (2001). Protein kinase C regulates the phosphorylation and cellular localization of occludin. *The Journal of Biological Chemistry*, *276*, 38480–38486.

Anttila, V., Stefansson, H., Kallela, M., Todt, U., Terwindt, G. M., Calafato, M. S., et al. (2010). Genome-wide association study of migraine implicates a common susceptibility variant on 8q22.1. *Nature Genetics*, *42*, 869–873.

Ash, S. C., Yang, D. Q., & Britt, D. E. (2008). LYRIC/AEG-1/MTDH/LYRIC overexpression modulates BCCIPalpha protein levels in prostate tumor cells. *Biochemical and Biophysical Research Communications*, *371*, 333–338.

Benyahia, B., Huguet, S., Decleves, X., Mokhtari, K., Criniere, E., Bernaudin, J. F., et al. (2004). Multidrug resistance-associated protein MRP1 expression in human gliomas: Chemosensitization to vincristine and etoposide by indomethacin in human glioma cell lines overexpressing MRP1. *Journal of Neuro-Oncology*, *66*, 65–70.

Bernardo, M. V., Yelo, E., Gimeno, L., Campillo, J. A., & Parrado, A. (2007). Identification of apoptosis-related PLZF target genes. *Biochemical and Biophysical Research Communications, 359*, 317–322.

Bhutia, S. K., Kegelman, T. P., Das, S. K., Azab, B., Su, Z. Z., Lee, S. G., et al. (2010). Astrocyte elevated gene-1 induces protective autophagy. *Proceedings of the National Academy of Sciences of the United States of America, 107*, 22243–22248.

Blanco, M. A., Aleckovic, M., Hua, Y., Li, T., Wei, Y., Xu, Z., et al. (2011). Identification of staphylococcal nuclease domain-containing 1 (SND1) as a Metadherin-interacting protein with metastasis-promoting functions. *The Journal of Biological Chemistry, 286*, 19982–19992.

Britt, D. E., Yang, D. F., Yang, D. Q., Flanagan, D., Callanan, H., Lim, Y. P., et al. (2004). Identification of a novel protein, LYRIC, localized to tight junctions of polarized epithelial cells. *Experimental Cell Research, 300*, 134–148.

Bronger, H., Konig, J., Kopplow, K., Steiner, H. H., Ahmadi, R., Herold-Mende, C., et al. (2005). ABCC drug efflux pumps and organic anion uptake transporters in human gliomas and the blood-tumor barrier. *Cancer Research, 65*, 11419–11428.

Brown, D. M., & Ruoslahti, E. (2004). Metadherin, a cell surface protein in breast tumors that mediates lung metastasis. *Cancer Cell, 5*, 365–374.

Brunet, A., Bonni, A., Zigmond, M. J., Lin, M. Z., Juo, P., Hu, L. S., et al. (1999). Akt promotes cell survival by phosphorylating and inhibiting a Forkhead transcription factor. *Cell, 96*, 857–868.

Buchan, J. R., & Parker, R. (2009). Eukaryotic stress granules: The ins and outs of translation. *Molecular Cell, 36*, 932–941.

Carnemolla, A., Fossale, E., Agostoni, E., Michelazzi, S., Calligaris, R., De Maso, L., et al. (2009). Rrs1 is involved in endoplasmic reticulum stress response in Huntington disease. *The Journal of Biological Chemistry, 284*, 18167–18173.

Carpenter, S., Carlson, T., Dellacasagrande, J., Garcia, A., Gibbons, S., Hertzog, P., et al. (2009). TRIL, a functional component of the TLR4 signaling complex, highly expressed in brain. *The Journal of Immunology, 183*, 3989–3995.

Chen, X., Dong, K., Long, M., Lin, F., Wang, X., Wei, J., et al. (2012). Serum anti-AEG-1/MTDH/LYRIC auto-antibody is a potential novel biomarker for malignant tumors. *Oncology Letters, 4*, 319–323.

Dephoure, N., Zhou, C., Villen, J., Beausoleil, S. A., Bakalarski, C. E., Elledge, S. J., et al. (2008). A quantitative atlas of mitotic phosphorylation. *Proceedings of the National Academy of Sciences of the United States of America, 105*, 10762–10767.

Dolecek, T. A., Propp, J. M., Stroup, N. E., & Kruchko, C. (2012). CBTRUS Statistical Report: Primary Brain and Central Nervous System Tumors Diagnosed in the United States in 2005–2009. *Neuro-Oncol, 14*(suppl 5), v1–v49.

El Mjiyad, N., Caro-Maldonado, A., Ramirez-Peinado, S., & Munoz-Pinedo, C. (2011). Sugar-free approaches to cancer cell killing. *Oncogene, 30*, 253–264.

Emdad, L., Lee, S. G., Su, Z. Z., Jeon, H. Y., Boukerche, H., Sarkar, D., et al. (2009). Astrocyte elevated gene-1 (AEG-1/MTDH/LYRIC) functions as an oncogene and regulates angiogenesis. *Proceedings of the National Academy of Sciences of the United States of America, 106*, 21300–21305.

Emdad, L., Sarkar, D., Lee, S. G., Su, Z. Z., Yoo, B. K., Dash, R., et al. (2010). Astrocyte elevated gene-1: A novel target for human glioma therapy. *Molecular Cancer Therapeutics, 9*, 79–88.

Emdad, L., Sarkar, D., Su, Z. Z., Randolph, A., Boukerche, H., Valerie, K., et al. (2006). Activation of the nuclear factor kappaB pathway by astrocyte elevated gene-1: Implications for tumor progression and metastasis. *Cancer Research, 66*, 1509–1516.

Engeland, C. E., Oberwinkler, H., Schümann, M., Krause, E., Müller, G. A., & Kräusslich, H. G. (2011). The cellular protein lyric interacts with HIV-1 Gag. *Journal of Virology, 85*, 13322–13332.

Fire, A., Xu, S., Montgomery, M. K., Kostas, S. A., Driver, S. E., & Mello, C. C. (1998). Potent and specific genetic interference by double-stranded RNA in Caenorhabditis elegans. *Nature, 391*, 806–811.

Friedman, H. S., Prados, M. D., Wen, P. Y., Mikkelsen, T., Schiff, D., Abrey, L. E., et al. (2009). Bevacizumab alone and in combination with irinotecan in recurrent glioblastoma. *Journal of Clinical Oncology: Official Journal of the American Society of Clinical Oncology, 27*, 4733–4740.

Fujishiro, M., Gotoh, Y., Katagiri, H., Sakoda, H., Ogihara, T., Anai, M., et al. (2001). MKK6/3 and p38 MAPK pathway activation is not necessary for insulin-induced glucose uptake but regulates glucose transporter expression. *The Journal of Biological Chemistry, 276*, 19800–19806.

Fricker, M., Hollinshead, M., White, N., & Vaux, D. (1997). Interphase nuclei of many mammalian cell types contain deep, dynamic, tubular membrane-bound invaginations of the nuclear envelope. *The Journal of Cellular Biology, 136*, 531–544.

Ghafouri, M., Amini, S., Khalili, K., & Sawaya, B. E. (2006). HIV-1 associated dementia: Symptoms and causes. *Retrovirology, 3*, 28.

Gravdal, K., Halvorsen, O. J., Haukaas, S. A., & Akslen, L. A. (2007). A switch from E-cadherin to N-cadherin expression indicates epithelial to mesenchymal transition and is of strong and independent importance for the progress of prostate cancer. *Clinical Cancer Research, 13*, 7003–7011.

Haas, B. R., & Sontheimer, H. (2010). Inhibition of the Sodium-Potassium-Chloride Cotransporter Isoform-1 reduces glioma invasion. *Cancer Research, 70*, 5597–5606.

Hazan, R. B., Phillips, G. R., Qiao, R. F., Norton, L., & Aaronson, S. A. (2000). Exogenous expression of N-cadherin in breast cancer cells induces cell migration, invasion, and metastasis. *The Journal of Cell Biology, 148*, 779–790.

Hoelzinger, D. B., Demuth, T., & Berens, M. E. (2007). Autocrine factors that sustain glioma invasion and paracrine biology in the brain microenvironment. *Journal of the National Cancer Institute, 99*, 1583–1593.

Hu, G., Chong, R. A., Yang, Q., Wei, Y., Blanco, M. A., Li, F., et al. (2009). MTDH activation by 8q22 genomic gain promotes chemoresistance and metastasis of poor-prognosis breast cancer. *Cancer Cell, 15*, 9–20.

Hutvagner, G., & Simard, M. J. (2008). Argonaute proteins: Key players in RNA silencing. *Nature Reviews. Molecular Cell Biology, 9*, 22–32.

Isozaki, Y., Hoshino, I., Nohata, N., Kinoshita, T., Akutsu, Y., Hanari, N., et al. (2012). Identification of novel molecular targets regulated by tumor suppressive miR-375 induced by histone acetylation in esophageal squamous cell carcinoma. *International Journal of Oncology, 41*, 985–994.

Jeon, H. Y., Choi, M., Howlett, E. L., Vozhilla, N., Yoo, B. K., Lloyd, J. A., et al. (2010). Expression patterns of astrocyte elevated gene-1 (AEG-1/MTDH/LYRIC) during development of the mouse embryo. *Gene Expression Patterns, 10*, 361–367.

Jian-bo, X., Hui, W., Yu-long, H., Chang-hua, Z., Long-juan, Z., Shi-rong, C., et al. (2011). Astrocyte-elevated gene-1 overexpression is associated with poor prognosis in gastric cancer. *Medical Oncology, 28*, 455–462.

Johnson, D. R., & Chang, S. M. (2012). Recent medical management of glioblastoma. *Advances in Experimental Medicine and Biology, 746*, 26–40.

Kang, D. C., Su, Z. Z., Sarkar, D., Emdad, L., Volsky, D. J., & Fisher, P. B. (2005). Cloning and characterization of HIV-1-inducible astrocyte elevated gene-1, AEG-1/MTDH/LYRIC. *Gene, 353*, 8–15.

Kaur, N., Chettiar, S., Rathod, S., Rath, P., Muzumdar, D., Shaikh, M. L., et al. (2013). Wnt3a mediated activation of Wnt/beta-catenin signaling promotes tumor progression in glioblastoma. *Molecular and Cellular Neurosciences, 54C*, 44–57.

Khuda, I. I., Koide, N., Noman, A. S., Dagvadorj, J., Tumurkhuu, G., Naiki, Y., et al. (2009). Astrocyte elevated gene-1 (AEG-1/MTDH/LYRIC) is induced by lipopolysaccharide as toll-like receptor 4 (TLR4) ligand and regulates TLR4 signalling. *Immunology, 128*, e700–e706.

Kikuno, N., Shiina, H., Urakami, S., Kawamoto, K., Hirata, H., Tanaka, Y., et al. (2007). Knockdown of astrocyte-elevated gene-1 inhibits prostate cancer progression through upregulation of FOXO3a activity. *Oncogene, 26*, 7647–7655.

Kirches, E., Oda, Y., Von Bossanyi, P., Diete, S., Schneider, T., Warich-Kirches, M., et al. (1997). Mdr1 mRNA expression differs between grade III astrocytomas and glioblastomas. *Clinical Neuropathology, 16*, 34–36.

Kirchhausen, T. (2000). Clathrin. *Annual Review of Biochemistry, 69*, 699–727.

Kochanek, D. M., & Wells, D. G. (2013). CPEB1 regulates the expression of MTDH/AEG-1/MTDH/LYRIC and glioblastoma cell migration. *Molecular Cancer Research, 11*, 149–160.

Kong, X., Moran, M. S., Zhao, Y., & Yang, Q. (2012). Inhibition of metadherin sensitizes breast cancer cells to AZD6244. *Cancer Biology & Therapy, 13*, 43–49.

Lee, S. G., Jeon, H. Y., Su, Z. Z., Richards, J. E., Vozhilla, N., Sarkar, D., et al. (2009). Astrocyte elevated gene-1 contributes to the pathogenesis of neuroblastoma. *Oncogene, 28*, 2476–2484.

Lee, S. G., Su, Z. Z., Emdad, L., Sarkar, D., & Fisher, P. B. (2006). Astrocyte elevated gene-1 (AEG-1/MTDH/LYRIC) is a target gene of oncogenic Ha-ras requiring phosphatidylinositol 3-kinase and c-Myc. *Proceedings of the National Academy of Sciences of the United States of America, 103*, 17390–17395.

Lee, S. G., Su, Z. Z., Emdad, L., Sarkar, D., Franke, T. F., & Fisher, P. B. (2008). Astrocyte elevated gene-1 activates cell survival pathways through PI3K-Akt signaling. *Oncogene, 27*, 1114–1121.

Li, C., Li, R., Song, H., Wang, D., Feng, T., Yu, X., et al. (2011). Significance of AEG-1/MTDH/LYRIC expression in correlation with VEGF, microvessel density and clinicopathological characteristics in triple-negative breast cancer. *Journal of Surgical Oncology, 103*, 184–192.

Li, J., Yang, L., Song, L., Xiong, H., Wang, L., Yan, X., et al. (2009). Astrocyte elevated gene-1 is a proliferation promoter in breast cancer via suppressing transcriptional factor FOXO1. *Oncogene, 28*, 3188–3196.

Ligthart, L., de Vries, B., Smith, A. V., Ikram, M. A., Amin, N., Hottenga, J. J., et al. (2011). Meta-analysis of genome-wide association for migraine in six population-based European cohorts. *European Journal of Human Genetics, 19*, 901–907.

Lino, M. M., & Merlo, A. (2011). PI3Kinase signaling in glioblastoma. *Journal of Neuro-Oncology, 103*, 417–427.

Liu, I. J., Chiu, C. Y., Chen, Y. C., & Wu, H. C. (2011). Molecular mimicry of human endothelial cell antigen by autoantibodies to nonstructural protein 1 of dengue virus. *The Journal of Biological Chemistry, 286*, 9726–9736.

Liu, H. Y., Liu, C. X., Han, B., Zhang, X. Y., & Sun, R. P. (2012). AEG-1/MTDH/LYRIC is associated with clinical outcome in neuroblastoma patients. *Cancer Biomarkers, 11*, 115–121.

Liu, H., Song, X., Liu, C., Xie, L., Wei, L., & Sun, R. (2009). Knockdown of astrocyte elevated gene-1 inhibits proliferation and enhancing chemo-sensitivity to cisplatin or doxorubicin in neuroblastoma cells. *Journal of Experimental & Clinical Cancer Research, 28*, 19.

Liu, L., Wu, J., Ying, Z., Chen, B., Han, A., Liang, Y., et al. (2010). Astrocyte elevated gene-1 upregulates matrix metalloproteinase-9 and induces human glioma invasion. *Cancer Research, 70*, 3750–3759.

Louis, D. N. (2006). Molecular pathology of malignant gliomas. *Annual Review of Pathology, 1*, 97–117.

Lu, H., Guo, X., Meng, X., Liu, J., Allen, C., Wray, J., et al. (2005). The BRCA2-interacting protein BCCIP functions in RAD51 and BRCA2 focus formation and homologous recombinational repair. *Molecular and Cellular Biology, 25*, 1949–1957.

Lu, C., & Shervington, A. (2008). Chemoresistance in gliomas. *Molecular and Cellular Biochemistry, 312*, 71–80.

Maragakis, N. J., & Rothstein, J. D. (2006). Mechanisms of disease: Astrocytes in neurodegenerative disease. *Nature Clinical Practice. Neurology, 2*, 679–689.

Martinez-Outschoorn, U. E., Lin, Z., Ko, Y. H., Goldberg, A. F., Flomenberg, N., Wang, C., et al. (2011). Understanding the metabolic basis of drug resistance: Therapeutic induction of the Warburg effect kills cancer cells. *Cell Cycle, 10*, 2521–2528.

Martinez-Outschoorn, U. E., Pavlides, S., Howell, A., Pestell, R. G., Tanowitz, H. B., Sotgia, F., et al. (2011). Stromal-epithelial metabolic coupling in cancer: Integrating autophagy and metabolism in the tumor microenvironment. *The International Journal of Biochemistry & Cell Biology, 43*, 1045–1051.

Martinez-Outschoorn, U. E., Whitaker-Menezes, D., Pavlides, S., Chiavarina, B., Bonuccelli, G., Casey, T., et al. (2010). The autophagic tumor stroma model of cancer or "battery-operated tumor growth": A simple solution to the autophagy paradox. *Cell Cycle, 9*, 4297–4306.

Mazurek, S., Boschek, C. B., Hugo, F., & Eigenbrodt, E. (2005). Pyruvate kinase type M2 and its role in tumor growth and spreading. *Seminars in Cancer Biology, 15*, 300–308.

McArthur, J. C., Steiner, J., Sacktor, N., & Nath, A. (2010). Human immunodeficiency virus-associated neurocognitive disorders: Mind the gap. *Annals of Neurology, 67*, 699–714.

Meng, X., Brachova, P., Yang, S., Xiong, Z., Zhang, Y., Thiel, K. W., et al. (2011). Knockdown of MTDH sensitizes endometrial cancer cells to cell death induction by death receptor ligand TRAIL and HDAC inhibitor LBH589 co-treatment. *PLoS One, 6*, e20920.

Meng, X., Fan, J., & Shen, Z. (2007). Roles of BCCIP in chromosome stability and cytokinesis. *Oncogene, 26*, 6253–6260.

Meng, X., Zhu, D., Yang, S., Wang, X., Xiong, Z., Zhang, Y., et al. (2012). Cytoplasmic Metadherin (MTDH) provides survival advantage under conditions of stress by acting as RNA-binding protein. *The Journal of Biological Chemistry, 287*, 4485–4491.

Mentlein, R., Hattermann, K., & Held-Feindt, J. (2012). Lost in disruption: Role of proteases in glioma invasion and progression. *Biochimica et Biophysica Acta, 1825*, 178–185.

Moser, J. J., & Fritzler, M. J. (2010). The microRNA and messengerRNA profile of the RNA-induced silencing complex in human primary astrocyte and astrocytoma cells. *PLoS One, 5*, e13445.

Nakai, E., Park, K., Yawata, T., Chihara, T., Kumazawa, A., Nakabayashi, H., et al. (2009). Enhanced MDR1 expression and chemoresistance of cancer stem cells derived from glioblastoma. *Cancer Investigation, 27*, 901–908.

Nakamura, N., Ramaswamy, S., Vazquez, F., Signoretti, S., Loda, M., & Sellers, W. R. (2000). Forkhead transcription factors are critical effectors of cell death and cell cycle arrest downstream of PTEN. *Molecular and Cellular Biology, 20*, 8969–8982.

Niethammer, A. G., Xiang, R., Becker, J. C., Wodrich, H., Pertl, U., Karsten, G., et al. (2002). A DNA vaccine against VEGF receptor 2 prevents effective angiogenesis and inhibits tumor growth. *Nature Medicine, 8*, 1369–1375.

Noch, E., Bookland, M., & Khalili, K. (2011). Astrocyte-elevated gene-1 (AEG-1/MTDH/LYRIC) induction by hypoxia and glucose deprivation in glioblastoma. *Cancer Biology & Therapy, 11*, 32–39.

Noch, E., & Khalili, K. (2009). Molecular mechanisms of necrosis in glioblastoma: The role of glutamate excitotoxicity. *Cancer Biology & Therapy, 8*, 1791–1797.

Nohata, N., Hanazawa, T., Kikkawa, N., Mutallip, M., Sakurai, D., Fujimura, L., et al. (2011). Tumor suppressive microRNA-375 regulates oncogene AEG-1/MTDH/LYRIC/MTDH in head and neck squamous cell carcinoma (HNSCC). *Journal of Human Genetics*, *56*, 595–601.

Olsen, J. V., Blagoev, B., Gnad, F., Macek, B., Kumar, C., Mortensen, P., et al. (2006). Global, in vivo, and site-specific phosphorylation dynamics in signaling networks. *Cell*, *127*, 635–648.

Olsen, J. V., Vermeulen, M., Santamaria, A., Kumar, C., Miller, M. L., Jensen, L. J., et al. (2010). Quantitative phosphoproteomics reveals widespread full phosphorylation site occupancy during mitosis. *Science Signaling*, *3*, ra3.

Pertl, U., Wodrich, H., Ruehlmann, J. M., Gillies, S. D., Lode, H. N., & Reisfeld, R. A. (2003). Immunotherapy with a posttranscriptionally modified DNA vaccine induces complete protection against metastatic neuroblastoma. *Blood*, *101*, 649–654.

Poltorak, A., He, X., Smirnova, I., Liu, M. Y., Van Huffel, C., Du, X., et al. (1998). Defective LPS signaling in C3H/HeJ and C57BL/10ScCr mice: Mutations in Tlr4 gene. *Science*, *282*, 2085–2088.

Purow, B., & Schiff, D. (2009). Advances in the genetics of glioblastoma: Are we reaching critical mass? *Nature Reviews. Neurology*, *5*, 419–426.

Qian, B. J., Yan, F., Li, N., Liu, Q. L., Lin, Y. H., Liu, C. M., et al. (2011). MTDH/AEG-1/MTDH/LYRIC-based DNA vaccine suppresses lung metastasis and enhances chemosensitivity to doxorubicin in breast cancer. *Cancer Immunology, Immunotherapy*, *60*, 883–893.

Ransom, C. B., O'Neal, J. T., & Sontheimer, H. (2001). Volume-activated chloride currents contribute to the resting conductance and invasive migration of human glioma cells. *The Journal of Neuroscience: The Official Journal of the Society for Neuroscience*, *21*, 7674–7683.

Raza, S. M., Lang, F. F., Aggarwal, B. B., Fuller, G. N., Wildrick, D. M., & Sawaya, R. (2002). Necrosis and glioblastoma: A friend or a foe? A review and a hypothesis. *Neurosurgery*, *51*, 2–12, discussion 12–13.

Rice, J., Ottensmeier, C. H., & Stevenson, F. K. (2008). DNA vaccines: Precision tools for activating effective immunity against cancer. *Nature Reviews. Cancer*, *8*, 108–120.

Rich, J. N., Guo, C., McLendon, R. E., Bigner, D. D., Wang, X. F., & Counter, C. M. (2001). A genetically tractable model of human glioma formation. *Cancer Research*, *61*, 3556–3560.

Roeb, W., Boyer, A., Cavenee, W. K., & Arden, K. C. (2007). PAX3-FOXO1 controls expression of the p57Kip2 cell-cycle regulator through degradation of EGR1. *Proceedings of the National Academy of Sciences of the United States of America*, *104*, 18085–18090.

Roussel, B. D., Kruppa, A. J., Miranda, E., Crowther, D. C., Lomas, D. A., & Marciniak, S. J. (2013). Endoplasmic reticulum dysfunction in neurological disease. *Lancet Neurology*, *12*, 105–118.

Rubenstein, B. M., & Kaufman, L. J. (2008). The role of extracellular matrix in glioma invasion: A cellular Potts model approach. *Biophysical Journal*, *95*, 5661–5680.

Sabri, F., Titanji, K., De Milito, A., & Chiodi, F. (2003). Astrocyte activation and apoptosis: Their roles in the neuropathology of HIV infection. *Brain Pathology*, *13*, 84–94.

Sakakibara, A., Furuse, M., Saitou, M., Ando-Akatsuka, Y., & Tsukita, S. (1997). Possible involvement of phosphorylation of occludin in tight junction formation. *The Journal of Cell Biology*, *137*, 1393–1401.

Sampson, J. H., Heimberger, A. B., Archer, G. E., Aldape, K. D., Friedman, A. H., Friedman, H. S., et al. (2010). Immunologic escape after prolonged progression-free survival with epidermal growth factor receptor variant III peptide vaccination in patients with newly diagnosed glioblastoma. *Journal of Clinical Oncology: Official Journal of the American Society of Clinical Oncology*, *28*, 4722–4729.

Sarkar, D., Park, E. S., Emdad, L., Lee, S. G., Su, Z. Z., & Fisher, P. B. (2008). Molecular basis of nuclear factor-kappaB activation by astrocyte elevated gene-1. *Cancer Research*, *68*, 1478–1484.

Schaich, M., Kestel, L., Pfirrmann, M., Robel, K., Illmer, T., Kramer, M., et al. (2009). A MDR1 (ABCB1) gene single nucleotide polymorphism predicts outcome of temozolomide treatment in glioblastoma patients. *Annals of Oncology: Official Journal of the European Society for Medical Oncology*, *20*, 175–181.

Scherer, H. (1938). Structural development in gliomas. *American Journal of Cancer*, *34*, 333–351.

Schulz, M., Brandner, S., Eberhagen, C., Eckardt-Schupp, F., Larsen, M. R., & Andrae, U. (2013). Quantitative phosphoproteomic analysis of early alterations in protein phosphorylation by 2,3,7,8-tetrachlorodibenzo-p-dioxin. *Journal of Proteome Research*, *12*, 866–882.

Seoane, J., Le, H. V., Shen, L., Anderson, S. A., & Massagué, J. (2004). Integration of Smad and forkhead pathways in the control of neuroepithelial and glioblastoma cell proliferation. *Cell*, *117*, 211–223.

Sotgia, F., Martinez-Outschoorn, U. E., Pavlides, S., Howell, A., Pestell, R. G., & Lisanti, M. P. (2011). Understanding the Warburg effect and the prognostic value of stromal caveolin-1 as a marker of a lethal tumor microenvironment. *Breast Cancer Research*, *13*, 213.

Spiegl-Kreinecker, S., Buchroithner, J., Elbling, L., Steiner, E., Wurm, G., Bodenteich, A., et al. (2002). Expression and functional activity of the ABC-transporter proteins P-glycoprotein and multidrug-resistance protein 1 in human brain tumor cells and astrocytes. *Journal of Neuro-Oncology*, *57*, 27–36.

Stupp, R., Mason, W. P., van den Bent, M. J., Weller, M., Fisher, B., Taphoorn, M. J., et al. (2005). Radiotherapy plus concomitant and adjuvant temozolomide for glioblastoma. *The New England Journal of Medicine*, *352*, 987–996.

Su, Z. Z., Chen, Y., Kang, D. C., Chao, W., Simm, M., Volsky, D. J., et al. (2003). Customized rapid subtraction hybridization (RaSH) gene microarrays identify overlapping expression changes in human fetal astrocytes resulting from human immunodeficiency virus-1 infection or tumor necrosis factor-alpha treatment. *Gene*, *306*, 67–78.

Sutherland, H. G., Lam, Y. W., Briers, S., Lamond, A. I., & Bickmore, W. A. (2004). 3D3/lyric: A novel transmembrane protein of the endoplasmic reticulum and nuclear envelope, which is also present in the nucleolus. *Experimental Cell Research*, *294*, 94–105.

Thirkettle, H. J., Girling, J., Warren, A. Y., Mills, I. G., Sahadevan, K., Leung, H., et al. (2009). LYRIC/AEG-1/MTDH/LYRIC is targeted to different subcellular compartments by ubiquitinylation and intrinsic nuclear localization signals. *Clinical Cancer Research*, *15*, 3003–3013.

Thirkettle, H. J., Mills, I. G., Whitaker, H. C., & Neal, D. E. (2009). Nuclear LYRIC/AEG-1/MTDH/LYRIC interacts with PLZF and relieves PLZF-mediated repression. *Oncogene*, *28*, 3663–3670.

Thuringer, D., Hammann, A., Benikhlef, N., Fourmaux, E., Bouchot, A., Wettstein, G., et al. (2011). Transactivation of the epidermal growth factor receptor by heat shock protein 90 via Toll-like receptor 4 contributes to the migration of glioblastoma cells. *The Journal of Biological Chemistry*, *286*, 3418–3428.

Vartak-Sharma, N., & Ghorpade, A. (2012). Astrocyte elevated gene-1 regulates astrocyte responses to neural injury: Implications for reactive astrogliosis and neurodegeneration. *Journal of Neuroinflammation*, *9*, 195.

Villen, J., Beausoleil, S. A., Gerber, S. A., & Gygi, S. P. (2007). Large-scale phosphorylation analysis of mouse liver. *Proceedings of the National Academy of Sciences of the United States of America*, *104*, 1488–1493.

Wang, J., Wang, Q., Cui, Y., Liu, Z. Y., Zhao, W., Wang, C. L., et al. (2012). Knockdown of cyclin D1 inhibits proliferation, induces apoptosis, and attenuates the invasive capacity of human glioblastoma cells. *Journal of Neuro-Oncology*, *106*, 473–484.

Warburg, O. (1956). On the origin of cancer cells. *Science (New York, NY)*, *123*, 309–314.

Wick, W., Platten, M., & Weller, M. (2001). Glioma cell invasion: Regulation of metalloproteinase activity by TGF-beta. *Journal of Neuro-Oncology*, *53*, 177–185.

Xia, Z., Zhang, N., Jin, H., Yu, Z., Xu, G., & Huang, Z. (2010). Clinical significance of astrocyte elevated gene-1 expression in human oligodendrogliomas. *Clinical Neurology and Neurosurgery*, *112*, 413–419.

Yang, Y., Wu, J., Guan, H., Cai, J., Fang, L., Li, J., et al. (2012). MiR-136 promotes apoptosis of glioma cells by targeting AEG-1/MTDH/LYRIC and Bcl-2. *FEBS Letters*, *586*, 3608–3612.

Yoo, B. K., Chen, D., Su, Z. Z., Gredler, R., Yoo, J., Shah, K., et al. (2010). Molecular mechanism of chemoresistance by astrocyte elevated gene-1. *Cancer Research*, *70*, 3249–3258.

Yoo, B. K., Emdad, L., Lee, S. G., Su, Z. Z., Santhekadur, P., Chen, D., et al. (2011). Astrocyte elevated gene-1 (AEG-1/MTDH/LYRIC): A multifunctional regulator of normal and abnormal physiology. *Pharmacology & Therapeutics*, *130*, 1–8.

Yoo, B. K., Emdad, L., Su, Z. Z., Villanueva, A., Chiang, D. Y., Mukhopadhyay, N. D., et al. (2009). Astrocyte elevated gene-1 regulates hepatocellular carcinoma development and progression. *The Journal of Clinical Investigation*, *119*, 465–477.

Yoo, B. K., Gredler, R., Vozhilla, N., Su, Z. Z., Chen, D., Forcier, T., et al. (2009). Identification of genes conferring resistance to 5-fluorouracil. *Proceedings of the National Academy of Sciences of the United States of America*, *106*, 12938–12943.

Yoo, B. K., Santhekadur, P. K., Gredler, R., Chen, D., Emdad, L., Bhutia, S., et al. (2011). Increased RNA-induced silencing complex (RISC) activity contributes to hepatocellular carcinoma. *Hepatology*, *53*, 1538–1548.

Yuan, C., Li, X., Yan, S., Yang, Q., Liu, X., & Kong, B. (2012). The MTDH (−470G > A) polymorphism is associated with ovarian cancer susceptibility. *PLoS One*, *7*, e51561.

Zhao, Y., Kong, X., Li, X., Yan, S., Yuan, C., Hu, W., et al. (2011). Metadherin mediates lipopolysaccharide-induced migration and invasion of breast cancer cells. *PLoS One*, *6*, e29363.

Zhou, Z., Deng, H., Yan, W., Huang, H., Deng, Y., Li, Y., et al. (2012). Expression of metadherin/AEG-1/MTDH/LYRIC gene is positively related to orientation chemotaxis and adhesion of human hepatocellular carcinoma cell lines of different metastatic potentials. *Journal of Huazhong University of Science and Technology. Medical Sciences*, *32*, 353–357.

Zhu, K., Dai, Z., Pan, Q., Wang, Z., Yang, G. H., Yu, L., et al. (2011). Metadherin promotes hepatocellular carcinoma metastasis through induction of epithelial-mesenchymal transition. *Clinical Cancer Research*, *17*, 7294–7302.

CHAPTER SEVEN

AEG-1/MTDH/LYRIC in Liver Cancer

Devanand Sarkar[1]
Department of Human and Molecular Genetics, VCU Institute of Molecular Medicine, VCU Massey Cancer Center, Virginia Commonwealth University,School of Medicine, Richmond, Virginia,USA
[1]Corresponding author: e-mail address: dsarkar@vcu.edu

Contents

Abstract

Hepatocellular carcinoma (HCC) is a highly virulent malignancy with diverse etiology. Identification of a common mediator of aggressive progression of HCC would be extremely beneficial not only for diagnostic/prognostic purposes but also for developing targeted therapies. *AEG-1/MTDH/LYRIC* gene is amplified in human HCC patients, and overexpression of AEG-1/MTDH/LYRIC has been identified in a high percentage of both hepatitis B virus and hepatitis C virus positive HCC cases, suggesting its key role in regulating hepatocarcinogenesis. Important insights into the molecular mechanisms mediating oncogenic properties of AEG-1/MTDH/LYRIC, especially regulating chemoresistance, angiogenesis, and metastasis, have been obtained from studies using HCC model. Additionally, analysis of HCC model has facilitated the identification of AEG-1/MTDH/LYRIC downstream genes and interacting proteins, thereby unraveling novel players regulating HCC development and progression leading to the development of novel interventional strategies. Characterization of a hepatocyte-specific AEG-1/MTDH/LYRIC transgenic mouse (Alb/AEG-1) has revealed novel aspects of AEG-1/MTDH/LYRIC

Advances in Cancer Research, Volume 120
ISSN 0065-230X
http://dx.doi.org/10.1016/B978-0-12-401676-7.00007-3

function in *in vivo* contexts. Combination of AEG-1/MTDH/LYRIC inhibition and chemotherapy has documented significant efficacy in abrogating human HCC xenografts in nude mice indicating the need for developing effective AEG-1/MTDH/LYRIC inhibition strategies to obtain objective response and survival benefits in terminal HCC patients.

1. HEPATOCELLULAR CARCINOMA: ETIOLOGY, EPIDEMIOLOGY, AND PATHOGENESIS

Hepatocellular carcinoma (HCC) is an epithelial tumor arising from the primary resident cells of the liver, the hepatocytes (El-Serag & Rudolph, 2007). HCC is the most common type of liver cancer accounting for >80% of all cancers in the liver (El-Serag & Rudolph, 2007). It is one of the five most common cancers and the third most common cause of cancer-related deaths worldwide (El-Serag, 2011). The epidemiology of HCC shows two major geographical patterns. In central and south-east Asia, sub-Saharan Africa, and the Amazon basin, it is the most common cancer (El-Serag, 2011). Globally HCC causes 662,000 deaths per year, about half of which is in China. The etiology is predominantly viral hepatitis, either Hepatitis B virus (HBV) or Hepatitis C virus (HCV). In China, ~90% HCC cases are HBV positive, while in Japan 70% of HCC cases are associated with HCV (Umemura, Ichijo, Yoshizawa, Tanaka, & Kiyosawa, 2009). Contamination of food with fungus, such as *Aspergillus flavus* generating aflatoxin, is another major cause of HCC, especially in Africa (Bressac, Kew, Wands, & Ozturk, 1991). The incidence of HCC is two to four times higher in males than in females, and the age of onset is between 30 and 50 years. The early onset of the disease indicates hepatitis infection earlier in life, such as at birth.

In North America and Western Europe, HCC is a rare cancer. However, the incidence of HCC is rising in Western countries. The incidence of HCC in the United States has increased from 1.6 to 4.9 (more than 200% increase) cases per 100,000 of population from 1975 to 2005 (Yang & Roberts, 2010). In the United States, the number of new cases of HCC in 2012 was estimated to be 28,720, out of which 20,550 were expected to die (Siegel, Naishadham, & Jemal, 2012). HCV and HBV constitute 48% and 16% of cases of HCC in the United States, respectively. Chronic alcoholism leading to cirrhosis and metabolic syndromes, such as nonalcoholic fatty liver disease usually associated with obesity and type 2 diabetes, are the most common causes in the remaining cases (Calle, Rodriguez, Walker-Thurmond, &

Thun, 2003; Donato et al., 2002; El-Serag, Tran, & Everhart, 2004; Starley, Calcagno, & Harrison, 2010). Other rare causes of HCC include hemochromatosis, α1-antitrypsin deficiency, autoimmune hepatitis, porphyrias, and Wilson disease (Dragani, 2010; El-Serag & Rudolph, 2007).

The pathogenesis of HCC usually follows two patterns, cirrhotic and noncirrhotic (Farazi & DePinho, 2006). Viral hepatitis triggers an immune response recruiting T-lymphocytes (Rehermann & Nascimbeni, 2005). Consequently, there is continuous hepatocyte necrosis, inflammation, regeneration, and fibrotic repair leading to cirrhosis. The endless cycle of necrosis/regeneration to restore the tissue architecture of the liver might initiate dysfunctional telomeres and genomic instability which, in turn, give rise to dysplastic lesions that progress to HCC. This scenario is more common for HCV infection and chronic alcoholism. The scenario might be different for HBV, which, unlike HCV, might be integrated into the genome resulting in microdeletion of genes (such as tumor suppressor genes), or the viral enhancer elements might upregulate genes, such as telomerase reverse transcriptase, platelet-derived growth factor receptor-β, and mitogen-activated protein kinase 1 (MAPK1), which favor limitless proliferation potential (Murakami et al., 2005). In HBV cases, HCC thus might develop in the absence of cirrhosis. However, chronic inflammation-induced cirrhotic changes are the predominant modality in HBV cases as well.

HCC might be either nodular or infiltrative macroscopically (Llovet, Bru, & Bruix, 1999). The nodular type might contain a large solitary mass or there might be multiple nodules. The nodules are not encapsulated but well circumscribed. The diffuse type is not well circumscribed and might infiltrate into the portal or hepatic vein. Microscopically, the tumor might be either well differentiated, in which the tumor cells resemble hepatocytes and form trabeculae, cords, and nests, or poorly differentiated, in which the malignant cells are pleomorphic, anaplastic, and giant. The tumor is usually compact with little stroma and contains a central necrotic core because of inadequate vascularization and hypoxia.

The incidence and mortality of HCC run parallel. HCC is a tumor with rapid growth and early vascular invasion (Llovet, Burroughs, & Bruix, 2003). It is also highly resistant to standard chemotherapy (Llovet et al., 2003; Pang et al., 2008; Poon et al., 2001). The treatment options for HCC depend upon the stages and grades of the disease (Llovet et al., 1999). With localized disease, surgical resection, radiofrequency ablation, and liver transplantations are the treatments of choice (Georgiades, Hong, & Geschwind, 2008; O'Neil & Venook, 2007). However, most

HCC patients present with advanced symptomatic tumors with underlying cirrhotic changes that are not amenable to surgical resection or transplantation. Transarterial chemoembolization and systemic therapy with doxorubicin alone or a combination of cisplatin, interferon, doxorubicin, and 5-fluorouracil (PIAF) are being used for advanced disease with moderate improvement in overall survival (OS) duration varying between 6.8 and 8.6 months (Leung et al., 1999; Llovet & Bruix, 2003; Llovet et al., 2002; Yeo et al., 2005). Sorafenib, an inhibitor of c-Raf and B-Raf kinases as well as of vascular endothelial growth factor receptor family, has been approved by FDA for unresectable advanced HCC (Llovet et al., 2008). While the median survival for placebo-treated patients was approximately 7.9 months, sorafenib-treated patients survived 10.7 months. VEGF pathway inhibitor bevacizumab, either alone or in combination with chemotherapy, also demonstrates very limited response (O'Neil & Venook, 2007; Zhu et al., 2006). In view of this dismal scenario, understanding the molecular pathogenesis of HCC and developing targeted and effective treatments are mandatory to significantly increase the survival interval and ameliorate the sufferings of the patients.

2. HCC: THE MOLECULAR ABNORMALITIES

Multiple etiologies have been linked to HCC, and therefore, no consistent genomic abnormalities have been attributed to this disease. Chronic HBV infection results in HCC by integration of HBV DNA into the genome leading to chromosomal instability and by HBV x protein (HBx) that activates a plethora of proto-oncogenes and signaling pathways associated with HCC (Brechot, Pourcel, Louise, Rain, & Tiollais, 1980; Feitelson & Duan, 1997). Chronic HCV infection causes HCC via core HCV proteins, NS3 and NS5A that inhibit the cyclin-dependent kinase inhibitor p21 and interact with p53 (Kwun, Jung, Ahn, Lee, & Jang, 2001; Majumder, Ghosh, Steele, Ray, & Ray, 2001). Mutations in numerous proto-oncogenes and tumor suppressor genes, such as p53, p73, Rb, APC, DLC-1, DLC-2, PTEN, SOCS1, GSTP1, HCCS1, Smad2/4, AXIN1, IGF-2, β-catenin, c-myc, and cyclin D1, have been detected in HCC (Boyault et al., 2007; Mann et al., 2007; Teufel et al., 2007). Oncogenomic approaches have identified cIAP1, an inhibitor of apoptosis, and Yap, a transcription factor, as candidate oncogenes for HCC (Zender et al., 2006).

The major signaling pathways activated in HCC are (i) MAPK that includes cascades of phosphorylation of ras, raf, mitogen-activated protein extracellular kinase (MEK), and extracellular signal-regulated kinase (ERK). Activation of this pathway has been well documented in HCC cell lines, *in vivo* HCC models, and human HCC specimens (Min, He, & Hui, 2011). (ii) Phosphotidyl-inositol-3-kinase (PI3K)/Akt/mTOR pathway (Llovet & Bruix, 2008). Activation of PI3K by growth factor receptors activates Akt that phosphorylates and inactivates proapoptotic proteins, such as Bad and caspase-9. Downstream of Akt is mTOR, a key regulator of cell translation machinery through two effector proteins, the eukaryotic initiation factor 4E-binding protein (4E-BP1) and the 40s ribosomal protein S6 kinase (p70s6k). These proteins regulate the translation of mRNAs of important genes regulating cell proliferation and angiogenesis, such as c-myc, cyclin D1, and HIF-1α. (iii) NF-κB pathway that might be activated by viral infection (Pikarsky et al., 2004). Persistent activation of NF-κB in the premalignant stage confers a survival advantage to hepatocytes that have acquired oncogenic mutations, thus favoring malignant transformation. (iv) Wnt/β-catenin signaling pathway (Ishizaki et al., 2004). The activation of multiple signaling pathways in different HCC makes it difficult to develop effective alternative therapies using small molecules. Identification of a key molecule that contributes to the activation of majority of these pathways would provide an important target for therapeutic intervention for HCC.

3. AEG-1/MTDH/LYRIC IS OVEREXPRESSED IN HCC

AEG-1/MTDH/LYRIC was initially identified as a HIV-1-inducible gene in primary human fetal astrocytes (Kang et al., 2005; Su et al., 2002). However, subsequent expression analysis documented that AEG-1 is overexpressed in a diverse array of cancers and AEG-1/MTDH/LYRIC is a downstream gene of oncogenic Ha-ras pathway, being transcriptionally regulated by c-Myc upon Ha-ras and PI3K activation, and AEG-1/MTDH/LYRIC cooperates with Ha-ras in promoting its transformation activity (Kang et al., 2005; Lee, Su, Emdad, Sarkar, & Fisher, 2006). Forced overexpression of AEG-1/MTDH/LYRIC increased proliferation and invasion, and activated NF-κB pathway by directly interacting with p65 subunit of NF-κB (Emdad et al., 2006; Sarkar et al., 2008). Additionally, AEG-1/MTDH/LYRIC was identified as a potent regulator of lung metastasis by breast cancer cells (Brown & Ruoslahti, 2004).

These initial studies prompted further analysis of AEG-1/MTDH/LYRIC in the context of HCC. Several studies analyzed AEG-1/MTDH/LYRIC expression and clinical correlation in human HCC patients and explored the potential molecular mechanism of AEG-1/MTDH/LYRIC overexpression. In the first study by Yoo et al., the expression of AEG-1/MTDH/LYRIC was analyzed by Western blotting in primary rat hepatocytes and human HCC cell lines HepG3, QGY-7703, SNU-423, Hep3B, Huh7, Sk-Hep-1, and Focus (Yoo, Emdad, et al., 2009). Among these cells, HepG3 do not form tumors in nude mice, while QGY-7703 HCC cells form aggressive tumors. Very low level of AEG-1/MTDH/LYRIC expression was detected in primary rat hepatocytes compared to all the human HCC cells. Interestingly, AEG-1/MTDH/LYRIC expression was higher in QGY-7703 cells compared to HepG3 cells. These findings were confirmed by immunohistochemistry in tissue microarrays (TMAs) containing 86 primary HCC, 23 metastatic HCC, and 9 normal adjacent liver samples. Very little to no AEG-1/MTDH/LYRIC immunostaining was detected in the nine normal liver samples while significant AEG-1/MTDH/LYRIC staining was observed in HCC samples. AEG-1/MTDH/LYRIC expression was detected predominantly in the perinuclear region. Among the 109 HCC samples, only 7 scored negative for AEG-1/MTDH/LYRIC and the remaining 102 (93.58%) showed variable levels of AEG-1/MTDH/LYRIC that could be correlated with the stages of the disease based on the BCLC staging system. Expression of AEG-1/MTDH/LYRIC gradually increased with the stages from I to IV as well as with the grades of differentiation from well-differentiated to poorly differentiated, and a statistically significant correlation was obtained between AEG-1/MTDH/LYRIC expression level and the stage of HCC.

To interrogate the molecular mechanism of AEG-1/MTDH/LYRIC overexpression, AEG-1/MTDH/LYRIC mRNA expression was analyzed using a gene expression microarray (Affymetrix U133 plus 2.0) across 132 human samples in various stages of human hepatocarcinogenesis: normal liver ($n=10$), cirrhotic tissue ($n=13$), low-grade dysplastic nodules ($n=10$), high-grade dysplastic nodules ($n=8$), and HCC ($n=91$). Expression of AEG-1/MTDH/LYRIC in HCV-related HCC was significantly increased in comparison to normal liver and cirrhotic tissue. Mean upregulation in comparison to normal liver and cirrhosis were 1.7 (t-test, $P=0.04$)- and 1.65 (t-test, $P<0.001$)-fold increase, respectively.

Next, it was examined how many tumor samples had DNA copy gains at the AEG-1/MTDH/LYRIC locus located in chromosome 8q. To do so, the average copy number for the 52 SNP array probes within 250 kb on either side of AEG-1/MTDH/LYRIC was calculated. For a copy number cutoff >3, 27 of 103 tumors showed gains of chromosome 8q (26%). Nine of these same tumors had a copy number cutoff >4 (8.7%). Analysis of the pairwise correlation between copy number at the AEG-1/MTDH/LYRIC locus and the log-base-2 expression of every transcript on the U133 Plus 2.0 array, regardless of its position in the genome, revealed a significant statistical correlation between AEG-1/MTDH/LYRIC copy number and expression level ($r=0.723$; permutation $P<0.004$). AEG-1/MTDH/LYRIC was the 19th most significant gene on this candidate gene list in chromosome 8. Finally, the list of significantly overexpressed or underexpressed genes associated with AEG-1/MTDH/LYRIC copy number was analyzed using the Significance Analysis of Microarrays package. Among the 91 tumors with expression data, 24 had copy gains >3 and 8 had copy gains >4. AEG-1/MTDH/LYRIC was among the list of significantly overexpressed genes using copy number cutoffs >3 or >4 (FDR q-value <0.002). These findings were strengthened by fluorescence *in situ* hybridization (FISH) for AEG-1/MTDH/LYRIC in human HCC TMA (CF and DS, unpublished result). Copy number gains involving AEG-1/MTDH/LYRIC were demonstrated in 32% of the core HCC tissue samples, with amplification of AEG-1/MTDH/LYRIC detected in 4 cores (7%) and low-level gain of AEG-1/MTDH/LYRIC present in an additional 15 cores (25%). Of the remaining cores, 67% exhibited multiple copies (polysomy) of chromosome 8; cells with trisomy 8 were universally present in these samples, together with variable numbers of cells with four or more copies of chromosome 8. In summary, AEG-1/MTDH/LYRIC is significantly overexpressed in HCC when compared to normal liver. This overexpression is associated with elevated copy numbers of AEG-1/MTDH/LYRIC, predominantly due to gains of large regions of chromosome 8q.

Following the initial observations, two additional studies have analyzed AEG-1/MTDH/LYRIC expression in HCC samples and correlated with patients' prognosis (Gong et al., 2012; Zhu et al., 2011). In the study by Zhu et al., AEG-1/MTDH/LYRIC expression was assessed in TMA of 323 HCC patients. The immunohistochemistry results showed that AEG-1/MTDH/LYRIC was primarily located in the membrane and the cytoplasm. Most of the tumor tissues expressed significantly higher levels of AEG-1/MTDH/LYRIC than adjacent nontumorous tissues, with AEG-1/MTDH/LYRIC^{High} accounting for 54.2% (175 of 323) of all the patients. The Pearson χ^2 test indicated that AEG-1/MTDH/LYRIC expression was closely associated with microvascular invasion ($P<0.001$), pathologic satellites ($P=0.007$), tumor differentiation ($P=0.002$), and TNM stage ($P=0.001$). These results suggest that tumors with more microvascular invasion or pathologic satellites, poorer differentiation, and TNM stages II and III are prone to exhibit higher AEG-1/MTDH/LYRIC expression. Expression of AEG-1/MTDH/LYRIC did not correlate with other clinicopathologic characteristics such as age, gender, liver cirrhosis, serum alpha-fetoprotein, tumor diameter, tumor encapsulation, or BCLC stage.

As of the last follow-up in March 2009, 54.2% (175 of 323) of the patients had suffered from recurrence and 51.1% (165 of 323) had died with local or distant recurrence. The 1-, 3-, and 5-year OS and cumulative recurrence rates in the whole cohort were 85.4% and 25.4%, 62.2% and 50.2%, 50.7% and 59.7%, respectively. Furthermore, the 1-, 3-, 5-year OS rates in the AEG-1/MTDH/LYRIC^{High} group were significantly lower than those in the AEG-1/MTDH/LYRIC^{Low} group (83.0% vs. 89.7%, 52.0% vs. 75.3%, 37.4% vs. 66.9%, respectively); the 1-, 3-, 5-year cumulative recurrence rates were markedly higher in the AEG-1/MTDH/LYRIC^{High} group than those in the AEG-1/MTDH/LYRIC^{Low} group (32.4% vs. 16.8%, 61.2% vs. 38.2%, 70.7% vs. 47.8%, respectively). Univariate and multivariate analyses revealed that along with tumor diameter, encapsulation, microvascular invasion, and TNM stage, AEG-1/MTDH/LYRIC was an independent prognostic factor for both OS (HR$=$1.870, $P<0.001$) and recurrence (HR$=$1.695, $P<0.001$).

The Gong et al. study showed that AEG-1/MTDH/LYRIC expression levels were elevated in HBV-related HCC tissues compared to normal liver tissues. There was a trend for gradually increased AEG-1/MTDH/LYRIC expression from normal liver tissue to hepatitis B and HBV-related HCC tissues. Furthermore, a statistical analysis revealed that AEG-1/MTDH/LYRIC expression significantly correlated with the American Joint Committee on Cancer (AJCC, seventh edition) stage ($P=0.020$), T classification ($P=0.007$), N classification ($P=0.044$), vascular invasion ($P=0.006$), and histological differentiation ($P=0.020$) in the HBV-related HCC patients. In addition, patients with high AEG-1/MTDH/LYRIC levels had shorter survival times compared to those with low AEG-1/MTDH/LYRIC expression ($P=0.001$).

miR-375 that targets AEG-1/MTDH/LYRIC has been shown to be downregulated in human HCC. Analysis of 60 pairs of HCC and matched adjacent nontumor tissues revealed that miR-375 was significantly suppressed in HCC tissues, and the abundance of miR-375 was inversely correlated with that of AEG-1/MTDH/LYRIC. Transcription of AEG-1/MTDH/LYRIC is induced by direct binding of c-myc to AEG-1/MTDH/LYRIC promoter upon activation of Ha-ras/PI3K/Akt signaling. Since PI3K/Akt is activated and c-myc is overexpressed in human HCC, they might also regulate AEG-1/MTDH/LYRIC transcription in HCC. Thus, multiple mechanisms, such as genomic amplification as well as transcriptional and posttranscriptional regulation, might be responsible for AEG-1/MTDH/LYRIC overexpression in human HCC.

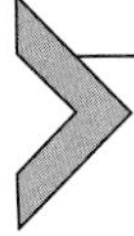

4. BIOLOGICAL CONSEQUENCE OF AEG-1/MTDH/LYRIC OVEREXPRESSION IN HCC

The availability of HepG3 cells, poorly aggressive HCC cells with low level of AEG-1/MTDH/LYRIC, and QGY-7703 cells, strongly aggressive HCC cells with high level of AEG-1/MTDH/LYRIC, facilitates the analysis of AEG-1/MTDH/LYRIC function by "gain-of-function" and "loss-of-function" approaches. Stable clones of HepG3 cells (Hep-AEG-1-8 and Hep-AEG-1-14) overexpressing AEG-1/MTDH/LYRIC demonstrated increased proliferation, colony formation, anchorage-independent growth, and Matrigel invasion *in vitro* and formed highly aggressive, metastatic, and angiogenic tumors in nude mice compared to the control clones (Hep-pc-4) (Yoo, Emdad, et al., 2009). Hep-AEG-1 clones expressed higher levels of proangiogenic factors, such as FGFα, VEGF, and pIGF compared to Hep-pc-4 clone. As a corollary, an adenovirus expressing AEG-1/MTDH/LYRIC siRNA significantly abrogated tumor formation by QGY-7703 cells in nude mice (Yoo, Emdad, et al., 2009). Stable clones of LM3 and 97H human HCC cells expressing AEG-1/MTDH/LYRIC shRNA demonstrated significant abrogation of migratory ability, as revealed by scratch assay and transwell migration assay, and upon orthotopic implantation in the liver of nude mice, these clones developed significantly less number of pulmonary and abdominal metastases (Zhu et al., 2011). These knockdown clones exhibited a switch in the EMT phenotype characterized by upregulation of E-cadherin, downregulation of N-cadherin and Snail, and increased cytoplasmic accumulation of β-catenin (Zhu et al., 2011).

Analysis of signal transduction pathway revealed activation of MEK/ERK, p38 MAPK, Akt, and NF-κB pathways in Hep-AEG-1 clones compared to control Hep-pc-4 clone (Yoo, Emdad, et al., 2009). Inhibition of

MEK/ERK and p38 MAPK pathways by their specific inhibitors PD98059 and SB203580, respectively, did not significantly affect increased proliferation conferred by AEG-1/MTDH/LYRIC; however, it markedly inhibited AEG-1/MTDH/LYRIC-induced Matrigel inhibition and anchorage-independent growth, indicating that MEK/ERK and p38 MAPK pathways might mediate more aggressive phenotype conferred by AEG-1/MTDH/LYRIC.

5. AEG-1/MTDH/LYRIC DOWNSTREAM GENES

To identify the downstream genes mediating the effects of AEG-1/MTDH/LYRIC in HCC cells, an Affymetrix oligonucleotide microarray (Human U133 plus 2.0) was performed between Hep-pc-4 and Hep-AEG-1-14 clones (Yoo, Emdad, et al., 2009). With a 1.5-fold cutoff, expressions of 5180 different oligonucleotides, that include ESTs and multiple oligonucleotides belonging to the same gene, were modulated in Hep-AEG-1-14 clone compared to Hep-pc-4 clone. One cluster of genes that were significantly modulated belongs to the Wnt signaling pathway. LEF-1, the transcription factor activated by Wnt signaling, was induced by 12.35-fold, while two negative regulators of Wnt signaling CTBP2 and APC were downregulated by 33.76- and 2.32-fold, respectively, in Hep-AEG-1-14 clone compared to Hep-pc-4 clone.

The second cluster of genes that were upregulated in Hep-AEG-1-14 clone is associated with chemoresistance. These genes include drug-metabolizing enzymes, such as dihydropyrimidine dehydrogenase (DPYD), principal enzyme inactivating 5-fluorouracil (5-FU), cytochrome P4502B6 (CYP2B6), involved in metabolism of multiple drugs, and dihydrodiol dehydrogenase (AKR1C2), conferring resistance to doxorubicin and cisplatin. The ATP-binding cassette transporter, ABCC11/MRP8, that causes efflux of multiple chemotherapeutics including 5-FU, was significantly induced in Hep-AEG-1-14 clone. Expression of LSF/TFCP2/LBP1-1c which activates the transcription of thymidylate synthase (TS), target of 5-FU, was significantly upregulated in Hep-AEG-1-14 clone.

Genes associated with invasion, such as claudin 4 and tetraspanin 8 (TSPAN8), were upregulated, and transgelin (TAGLN, a suppressor of MMP-9) was downregulated significantly in Hep-AEG-1-14 clone. IGFBP7, a secreted protein involved in senescence induction, was markedly downregulated in Hep-AEG-1-14 clone. Pyruvate kinase, a key enzyme of the glycolytic pathway, was also upregulated in Hep-AEG-1-14 clone. The up- or downregulation of majority of these genes by AEG-1/MTDH/LYRIC were confirmed by Taqman quantitative PCR and correlated well with the findings of microarray. Immunohistochemical analysis of tissue sections of matched normal liver and HCC samples revealed concomitant upregulation of AEG-1/MTDH/LYRIC, LEF-1, LSF, and DPYD and downregulation of IGFBP7 in HCC in 13 out of 18 patients further strengthening the potential regulation of these genes by AEG-1/MTDH/LYRIC.

5.1. LEF-1 and Wnt/β-catenin signaling

Activation of Wnt/β-catenin signaling plays an important role in hepatocarcinogenesis (Ishizaki et al., 2004; Thompson & Monga, 2007). The marked induction in LEF-1 and concomitant downregulation of CTBP2 prompted further analysis of AEG-1/MTDH/LYRIC-induced activation of Wnt/β-catenin pathway (Yoo, Emdad, et al., 2009). LEF-1 and its downstream target c-Myc were significantly upregulated, and LEF-1-regulated luciferase reporter activity was significantly higher in Hep-AEG-1 clones compared to Hep-pc-4 clone (Yoo, Emdad, et al., 2009). Knocking down LEF-1 by siRNA significantly abrogated AEG-1/MTDH/LYRIC-induced Matrigel invasion. Nuclear accumulation of β-catenin was significantly augmented in Hep-AEG-1 clones compared to Hep-pc-4 clone. β-Catenin is phosphorylated by GSK3β and undergoes proteasomal degradation while phosphorylation of GSK3β inactivates it and allows nuclear translocation of β-catenin. Since AEG-1/MTDH/LYRIC activates ERK42/44, it was hypothesized that ERK42/44 might phosphorylate GSK3β. The level of phosphorylated GSK3β was significantly higher in Hep-AEG-1-14 clone compared to Hep-pc-4 clone. As a consequence, the level of phosphorylated β-catenin was downregulated resulting in an increase in total β-catenin. Treatment with PD98059 decreased phosphorylated GSK3β level thus activating it, resulting in an increase in phosphorylated β-catenin level and a decrease in total β-catenin. These findings indicate that AEG-1/MTDH/LYRIC activates the Wnt signaling pathway by directly inducing LEF-1 level and indirectly by activating ERK42/44 thus facilitating nuclear translocation of β-catenin (Fig 7.1).

5.2. LSF, DPYD, and resistance to 5-FU

5-FU is a common chemotherapeutic for HCC. 5-FU is converted intracellularly into 5′-fluoro-2′-deoxyuridine by thymidine phosphorylase with

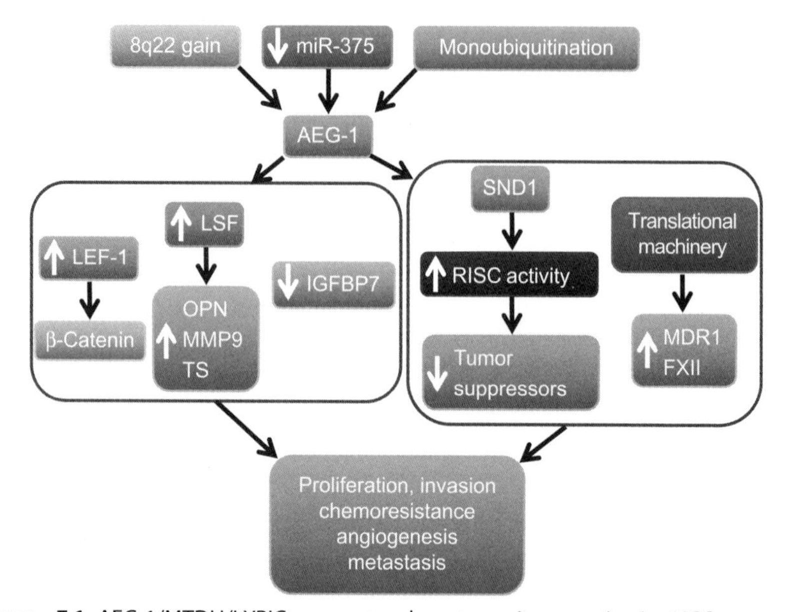

Figure 7.1 AEG-1/MTDH/LYRIC promotes hepatocarcinogenesis. In HCC genomic amplification (8q22 gain), downregulation of miRNA-375 and stabilization of AEG-1/MTDH/LYRIC protein by monoubiquitination have been identified as potential mechanisms of AEG-1/MTDH/LYRIC overexpression. Microarray analysis identified AEG-1/MTDH/LYRIC downstream genes (left box). AEG-1/MTDH/LYRIC increases LEF-1 with resultant activation of β-catenin pathway. AEG-1/MTDH/LYRIC induces the transcription factor late SV40 factor (LSF) which transcriptionally regulates osteopontin (OPN), matrix metalloproteinase-9 (MMP9), and thymidylate synthase (TS). AEG-1/MTDH/LYRIC downregulates insulin-like growth factor-binding protein-7 (IGFBP7), an inducer of senescence. Right box shows AEG-1/MTDH/LYRIC-interacting proteins identified in HCC cells. AEG-1/MTDH/LYRIC interacts with staphylococcal nuclease domain containing-1 (SND1) and together these two proteins increase RNA-induced silencing complex (RISC) activity that facilitates oncogenic miRNA-mediated degradation of tumor suppressor mRNAs. AEG-1/MTDH/LYRIC interacts with the translational machinery that potentially facilitates association of specific mRNAs, such as multidrug resistance 1 (MDR1) and coagulation factor XII (FXII), to polysome increasing their translation. The net effects of these events are augmentation of hallmarks of cancer. (See Page 6 in Color Section at the back of the book.)

subsequent phosphorylation by thymidine kinase into the active metabolite 5-fluoro-2′-deoxyuridine 5′-monophosphate (FdUMP) (Longley, Harkin, & Johnston, 2003). FdUMP inhibits TS which reduces the thymidine pool and increases the uracil pool leading to the inhibition of DNA synthesis. 5-FU is converted into its inactive metabolite fluoro-5,6-dihydrouracil (FUH_2) by

DPYD. TS and DPYD gene expression and/or activity are major determinants of the efficacy of 5-FU (Oguri et al., 2005; Yoshinare et al., 2003).

The transcription factor late SV40 factor (LSF), also known as TFCP2, functions as a transcriptional activator or repressor. It activates transcription of serum amyloid A3, IL-4, α-globin, α-A crystallin, TS, and PAX6 in different vertebrate species (Hansen, Owens, & Saxena, 2009; Santhekadur, Rajasekaran, et al., 2012; Veljkovic & Hansen, 2004). In cell-free extracts, it activates RNA polymerase II transcription by binding to basal promoter factor TFIIB. LSF also inhibits transcription of HIV LTR by binding to YY1 and histone deacetylase 1. A major cellular target of LSF is the TS gene, which encodes the rate-limiting enzyme in production of dTTP, required for DNA synthesis. LSF binds to the TS promoter and upregulates TS mRNA at the G_1/S transition (Powell, Rudge, Zhu, Johnson, & Hansen, 2000). Inhibition of LSF by a dominant-negative construct (LSFdn) inhibits TS induction and induces apoptosis while addition of thymidine in the medium protects the cells from inhibition of DNA synthesis and induction of apoptosis. As a consequence, LSF plays an important role in DNA synthesis and cell survival. However, even though LSF has been studied for more than 25 years, no study was performed to dissect its potential role in carcinogenesis.

Expression of DPYD and LSF and its target gene TS was significantly higher in Hep-AEG-1 clones compared to Hep-pc-4 clone (Yoo, Gredler, et al., 2009). Nuclear accumulation of LSF and LSF-dependent luciferase reporter activity were also significantly higher in Hep-AEG-1 clone compared to Hep-pc-4 clone. Knocking down AEG-1/MTDH/LYRIC by siRNA resulted in significant downregulation of DPYD, LSF, and TS level while an LSFdn also induced downregulation of TS (Yoo, Gredler, et al., 2009). AEG-1/MTDH/LYRIC-overexpressing cells showed significant resistance to 5-FU which could be overcome by inhibiting AEG-1/MTDH/LYRIC, LSF, or DPYD. A lentivirus delivering AEG-1/MTDH/LYRIC siRNA in combination with 5-FU markedly inhibited growth of QGY-7703 HCC cells xenotransplanted in athymic nude mice when compared to either agent alone (Yoo, Gredler, et al., 2009).

Several follow-up studies highlighted the importance of LSF in hepatocarcinogenesis. Analysis of 109 human HCC samples by immunohistochemistry documented the overexpression of LSF in 100 samples compared to the normal liver which was associated with polyploidy of chromosome 12, where LSF gene is located, in 68% cases (Yoo, Emdad, et al., 2010).

LSF expression is almost undetectable in normal hepatocytes, is very low in HepG3 cells, and is markedly high in most of the human HCC cell lines including QGY-7703. Forced overexpression of LSF in HepG3 cells increased proliferation, colony formation, Matrigel invasion, and anchorage-independent growth *in vitro* and resulted in highly aggressive, angiogenic, multiorgan metastatic tumors in nude mice upon subcutaneous xenograft assays, and tail vein injection (Yoo, Emdad, et al., 2010). Conversely, inhibition of LSF by a dominant-negative construct in QGY-7703 cells significantly abrogated *in vitro* proliferation, colony formation, Matrigel invasion, anchorage-independent growth, *in vivo* tumor growth, and metastasis. Affymetrix microarray identified many genes associated with invasion, angiogenesis, chemoresistance, and metastasis to be modulated by LSF, and the most robust induction was observed for osteopontin (OPN), a known mediator for tumor progression and metastasis (Bellahcene, Castronovo, Ogbureke, Fisher, & Fedarko, 2008). It was documented that LSF transcriptionally regulates OPN, and inhibition of OPN significantly abrogated the oncogenic functions of LSF (Yoo, Emdad, et al., 2010). OPN, induced by LSF, activates c-Met via interaction with CD44 receptor, and inhibition of c-Met significantly abrogated LSF-induced tumorigenesis and metastasis (Yoo, Gredler, et al., 2011). LSF transcriptionally upregulates MMP-9 that plays an important role in LSF-induced angiogenesis (Santhekadur, Gredler, et al., 2012). LSF also activates two important protumorigenic signaling, ERK and NF-κB (Yoo, Emdad, et al., 2010). Since the majority of HCC arises in the background of HBV or HCV infection, the activation of NF-κB by LSF indicates that LSF might also be involved in the inflammatory component of hepatocarcinogenesis. Indeed, a number of inflammatory cytokines induce LSF expression. Thus LSF promotes cell survival, chemoresistance, angiogenesis, and metastasis, all important hallmarks of cancer.

A high-throughput screening identified small-molecule inhibitors of LSF DNA binding and the prototype of these molecules, Factor quinolinone inhibitor 1 (FQI1), profoundly inhibited cell viability and induced apoptosis in human HCC cells without exerting harmful effects to normal immortal human hepatocytes and primary mouse hepatocytes (Grant et al., 2012). In nude mice xenograft studies, FQI1 markedly inhibited growth of human HCC xenografts as well as angiogenesis without exerting any toxicity. Thus, the original studies analyzing AEG-1/MTDH/LYRIC downstream genes identified a novel oncogene, its downstream pathways, and a class of novel anti-HCC therapeutics with promising efficiency.

5.3. IGFBP7: A tumor suppressor for HCC

IGFBP7, also known as mac25 or IGFBP-related protein-1 (IGFBP-rP1), is a secreted protein belonging to the IGFBP family. IGFBPs regulate the bioavailability of IGFs and insulin by physical binding and thereby limit access of IGFs or insulin to their corresponding receptors inhibiting the activity (Hwa, Oh, & Rosenfeld, 1999). IGFBP7 differs from the other six members of this family by lacking the C-terminus and having 100 times lower affinity for IGF-I but significantly higher affinity for insulin. IGFBP7 also exerts insulin/IGF-independent action where it induces senescence and apoptosis by inhibiting BRAF/MEK/ERK signaling cascade. IGFBP7 has been implicated to be a tumor suppressor for breast, prostate, colorectal cancers, and melanoma (Burger et al., 1998; Ruan et al., 2007; Sprenger, Damon, Hwa, Rosenfeld, & Plymate, 1999; Wajapeyee, Serra, Zhu, Mahalingam, & Green, 2008). IGFBP7 was identified as the most robustly downregulated gene by AEG-1/MTDH/LYRIC (Yoo, Emdad, et al., 2009). Using an immunohistochemical approach in TMA, high IGFBP7 expression was documented in normal human liver, while in 104 HCC patients, there was a statistically significant gradual decrease in IGFBP7 expression with the stages of HCC (Chen et al., 2011). Moreover, in each stage, IGFBP7 expression was much lower in poorly differentiated grades compared to moderately differentiated grades. FISH in TMA revealed loss of heterozygosity for IGFBP7 locus in 26% of HCC patients (Chen et al., 2011). IGFBP7 mRNA and protein expression was significantly lower in human HCC cells compared to THLE-3 cells, which are normal human hepatocytes immortalized by SV40 T/t Ag. Inverse correlation between AEG-1/MTDH/LYRIC and IGFBP7 expression was observed in normal liver and matched HCC samples in 13 out of 18 patients. A separate study analyzing 104 patients documented negative IGFBP7 staining in 35.6% HCC patients which statistically correlated with large tumor size, increased vascular invasion, poor OS, and disease-free survival rates (Tomimaru et al., 2012). Thus IGFBP7 downregulation might be a clinically relevant prognostic marker for aggressive HCC. Thus IGFBP7 downregulation might be a clinically relevant prognostic marker for aggressive HCC.

To analyze the role of IGFBP7 downregulation in mediating AEG-1/MTDH/LYRIC function, stable IGFBP7-overexpressing clones were established in Hep-AEG-1-14 background (Chen et al., 2011). In *in vitro* assays, these clones showed small but significant inhibition of proliferation. However, in subcutaneous xenograft assays, the IGFBP7-overexpressing clones demonstrate profound inhibition of tumor growth. This marked *in vivo* inhibitory effect of IGFBP7 could be attributed to its ability to induce senescence as revealed by senescence-associated β-galactosidase (SA β-gal) and senescence-associated heterochromatin foci assays, and inhibit angiogenesis as revealed by human vascular endothelial cell (HUVEC) differentiation and chicken chorioallantoic membrane (CAM) assays. Reduced activation of IGF-IR, Akt, and ERK was observed in IGFBP7-overexpressing clones. Thus, IGFBP7 downregulation might play an important role in promoting AEG-1/MTDH/LYRIC function. The molecular mechanism by which AEG-1/MTDH/LYRIC downregulates IGFBP7 remains to be determined.

The profound growth inhibitory properties of IGFBP7 suggest it to be a potential therapeutic for HCC. A replication incompetent adenovirus expressing IGFBP7 (Ad.IGFBP7) profoundly inhibited viability and induced apoptosis in multiple human HCC cell lines by inducing reactive oxygen species (ROS) and activating a DNA damage response (DDR) and p38 MAPK (Chen et al., 2013). In orthotopic xenograft models of human HCC in athymic nude mice, intravenous administration of Ad.IGFBP7 profoundly inhibited primary tumor growth and intrahepatic metastasis. In a nude mice subcutaneous model, xenografts from human HCC cells were established in both flanks and only left-sided tumors received intratumoral injection of Ad.IGFBP7. Growth of both left-sided injected tumors and right-sided uninjected tumors was markedly inhibited by Ad.IGFBP7 with profound suppression of angiogenesis. These findings indicate that Ad.IGFBP7 might be a potent therapeutic eradicating both primary HCC and distant metastasis and might be an effective treatment option for terminal HCC patients either alone or in combination with other agents.

6. AEG-1/MTDH/LYRIC-INTERACTING PROTEINS

Apart from having a transmembrane domain and three nuclear/nucleolar localization signals, AEG-1/MTDH/LYRIC protein does not have any known domains and motifs. However, AEG-1/MTDH/LYRIC is localized in the nucleus and nucleolus, in the ER/perinuclear region, abundantly in the cytoplasm, and also in the membrane. In the nucleus, multiple interacting partners of AEG-1/MTDH/LYRIC have been identified, such as p65 subunit of NF-κB, CBP, and PLZF, exerting diverse levels of transcriptional regulation (Emdad et al., 2006; Sarkar et al., 2008; Thirkettle, Mills, Whitaker, & Neal, 2009). These findings strongly suggest that depending upon its subcellular localization AEG-1/MTDH/LYRIC might interact with different protein complexes, thereby mediating the oncogenic functions of AEG-1/MTDH/LYRIC. With that end in view, the identification of AEG-1/MTDH/LYRIC-interacting protein was endeavored using two approaches. The first approach involved yeast two-hybrid (Y2H) screening. The N-terminal (a.a. 1–57) and C-terminal (a.a. 68–582) regions of AEG-1/MTDH/LYRIC that precedes and follows the transmembrane domain, respectively, were used as baits to separately screen a human liver cDNA library using the technology of Hybrigenics (http://www.hybrigenics-services.com) (Yoo, Santhekadur, et al., 2011). The C-terminal region showed autoactivator function, thereby complicating the assay. However, using selective medium containing 20 mM of 3-aminotriazole, the inhibitor of the reporter gene product, the assay could be optimized. Despite these efforts, only five known proteins with moderate confidence in the interaction were identified, namely, staphylococcal nuclease domain-containing 1 (SND1), chaperonin-containing TCP1, subunit 3 (gamma) (CCT3), extracellular matrix protein 2 (ECM2), inter-alpha (globulin) inhibitor H4 (ITIH4), and solute carrier family 22, member 1 (SLC22A1) (Yoo, Santhekadur, et al., 2011). A number of unknown proteins were also identified by this assay.

For the N-terminal interaction, 11 known proteins were identified. Arginase (ARG1) and 24-dehydrocholesterol reductase showed high and prostaglandin reductase 1 showed good confidence of interaction, respectively. The eight proteins showing moderate confidence of interaction were ATPase, class VI, type 11C (ATP11C), cytochrome P450, family 2, subfamily C (CYP2C8), ITIH3, MAP/microtubule affinity-regulating kinase 2, quinoid dihydropteridine reductase, retinoid X receptor, beta, tenascin N, and tripartite motif-containing 8. Three previously uncharacterized proteins were also identified as potential AEG-1/MTDH/LYRIC-interacting protein.

The relatively modest result of the Y2H screening prompted the employment of alternative strategy of coimmunoprecipitation coupled with mass spectrometry. Cell lysates from Hep-AEG-1-14 and Hep-pc-4 cells were subjected to immunoprecipitation using protein A agarose conjugated with anti-HA antibody (anti-HA agarose) (Yoo, Santhekadur, et al., 2011). The immunoprecipitates were eluted using HA peptide and were run in a SDS-PAGE gel. The gel was stained with Coomassie blue and the stained bands, that were present only in Hep-AEG-1-14 immunoprecipitates but not in Hep-pc-4 immunoprecipitates, were cut and were subjected to

LC-MS/MS analysis after in-gel trypsin digestion. A total of 182 potential AEG-1/MTDH/LYRIC-interacting proteins were thus identified. However, the most represented proteins were AEG-1/MTDH/LYRIC and SND1.

6.1. Staphylococcal nuclease domain-containing 1

The identification of SND1 as an AEG-1/MTDH/LYRIC-interacting protein by both genomic and proteomic approaches strongly suggested further exploration. SND1, also known as p100 coactivator or Tudor-SN, is a multifunctional protein modulating transcription, mRNA splicing, RNAi function, and mRNA stability (Caudy et al., 2003; Paukku, Kalkkinen, Silvennoinen, Kontula, & Lehtonen, 2008; Paukku, Yang, & Silvennoinen, 2003; Yang et al., 2002, 2007). In the cytoplasm, SND1 functions as a nuclease in the RNA-induced silencing complex (RISC) in which small RNAs (such as siRNAs or miRNAs) are complexed with ribonucleoproteins to ensue RNAi-mediated gene silencing (Caudy et al., 2003). Immunofluorescence analysis using human HCC cell lines and HCC tissue samples demonstrated AEG-1/MTDH/LYRIC and SND1 interaction in the cytoplasm which was confirmed by coimmunoprecipitation analysis (Yoo, Santhekadur, et al., 2011). It was documented that AEG-1/MTDH/LYRIC also interacts with Argonaute 2, the major nuclease in the RISC, and thus AEG-1/MTDH/LYRIC is a *bona fide* component of RISC. Using a luciferase-based reporter assay and AEG-1/MTDH/LYRIC or SND1 overexpressing and knockdown clones, it was documented that both AEG-1/MTDH/LYRIC and SND1 are required for optimum RISC activity. Immunohistochemical analysis of TMA containing 109 HCC samples documented higher SND1 expression in 81 cases (~74%) compared to the normal liver, and SND1 expression gradually increased with the stages and grades of the disease. As a corollary, higher RISC activity was observed in human HCC cells compared to normal immortal hepatocytes THLE-3 cells. Increased RISC activity, conferred by AEG-1/MTDH/LYRIC or SND1, resulted in increased degradation of tumor suppressor mRNAs that are the target of oncomiRs. These mRNAs include PTEN, target of miR-221 and miR-21; CDKN1C (p57), target of miR-221; CDKN1A (p21), target of miR-106b; SPRY2, target of miR-21; and TGFBR2, target of miR-93. Overexpression of AEG-1/MTDH/LYRIC or SND1 downregulates, while knockdown of AEG-1/MTDH/LYRIC or SND1 upregulates all these mRNA levels in HCC cells thus supporting the hypothesis that increased RISC activity might contribute to hepatocarcinogenesis by augmenting oncomiR-mediated degradation of tumor suppressor mRNAs.

Inhibition of SND1 enzymatic activity using 3′, 5′-deoxythymidine bisphosphate (pdTp) significantly abrogated proliferation and colony formation by Hep-AEG-1 clones, indicating that AEG-1/SND1 interaction is required to mediate AEG-1/MTDH/LYRIC function (Yoo, Santhekadur, et al., 2011). Stable overexpression of SND1 in Hep3B cells, expressing low level of SND1, increased *in vitro* proliferation and colony formation and *in vivo* tumor formation in nude mice (Yoo, Santhekadur, et al., 2011). Stable knockdown of SND1 by shRNA in QGY-7703 cells, expressing high level of SND1, significantly abrogated the aforementioned *in vitro* and *in vivo* phenotypes. Conditioned medium from Hep3B–SND1 cells augmented while that from QGY-SND1si cells significantly inhibited angiogenesis as analyzed by CAM assay and HUVEC differentiation assay (Santhekadur, Das, et al., 2012). A linear pathway was unraveled in which SND1-induced activation of NF-κB resulted in induction of miR-221 and subsequent induction of angiogenic factors angiogenin and CXCL16 (Santhekadur, Das, et al., 2012). Inhibition of either of these components resulted in significant inhibition of SND1-induced angiogenesis thus highlighting the importance of this molecular cascade in regulating SND1 function. As SND1 regulates NF-κB and miR-221, two important determinants of HCC controlling the aggressive phenotype, SND1 inhibition might be an effective strategy to counteract this fatal malady.

Apart from its role in mRNA degradation as a component of RISC, SND1 has also been documented to bind to 3′-UTR of specific mRNAs, such as angiotensin II type 1 receptor (AT1R), and increase translation by reducing decay (Paukku et al., 2008). It was documented that by increasing the level of AT1R by posttranscriptional stabilization of its mRNA, SND1 activates AT1R downstream signaling, such as activation of ERK and SMAD2, resulting in activation of TGFβ signaling that regulates EMT, migration, and invasion of human HCC cells (PKS and DS, unpublished results). A significant positive correlation between SND1 and AT1R expression level was observed in human HCC samples validating the regulation of AT1R by SND1. Angiotensin-converting enzyme inhibitors and AT1R blockers inhibit hepatic fibrosis and subsequent hepatocarcinogenesis. The results of this study unravel a novel mechanism of AT1R activation in

HCC, brought forth by SND1 overexpression, thereby strengthening the rationale of using AT1R blockers as adjuvant therapy for HCC.

6.2. Translational machinery and chemoresistance

One striking phenotype conferred by AEG-1/MTDH/LYRIC is chemoresistance. Upregulation of mRNA of DPYD, LSF, and TS by AEG-1/MTDH/LYRIC underlies the mechanism of 5-FU resistance (Yoo, Gredler, et al., 2009). However, AEG-1/MTDH/LYRIC employs a different mechanism to induce resistance to anthracycline compounds, such as doxorubicin. AEG-1/MTDH/LYRIC upregulated the expression of multidrug resistance gene 1 (MDR1/ABCB1) at the protein level but not at the mRNA level resulting in increased efflux and decreased accumulation of doxorubicin in HCC cells (Yoo, Chen, et al., 2010). Overexpression of AEG-1/MTDH/LYRIC, either forced or endogenous, provided significant resistance to doxorubicin which could be overcome either by knocking down AEG-1/MTDH/LYRIC with siRNA or by blocking MDR1 with siRNA or chemical inhibitors. A lentivirus expressing AEG-1/MTDH/LYRIC shRNA significantly augmented the efficacy of doxorubicin in inhibiting growth of human HCC cells in subcutaneous xenograft assays in athymic nude mice (Yoo, Chen, et al., 2010). It was demonstrated that AEG-1/MTDH/LYRIC facilitates association of MDR1 mRNA to polysomes thus facilitating MDR1 translation. The increased association of MDR1 mRNA to polysome could be blocked by LY294002, indicating that activation of PI3K/Akt pathway by AEG-1/MTDH/LYRIC is necessary to induce this effect. To provide support for this observation, a low dose of LY294002, which itself did not exert significant toxicity, significantly increased cytotoxicity of doxorubicin. Additionally, AEG-1/MTDH/LYRIC inhibited ubiquitination and proteasomal degradation of MDR1 protein resulting in further accumulation of MDR1 protein (Yoo, Chen, et al., 2010).

The observation that AEG-1/MTDH/LYRIC increases translation of MDR1 prompted further analysis of regulation of the translational machinery by AEG-1/MTDH/LYRIC. The phosphorylation and activation status of key cap-dependent translational initiation factors, namely, eukaryotic translation initiation factor 4E (eIF4E), eukaryotic translation initiation factor 4G (eIF4G), and 4E binding protein (4E-BP) that are regulated by the PI3K/Akt/mTOR pathway in Hep-pc-4 cells and Hep-AEG-1-14 cells untreated or treated with LY294002 were analyzed (Yoo, Chen, et al., 2010). Marked phosphorylation of eIF4G was observed in Hep-AEG-1-14 cells compared to Hep-pc-4 cells, which was not affected by LY294002 treatment. The phosphorylation of eIF4E and 4E-BP was not modulated by AEG-1/MTDH/LYRIC. Hypophosphorylated 4E-BP binds to the 5′ cap binding protein of eIF4E, thereby preventing its interaction with eIF4G and inhibiting translation. Phosphorylation of 4E-BP via the mTOR pathway releases 4E-BP from eIF4E resulting in the recruitment of eIF4G to the 5′ cap and thereby allowing translation initiation to proceed. As 4E-BP phosphorylation by AEG-1/MTDH/LYRIC was not detected, the phosphorylation of eIF4G and subsequent augmentation of translation by AEG-1/MTDH/LYRIC might be independent of the mTOR pathway. One possible mechanism by which AEG-1/MTDH/LYRIC facilitates translation initiation might be direct involvement of AEG-1/MTDH/LYRIC in the translational machinery. AEG-1/MTDH/LYRIC is a highly basic protein that might function as a scaffold facilitating binding of nucleic acids, including mRNA. Indeed, several eukaryotic translation initiation factors, translation elongation factors, and ribosomal proteins were identified as potential AEG-1/MTDH/LYRIC-interacting proteins, and AEG-1/MTDH/LYRIC was implicated to be a potential RNA-binding protein in other systems (Meng et al., 2012; Yoo, Santhekadur, et al., 2011). The mechanism by which AEG-1/MTDH/LYRIC increases translation of specific mRNAs remains to be elucidated.

7. THE ROLE OF AEG-1/MTDH/LYRIC IN HEPATOCARCINOGENESIS: NOVEL INSIGHTS FROM A MOUSE MODEL

The oncogenic properties of AEG-1/MTDH/LYRIC have been validated through numerous studies employing human patient samples, *in vitro* cell culture system, and nude mice xenograft models. When overexpressed in normal immortal human cells, AEG-1/MTDH/LYRIC significantly protects from serum starvation-induced apoptosis (Lee et al., 2008). Overexpression of AEG-1/MTDH/LYRIC in normal immortal cloned rat embryo fibroblasts results in aggressive tumor formation in nude mice (Emdad et al., 2009). These studies indicate that AEG-1/MTDH/LYRIC could confer transforming properties to normal immortal cells. However, whether AEG-1/MTDH/LYRIC alone might induce transformation and evoke an explicit carcinogenic phenotype was not clear. To better comprehend the role of AEG-1/MTDH/LYRIC in hepatocarcinogenesis and to decipher the underlying molecular mechanism(s) in an *in vivo* context, a transgenic mouse with hepatocyte-specific expression of AEG-1/MTDH/LYRIC (Alb/AEG-1) was generated (Srivastava et al., 2012). The hepatocyte-specific expression was engineered by using a mouse albumin promoter/enhancer element to drive human AEG-1/MTDH/LYRIC expression. Alb/AEG-1 mice did not develop spontaneous HCC when followed for a period of 1 year. However, when treated with a hepatocarcinogen, *N*-nitrosodiethylamine (DEN), at 28 weeks of age, only two out of 11 WT animals showed a few very small nodules in the liver, whereas all of the 17 Alb/AEG-1 mice livers harbored numerous nodules of different sizes. Histological analysis of the liver of WT mice showed a few dysplastic, hyperchromatic nuclei indicating that with time HCC would eventually develop. In Alb/AEG-1 mice, a marked increase in dysplastic, hyperchromatic nuclei was observed in both the nodules and the adjacent normal liver. The most striking feature was observed in the hepatic nodules of Alb/AEG-1 mice, showing profound steatotic phenotypes with large lipid droplets in the hepatocytes. A moderate level of steatosis was also observed in the adjacent normal liver in Alb/AEG-1 mice. There was a significant increase in hepatic enzymes, such as AST, ALT, and alkaline phosphatase, in the sera of Alb/AEG-1 mice versus the sera of WT mice. At 32 weeks of age, the WT mice developed hepatic nodules; however, the nodules developed in Alb/AEG-1 mice were markedly larger. These findings indicate that AEG-1/MTDH/LYRIC significantly accelerated the hepatocarcinogenic process in DEN-treated animals.

Affymetrix microarray analysis of DEN-treated WT and Alb/AEG-1 livers revealed upregulation of several genes that include HCC marker α-feto protein (Afp); invasion- and metastasis-associated genes Tspan8 and Lcn2; several genes associated with fat metabolism, such as, Scd2, Lpl, Apoa4, and Apoc2; and genes regulating angiogenesis, such as TFF3. Tspan8 was also found to be upregulated by AEG-1/MTDH/LYRIC in human HCC cells indicating a potential key role in mediating AEG-1/MTDH/LYRIC function.

Analysis of hepatocytes from WT and Alb/AEG-1 hepatocytes provided several important insights into the function of AEG-1/MTDH/LYRIC. Alb/AEG-1 hepatocytes demonstrated marked resistance to doxorubicin and 5-FU treatment when compared to their WT littermates. Primary mouse hepatocytes, cultured in the presence of growth factors, do not divide and show decreasing viability after ~4 days as they enter senescence. The viability of Alb/AEG-1 hepatocytes in complete growth media was significantly higher than that of WT hepatocytes, as monitored by standard MTT assay over a 7-day period. Upon removal of growth factors, the WT hepatocytes started losing viability within 1 day and by 3 days more than 50% of the cells were dead. In contrast, Alb/AEG-1 hepatocytes were significantly resistant to removal of growth factors, and even after 7 days in basal media, the cell viability was only reduced by 20%. These observations indicate that AEG-1/MTDH/LYRIC might autonomously activate growth factor-induced signaling and might inhibit pathways mediating senescence. Indeed, Alb/AEG-1 hepatocytes exhibited higher levels of activated (phosphorylated) ERK, Akt, and p38 MAPK and antiapoptotic proteins, Bcl-2 and Mcl-1 but not Bcl-x_L, when compared to the WT hepatocytes.

Very interestingly, Alb/AEG-1 hepatocytes displayed significant resistance to senescence. At day 7 following isolation, the WT hepatocytes became large and vacuolated and ~55% cells were positive for SA β-gal while only 3% Alb/AEG-1 hepatocytes stained positive for SA β-gal. WT hepatocytes had higher levels of ROS and displayed activation of a DDR, characterized by activation of ATM, ATR, CHK1, CHK2, and p53 and upregulation of p21. In contrast, the ROS level in Alb/AEG-1 hepatocytes was significantly lower and the DDR response was also markedly dampened. Thus, activation of growth factor signaling, upregulation of antiapoptotic proteins, and abrogation of ROS-induced DDR response and senescence might underlie the mechanism by which AEG-1/MTDH/LYRIC promotes a prosurvival response.

Conditioned media (CM) from Alb/AEG-1 hepatocytes induced a significant angiogenic response, as revealed by HUVEC differentiation and CAM assays, when compared to WT hepatocytes. Proteomic analysis of the CM identified upregulation of several components of the coagulation pathway, including Fibrinogen α and β chains, Factor XII (FXII), plasminogen, and prothrombin, that are known to play significant roles in cancer angiogenesis, metastasis, and invasion in Alb/AEG-1 hepatocytes compared to the WT hepatocytes. The role of FXII, showing 56-fold upregulation in Alb/AEG-1 hepatocytes over WT hepatocytes, and TFF3, identified by Affymetrix microarray, was analyzed by knocking them down by siRNA in Alb/AEG-1 hepatocytes and subjecting the CM to angiogenesis assays. Knocking down either FXII or TFF3 markedly abrogated the angiogenic response by AEG-1/MTDH/LYRIC although the effect of FXII siRNA was much more pronounced indicating that FXII might play a pivotal role in mediating AEG-1/MTDH/LYRIC-induced angiogenesis. FXII cross talks with EGFR-activating MAPK and Akt signaling to promote proliferation and differentiation of endothelial cells. HUVECs were treated with CM from Alb/AEG-1 hepatocytes transfected with either control siRNA or FXII siRNA. While CM from control siRNA-treated HUVEC maintained activation of EGFR, Akt, ERK, and p38 MAPK, absence of FXII in the CM from FXII knockdown cells significantly abrogated the activation of EGFR, Akt, ERK, and p38 MAPK in HUVEC further highlighting the importance of FXII in AEG-1/MTDH/LYRIC action. Compared to the changes in the protein level, FXII mRNA level showed little increase in Alb/AEG-1 hepatocytes over WT hepatocytes. It was demonstrated that similar to MDR1, FXII mRNA also showed increased association with polysomes upon AEG-1/MTDH/LYRIC overexpression. The level of miRNA-181a, that targets FXII, was similar between WT and Alb/AEG-1 hepatocytes, indicating that the translational, and not miRNA-mediated, regulation is the molecular mechanism by which AEG-1/MTDH/LYRIC increases FXII expression.

8. CONCLUSION

HCC is a highly radio- and chemoresistant cancer, and there is virtually no therapy for advanced HCC. The only FDA-approved targeted drug, the Raf kinase inhibitor sorafenib, provides a survival benefit of only 2.8 months in nonresectable HCC patients (Llovet et al., 2008). As such there is a dire need of developing new modalities of treatment providing lasting disease-free survival benefit to the patients. Our extensive studies in human cell lines and Alb/AEG-1 mice firmly establish that even though AEG-1/MTDH/LYRIC may not induce HCC it is required for subsequent progression of the disease. Thus, AEG-1/MTDH/LYRIC inhibition, in combination with other modalities of therapy, might be an effective way to counteract HCC. As AEG-1/MTDH/LYRIC itself does not catalyze any reaction and functions most likely as a scaffold protein facilitating formation of protein complexes, developing small molecule inhibitors for AEG-1/MTDH/LYRIC by a standard high-throughput library screening approach may not be tenable. Resolution of crystal structure of AEG-1/MTDH/LYRIC is thus mandatory to identify key regions facilitating protein–protein interaction and develop peptidomimetic inhibitors using fragment-based drug discovery approach (Bembenek, Tounge, & Reynolds, 2009; Chen et al., 2007). While AEG-1/MTDH/LYRIC is detected primarily in the nucleus in normal hepatocytes, in HCC cells it is detected mainly in the cytoplasm, suggesting that in hepatocytes AEG-1/MTDH/LYRIC might serve unique functions that are different from its role in HCC cells. Liver is a specialized organ intimately involved in metabolic and hormonal regulation. Whether AEG-1/MTDH/LYRIC plays any role in regulating the specialized functions of liver remains to be determined. An AEG-1/MTDH/LYRIC knockout mouse model is thus required to be developed to understand the role of this unique molecule in normal and diseased states.

REFERENCES

Bellahcene, A., Castronovo, V., Ogbureke, K. U., Fisher, L. W., & Fedarko, N. S. (2008). Small integrin-binding ligand N-linked glycoproteins (SIBLINGs): Multifunctional proteins in cancer. *Nature Reviews. Cancer*, *8*, 212–226.

Bembenek, S. D., Tounge, B. A., & Reynolds, C. H. (2009). Ligand efficiency and fragment-based drug discovery. *Drug Discovery Today*, *14*, 278–283.

Boyault, S., Rickman, D. S., de Reynies, A., Balabaud, C., Rebouissou, S., Jeannot, E., et al. (2007). Transcriptome classification of HCC is related to gene alterations and to new therapeutic targets. *Hepatology*, *45*, 42–52.

Brechot, C., Pourcel, C., Louise, A., Rain, B., & Tiollais, P. (1980). Presence of integrated hepatitis B virus DNA sequences in cellular DNA of human hepatocellular carcinoma. *Nature*, *286*, 533–535.

Bressac, B., Kew, M., Wands, J., & Ozturk, M. (1991). Selective G to T mutations of p53 gene in hepatocellular carcinoma from southern Africa. *Nature*, *350*, 429–431.

Brown, D. M., & Ruoslahti, E. (2004). Metadherin, a cell surface protein in breast tumors that mediates lung metastasis. *Cancer Cell*, *5*, 365–374.

Burger, A. M., Zhang, X., Li, H., Ostrowski, J. L., Beatty, B., Venanzoni, M., et al. (1998). Down-regulation of T1A12/mac25, a novel insulin-like growth factor binding protein related gene, is associated with disease progression in breast carcinomas. *Oncogene*, *16*, 2459–2467.

Calle, E. E., Rodriguez, C., Walker-Thurmond, K., & Thun, M. J. (2003). Overweight, obesity, and mortality from cancer in a prospectively studied cohort of U.S. adults. *The New England Journal of Medicine*, *348*, 1625–1638.

Caudy, A. A., Ketting, R. F., Hammond, S. M., Denli, A. M., Bathoorn, A. M., Tops, B. B., et al. (2003). A micrococcal nuclease homologue in RNAi effector complexes. *Nature*, *425*, 411–414.

Chen, D., Siddiq, A., Emdad, L., Rajasekaran, D., Gredler, R., Shen, X.-N., et al. (2013). Insulin-like growth factor binding protein-7 (IGFBP7): A promising gene therapeutic for hepatocellular carcinoma (HCC). *Molecular Therapy*, *21*(4), 758–766.

Chen, D., Yoo, B. K., Santhekadur, P. K., Gredler, R., Bhutia, S. K., Das, S. K., et al. (2011). Insulin-like growth factor-binding protein-7 functions as a potential tumor suppressor in hepatocellular carcinoma. *Clinical Cancer Research*, *17*, 6693–6701.

Chen, J., Zhang, Z., Stebbins, J. L., Zhang, X., Hoffman, R., Moore, A., et al. (2007). A fragment-based approach for the discovery of isoform-specific p38alpha inhibitors. *ACS Chemical Biology*, *2*, 329–336.

Donato, F., Tagger, A., Gelatti, U., Parrinello, G., Boffetta, P., Albertini, A., et al. (2002). Alcohol and hepatocellular carcinoma: The effect of lifetime intake and hepatitis virus infections in men and women. *American Journal of Epidemiology*, *155*, 323–331.

Dragani, T. A. (2010). Risk of HCC: Genetic heterogeneity and complex genetics. *Journal of Hepatology, 52*, 252–257.

El-Serag, H. B. (2011). Hepatocellular carcinoma. *The New England Journal of Medicine, 365*, 1118–1127.

El-Serag, H. B., & Rudolph, K. L. (2007). Hepatocellular carcinoma: Epidemiology and molecular carcinogenesis. *Gastroenterology, 132*, 2557–2576.

El-Serag, H. B., Tran, T., & Everhart, J. E. (2004). Diabetes increases the risk of chronic liver disease and hepatocellular carcinoma. *Gastroenterology, 126*, 460–468.

Emdad, L., Lee, S. G., Su, Z. Z., Jeon, H. Y., Boukerche, H., Sarkar, D., et al. (2009). Astrocyte elevated gene-1 (AEG-1) functions as an oncogene and regulates angiogenesis. *Proceedings of the National Academy of Sciences of the United States of America, 106*, 21300–21305.

Emdad, L., Sarkar, D., Su, Z. Z., Randolph, A., Boukerche, H., Valerie, K., et al. (2006). Activation of the nuclear factor kappaB pathway by astrocyte elevated gene-1: Implications for tumor progression and metastasis. *Cancer Research, 66*, 1509–1516.

Farazi, P. A., & DePinho, R. A. (2006). Hepatocellular carcinoma pathogenesis: From genes to environment. *Nature Reviews. Cancer, 6*, 674–687.

Feitelson, M. A., & Duan, L. X. (1997). Hepatitis B virus X antigen in the pathogenesis of chronic infections and the development of hepatocellular carcinoma. *The American Journal of Pathology, 150*, 1141–1157.

Georgiades, C. S., Hong, K., & Geschwind, J. F. (2008). Radiofrequency ablation and chemoembolization for hepatocellular carcinoma. *Cancer Journal, 14*, 117–122.

Gong, Z., Liu, W., You, N., Wang, T., Wang, X., Lu, P., et al. (2012). Prognostic significance of metadherin overexpression in hepatitis B virus-related hepatocellular carcinoma. *Oncology Reports, 27*, 2073–2079.

Grant, T. J., Bishop, J. A., Christadore, L. M., Barot, G., Chin, H. G., Woodson, S., et al. (2012). Antiproliferative small molecule inhibitors of transcription factor LSF reveal oncogene addiction to LSF in hepatocellular carcinoma. *Proceedings of the National Academy of Sciences of the United States of America, 109*, 4503–4508.

Hansen, U., Owens, L., & Saxena, U. H. (2009). Transcription factors LSF and E2Fs: Tandem cyclists driving G0 to S? *Cell Cycle, 8*, 2146–2151.

Hwa, V., Oh, Y., & Rosenfeld, R. G. (1999). The insulin-like growth factor-binding protein (IGFBP) superfamily. *Endocrine Reviews, 20*, 761–787.

Ishizaki, Y., Ikeda, S., Fujimori, M., Shimizu, Y., Kurihara, T., Itamoto, T., et al. (2004). Immunohistochemical analysis and mutational analyses of beta-catenin, Axin family and APC genes in hepatocellular carcinomas. *International Journal of Oncology, 24*, 1077–1083.

Kang, D. C., Su, Z. Z., Sarkar, D., Emdad, L., Volsky, D. J., & Fisher, P. B. (2005). Cloning and characterization of HIV-1-inducible astrocyte elevated gene-1, AEG-1. *Gene, 353*, 8–15.

Kwun, H. J., Jung, E. Y., Ahn, J. Y., Lee, M. N., & Jang, K. L. (2001). p53-dependent transcriptional repression of p21(waf1) by hepatitis C virus NS3. *The Journal of General Virology, 82*, 2235–2241.

Lee, S. G., Su, Z. Z., Emdad, L., Sarkar, D., & Fisher, P. B. (2006). Astrocyte elevated gene-1 (AEG-1) is a target gene of oncogenic Ha-ras requiring phosphatidylinositol 3-kinase and c-Myc. *Proceedings of the National Academy of Sciences of the United States of America, 103*, 17390–17395.

Lee, S. G., Su, Z. Z., Emdad, L., Sarkar, D., Franke, T. F., & Fisher, P. B. (2008). Astrocyte elevated gene-1 activates cell survival pathways through PI3K-Akt signaling. *Oncogene, 27*, 1114–1121.

Leung, T. W., Patt, Y. Z., Lau, W. Y., Ho, S. K., Yu, S. C., Chan, A. T., et al. (1999). Complete pathological remission is possible with systemic combination chemotherapy for inoperable hepatocellular carcinoma. *Clinical Cancer Research, 5*, 1676–1681.

Llovet, J. M., Bru, C., & Bruix, J. (1999). Prognosis of hepatocellular carcinoma: The BCLC staging classification. *Seminars in Liver Disease, 19*, 329–338.

Llovet, J. M., & Bruix, J. (2003). Systematic review of randomized trials for unresectable hepatocellular carcinoma: Chemoembolization improves survival. *Hepatology, 37*, 429–442.

Llovet, J. M., & Bruix, J. (2008). Molecular targeted therapies in hepatocellular carcinoma. *Hepatology, 48*, 1312–1327.

Llovet, J. M., Burroughs, A., & Bruix, J. (2003). Hepatocellular carcinoma. *Lancet, 362*, 1907–1917.

Llovet, J. M., Real, M. I., Montana, X., Planas, R., Coll, S., Aponte, J., et al. (2002). Arterial embolisation or chemoembolisation versus symptomatic treatment in patients with unresectable hepatocellular carcinoma: A randomised controlled trial. *Lancet, 359*, 1734–1739.

Llovet, J. M., Ricci, S., Mazzaferro, V., Hilgard, P., Gane, E., Blanc, J. F., et al. (2008). Sorafenib in advanced hepatocellular carcinoma. *The New England Journal of Medicine, 359*, 378–390.

Longley, D. B., Harkin, D. P., & Johnston, P. G. (2003). 5-fluorouracil: Mechanisms of action and clinical strategies. *Nature Reviews. Cancer, 3*, 330–338.

Majumder, M., Ghosh, A. K., Steele, R., Ray, R., & Ray, R. B. (2001). Hepatitis C virus NS5A physically associates with p53 and regulates p21/waf1 gene expression in a p53-dependent manner. *Journal of Virology, 75*, 1401–1407.

Mann, C. D., Neal, C. P., Garcea, G., Manson, M. M., Dennison, A. R., & Berry, D. P. (2007). Prognostic molecular markers in hepatocellular carcinoma: A systematic review. *European Journal of Cancer, 43*, 979–992.

Meng, X., Zhu, D., Yang, S., Wang, X., Xiong, Z., Zhang, Y., et al. (2012). Cytoplasmic Metadherin (MTDH) provides survival advantage under conditions of stress by acting as RNA-binding protein. *The Journal of Biological Chemistry, 287*, 4485–4491.

Min, L., He, B., & Hui, L. (2011). Mitogen-activated protein kinases in hepatocellular carcinoma development. *Seminars in Cancer Biology, 21*, 10–20.

Murakami, Y., Saigo, K., Takashima, H., Minami, M., Okanoue, T., Brechot, C., et al. (2005). Large scaled analysis of hepatitis B virus (HBV) DNA integration in HBV related hepatocellular carcinomas. *Gut, 54*, 1162–1168.

Oguri, T., Achiwa, H., Bessho, Y., Muramatsu, H., Maeda, H., Niimi, T., et al. (2005). The role of thymidylate synthase and dihydropyrimidine dehydrogenase in resistance to 5-fluorouracil in human lung cancer cells. *Lung Cancer, 49*, 345–351.

O'Neil, B. H., & Venook, A. P. (2007). Hepatocellular carcinoma: The role of the North American GI Steering Committee Hepatobiliary Task Force and the advent of effective drug therapy. *The Oncologist, 12*, 1425–1432.

Pang, R. W., Joh, J. W., Johnson, P. J., Monden, M., Pawlik, T. M., & Poon, R. T. (2008). Biology of hepatocellular carcinoma. *Annals of Surgical Oncology, 15*, 962–971.

Paukku, K., Kalkkinen, N., Silvennoinen, O., Kontula, K. K., & Lehtonen, J. Y. (2008). p100 increases AT1R expression through interaction with AT1R 3′-UTR. *Nucleic Acids Research, 36*, 4474–4487.

Paukku, K., Yang, J., & Silvennoinen, O. (2003). Tudor and nuclease-like domains containing protein p100 function as coactivators for signal transducer and activator of transcription 5. *Molecular Endocrinology, 17*, 1805–1814.

Pikarsky, E., Porat, R. M., Stein, I., Abramovitch, R., Amit, S., Kasem, S., et al. (2004). NF-kappaB functions as a tumour promoter in inflammation-associated cancer. *Nature, 431*, 461–466.

Poon, R. T., Fan, S. T., Lo, C. M., Ng, I. O., Liu, C. L., Lam, C. M., et al. (2001). Improving survival results after resection of hepatocellular carcinoma: A prospective study of 377 patients over 10 years. *Annals of Surgery, 234*, 63–70.

Powell, C. M., Rudge, T. L., Zhu, Q., Johnson, L. F., & Hansen, U. (2000). Inhibition of the mammalian transcription factor LSF induces S-phase-dependent apoptosis by downregulating thymidylate synthase expression. *The EMBO Journal*, *19*, 4665–4675.

Rehermann, B., & Nascimbeni, M. (2005). Immunology of hepatitis B virus and hepatitis C virus infection. *Nature Reviews. Immunology*, *5*, 215–229.

Ruan, W., Xu, E., Xu, F., Ma, Y., Deng, H., Huang, Q., et al. (2007). IGFBP7 plays a potential tumor suppressor role in colorectal carcinogenesis. *Cancer Biology & Therapy*, *6*, 354–359.

Santhekadur, P. K., Das, S. K., Gredler, R., Chen, D., Srivastava, J., Robertson, C., et al. (2012). Multifunction protein staphylococcal nuclease domain containing 1 (SND1) promotes tumor angiogenesis in human hepatocellular carcinoma through novel pathway that involves nuclear factor kappaB and miR-221. *The Journal of Biological Chemistry*, *287*, 13952–13958.

Santhekadur, P. K., Gredler, R., Chen, D., Siddiq, A., Shen, X. N., Das, S. K., et al. (2012). Late SV40 factor (LSF) enhances angiogenesis by transcriptionally up-regulating matrix metalloproteinase-9 (MMP-9). *The Journal of Biological Chemistry*, *287*, 3425–3432.

Santhekadur, P. K., Rajasekaran, D., Siddiq, A., Gredler, R., Chen, D., Schaus, S. E., et al. (2012). The transcription factor LSF: A novel oncogene for hepatocellular carcinoma. *American Journal of Cancer Research*, *2*, 269–285.

Sarkar, D., Park, E. S., Emdad, L., Lee, S. G., Su, Z. Z., & Fisher, P. B. (2008). Molecular basis of nuclear factor-kappaB activation by astrocyte elevated gene-1. *Cancer Research*, *68*, 1478–1484.

Siegel, R., Naishadham, D., & Jemal, A. (2012). Cancer statistics, 2012. *CA: A Cancer Journal for Clinicians*, *62*, 10–29.

Sprenger, C. C., Damon, S. E., Hwa, V., Rosenfeld, R. G., & Plymate, S. R. (1999). Insulin-like growth factor binding protein-related protein 1 (IGFBP-rP1) is a potential tumor suppressor protein for prostate cancer. *Cancer Research*, *59*, 2370–2375.

Srivastava, J., Siddiq, A., Emdad, L., Santhekadur, P., Chen, D., Gredler, R., et al. (2012). Astrocyte elevated gene-1 (AEG-1) promotes hepatocarcinogenesis: Novel insights from a mouse model. *Hepatology*, *56*(5), 1782–1791.

Starley, B. Q., Calcagno, C. J., & Harrison, S. A. (2010). Nonalcoholic fatty liver disease and hepatocellular carcinoma: A weighty connection. *Hepatology*, *51*, 1820–1832.

Su, Z. Z., Kang, D. C., Chen, Y., Pekarskaya, O., Chao, W., Volsky, D. J., et al. (2002). Identification and cloning of human astrocyte genes displaying elevated expression after infection with HIV-1 or exposure to HIV-1 envelope glycoprotein by rapid subtraction hybridization, RaSH. *Oncogene*, *21*, 3592–3602.

Teufel, A., Staib, F., Kanzler, S., Weinmann, A., Schulze-Bergkamen, H., & Galle, P. R. (2007). Genetics of hepatocellular carcinoma. *World Journal of Gastroenterology*, *13*, 2271–2282.

Thirkettle, H. J., Mills, I. G., Whitaker, H. C., & Neal, D. E. (2009). Nuclear LYRIC/AEG-1 interacts with PLZF and relieves PLZF-mediated repression. *Oncogene*, *28*, 3663–3670.

Thompson, M. D., & Monga, S. P. (2007). WNT/beta-catenin signaling in liver health and disease. *Hepatology*, *45*, 1298–1305.

Tomimaru, Y., Eguchi, H., Wada, H., Kobayashi, S., Marubashi, S., Tanemura, M., et al. (2012). IGFBP7 downregulation is associated with tumor progression and clinical outcome in hepatocellular carcinoma. *International Journal of Cancer*, *130*, 319–327.

Umemura, T., Ichijo, T., Yoshizawa, K., Tanaka, E., & Kiyosawa, K. (2009). Epidemiology of hepatocellular carcinoma in Japan. *Journal of Gastroenterology*, *44*(Suppl. 19), 102–107.

Veljkovic, J., & Hansen, U. (2004). Lineage-specific and ubiquitous biological roles of the mammalian transcription factor LSF. *Gene*, *343*, 23–40.

Wajapeyee, N., Serra, R. W., Zhu, X., Mahalingam, M., & Green, M. R. (2008). Oncogenic BRAF induces senescence and apoptosis through pathways mediated by the secreted protein IGFBP7. *Cell, 132*, 363–374.

Yang, J., Aittomaki, S., Pesu, M., Carter, K., Saarinen, J., Kalkkinen, N., et al. (2002). Identification of p100 as a coactivator for STAT6 that bridges STAT6 with RNA polymerase II. *The EMBO Journal, 21*, 4950–4958.

Yang, J. D., & Roberts, L. R. (2010). Hepatocellular carcinoma: A global view. *Nature Reviews. Gastroenterology & Hepatology*, 7, 448–458.

Yang, J., Valineva, T., Hong, J., Bu, T., Yao, Z., Jensen, O. N., et al. (2007). Transcriptional co-activator protein p100 interacts with snRNP proteins and facilitates the assembly of the spliceosome. *Nucleic Acids Research, 35*, 4485–4494.

Yeo, W., Mok, T. S., Zee, B., Leung, T. W., Lai, P. B., Lau, W. Y., et al. (2005). A randomized phase III study of doxorubicin versus cisplatin/interferon alpha-2b/doxorubicin/fluorouracil (PIAF) combination chemotherapy for unresectable hepatocellular carcinoma. *Journal of the National Cancer Institute, 97*, 1532–1538.

Yoo, B. K., Chen, D., Su, Z.-Z., Gredler, R., Yoo, J., Shah, K., et al. (2010). Molecular mechanism of chemoresistance by Astrocyte Elevated Gene-1 (AEG-1). *Cancer Research, 70*, 3249–3258.

Yoo, B. K., Emdad, L., Gredler, R., Fuller, C., Dumur, C. I., Jones, K. H., et al. (2010). Transcription factor Late SV40 Factor (LSF) functions as an oncogene in hepatocellular carcinoma. *Proceedings of the National Academy of Sciences of the United States of America, 107*, 8357–8362.

Yoo, B. K., Emdad, L., Su, Z. Z., Villanueva, A., Chiang, D. Y., Mukhopadhyay, N. D., et al. (2009). Astrocyte elevated gene-1 regulates hepatocellular carcinoma development and progression. *The Journal of Clinical Investigation, 119*, 465–477.

Yoo, B. K., Gredler, R., Chen, D., Santhekadur, P. K., Fisher, P. B., & Sarkar, D. (2011). c-Met activation through a novel pathway involving osteopontin mediates oncogenesis by the transcription factor LSF. *Journal of Hepatology, 55*, 1317–1324.

Yoo, B. K., Gredler, R., Vozhilla, N., Su, Z. Z., Chen, D., Forcier, T., et al. (2009). Identification of genes conferring resistance to 5-fluorouracil. *Proceedings of the National Academy of Sciences of the United States of America, 106*, 12938–12943.

Yoo, B. K., Santhekadur, P. K., Gredler, R., Chen, D., Emdad, L., Bhutia, S. K., et al. (2011). Increased RNA-induced silencing complex (RISC) activity contributes to hepatocelllular carcinoma. *Hepatology, 53*, 1538–1548.

Yoshinare, K., Kubota, T., Watanabe, M., Wada, N., Nishibori, H., Hasegawa, H., et al. (2003). Gene expression in colorectal cancer and in vitro chemosensitivity to 5-fluorouracil: A study of 88 surgical specimens. *Cancer Science, 94*, 633–638.

Zender, L., Spector, M. S., Xue, W., Flemming, P., Cordon-Cardo, C., Silke, J., et al. (2006). Identification and validation of oncogenes in liver cancer using an integrative oncogenomic approach. *Cell, 125*, 1253–1267.

Zhu, A. X., Blaszkowsky, L. S., Ryan, D. P., Clark, J. W., Muzikansky, A., Horgan, K., et al. (2006). Phase II study of gemcitabine and oxaliplatin in combination with bevacizumab in patients with advanced hepatocellular carcinoma. *Journal of Clinical Oncology, 24*, 1898–1903.

Zhu, K., Dai, Z., Pan, Q., Wang, Z., Yang, G. H., Yu, L., et al. (2011). Metadherin promotes hepatocellular carcinoma metastasis through induction of epithelial-mesenchymal transition. *Clinical Cancer Research, 17*, 7294–7302.

INDEX

Note: Page numbers followed by "*f*" indicate figures, and "*t*" indicate tables.

A

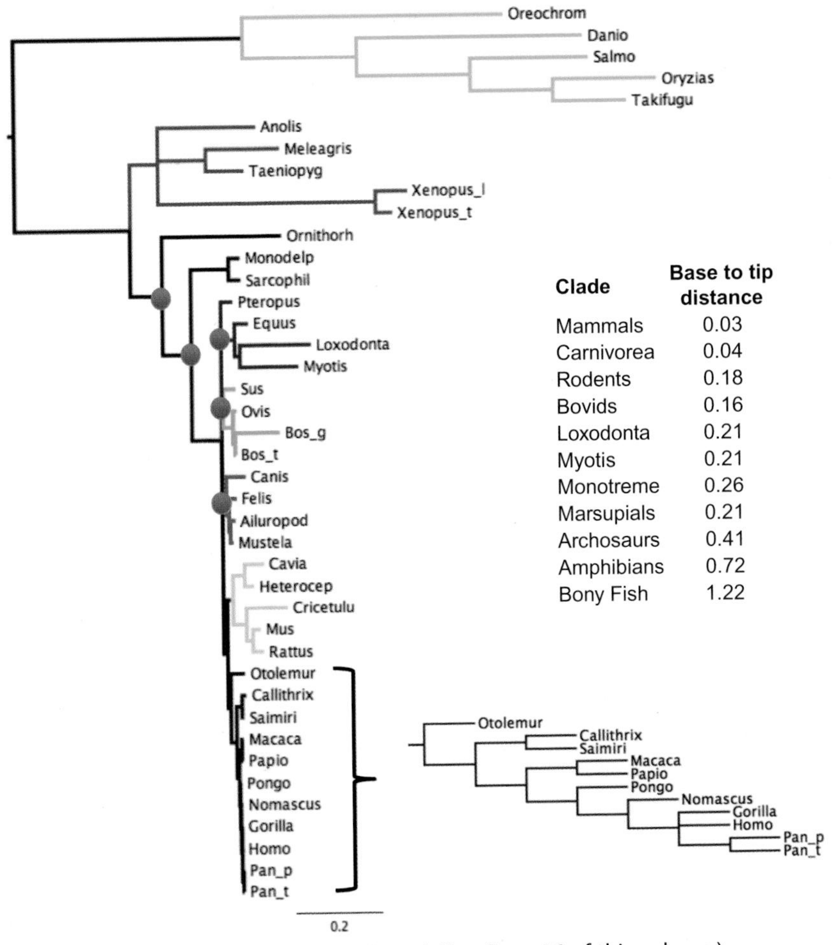

Figure 1.2, Seok-Geun Lee *et al.* (See Page 16 of this volume.)

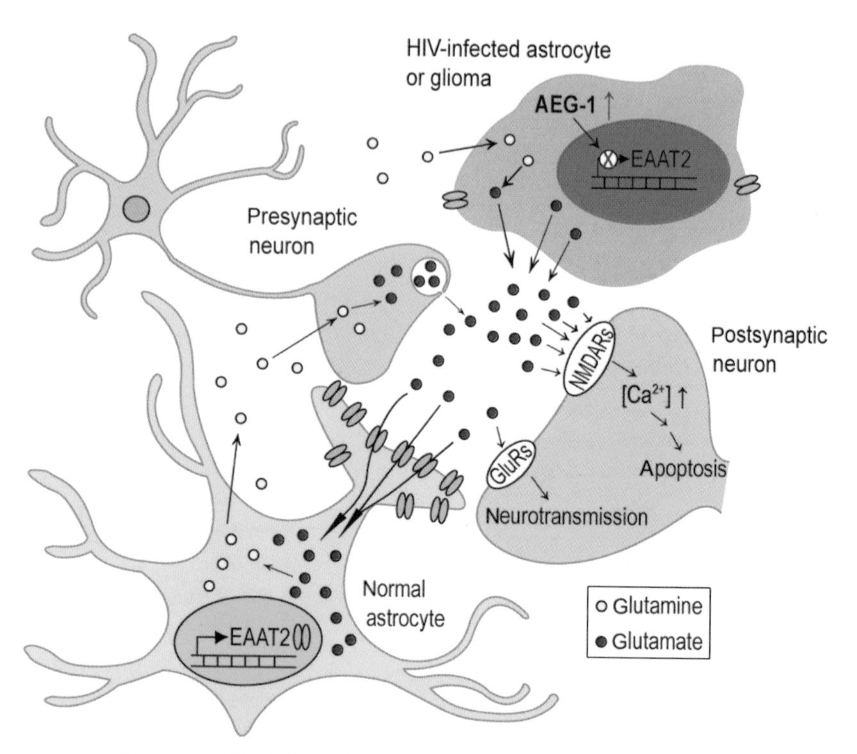

Figure 1.3, Seok-Geun Lee *et al.* (See Page 25 of this volume.)

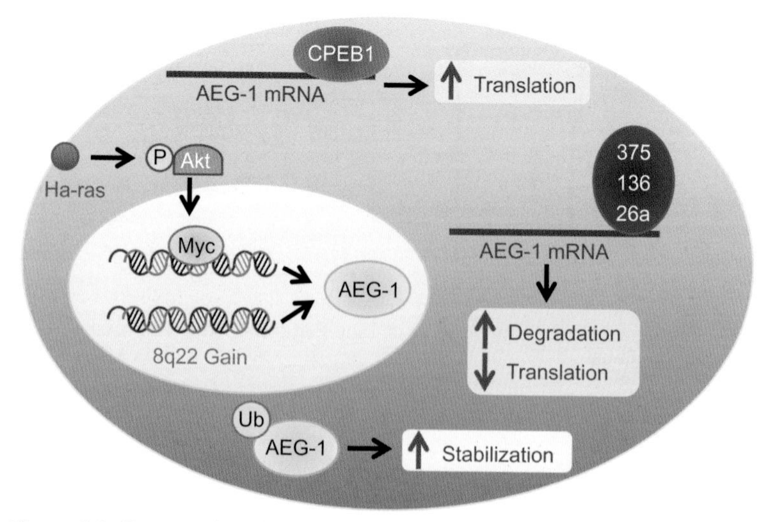

Figure 2.1, Devanand Sarkar and Paul B. Fisher (See Page 42 of this volume.)

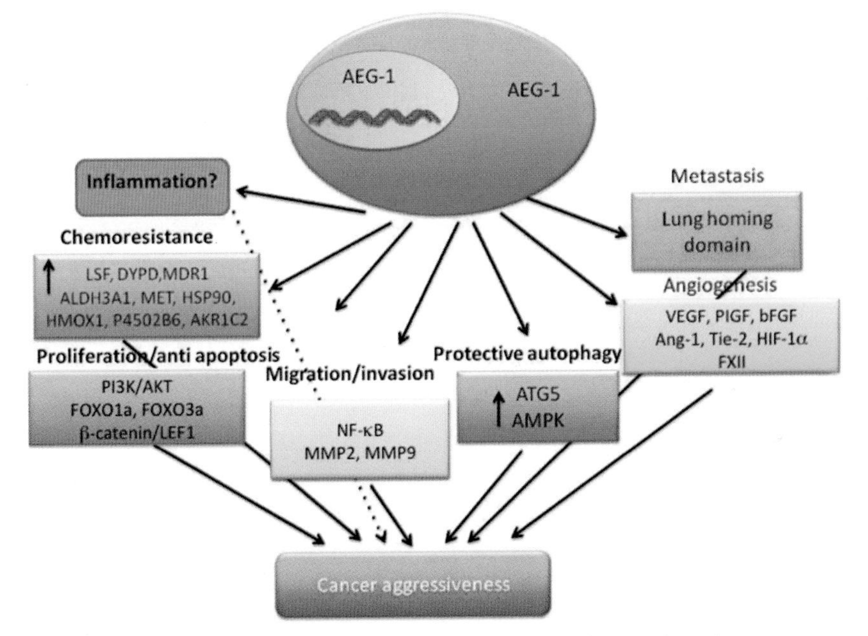

Figure 3.1, Luni Emdad *et al.* (See Page 77 of this volume.)

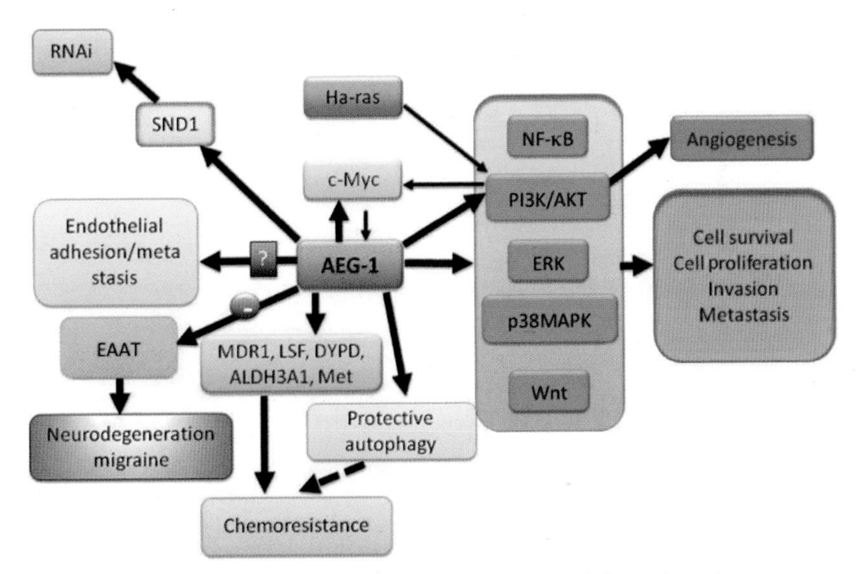

Figure 3.2, Luni Emdad *et al.* (See Page 78 of this volume.)

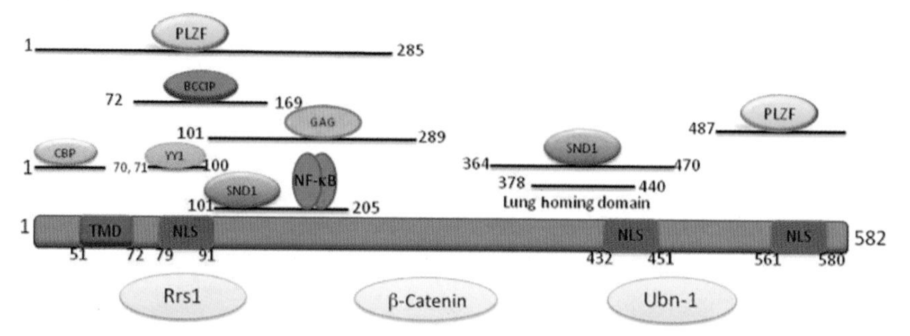

Figure 3.3, Luni Emdad *et al.* (See Page 90 of this volume.)

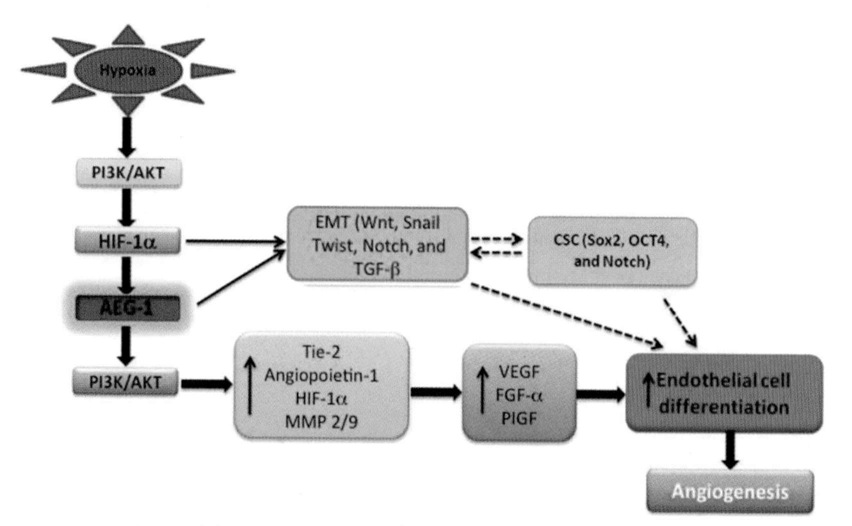

Figure 3.4, Luni Emdad *et al.* (See Page 101 of this volume.)

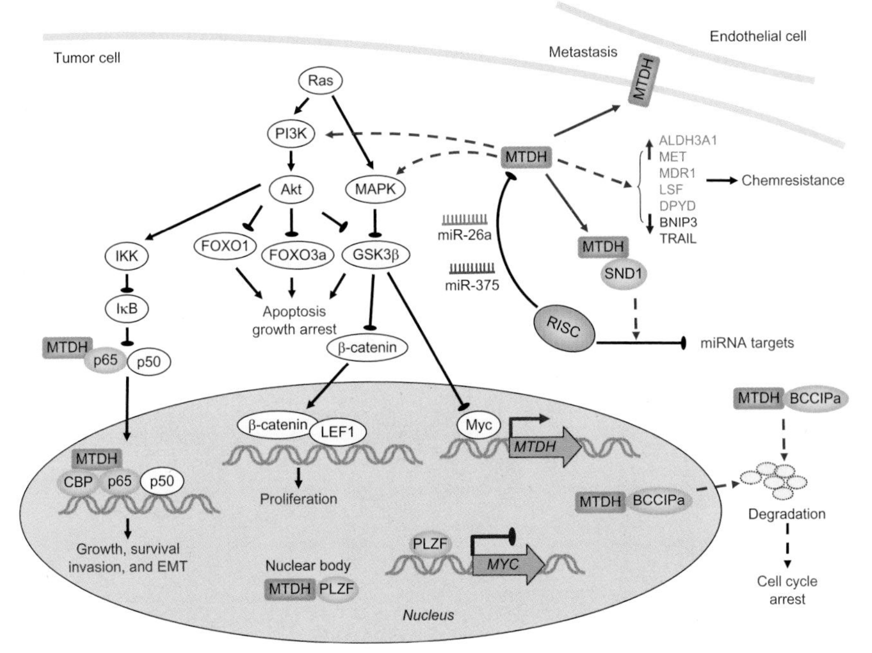

Figure 4.1, Liling Wan and Yibin Kang (See Page 125 of this volume.)

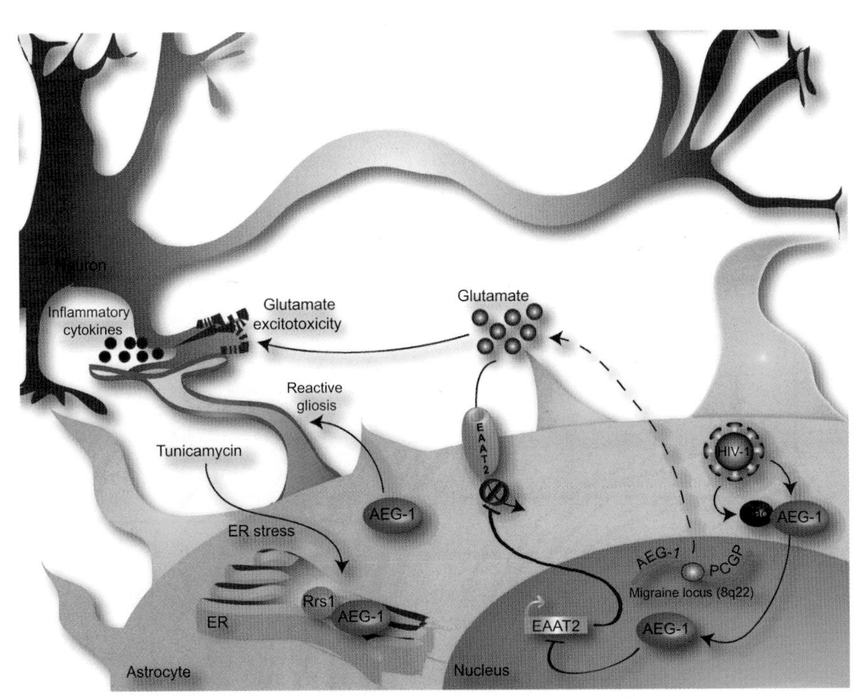

Figure 6.1, Evan K. Noch and Kamel Khalili (See Page 163 of this volume.)

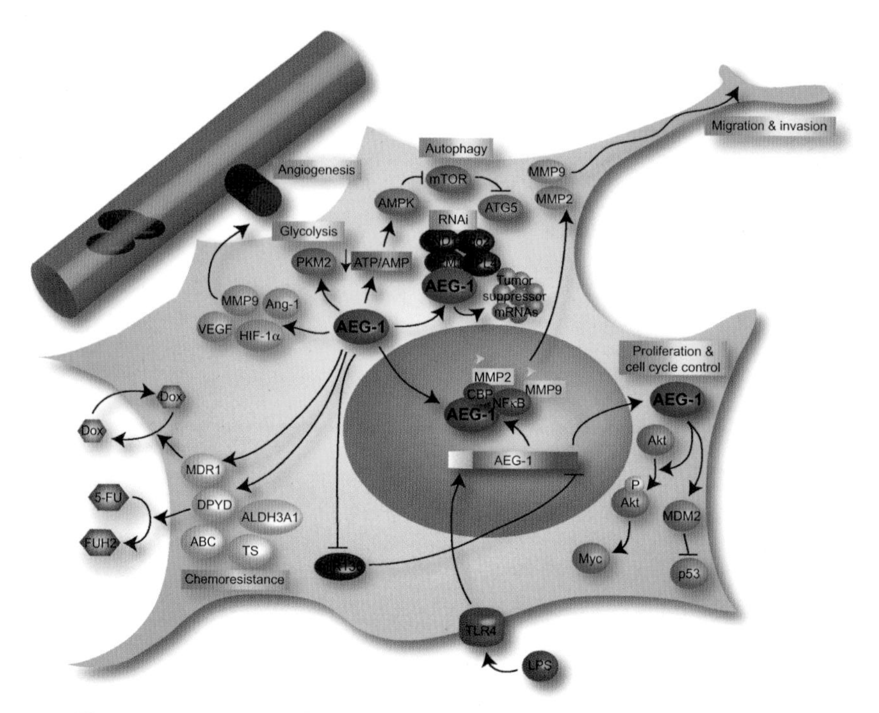

Figure 6.2, Evan K. Noch and Kamel Khalili (See Page 170 of this volume.)

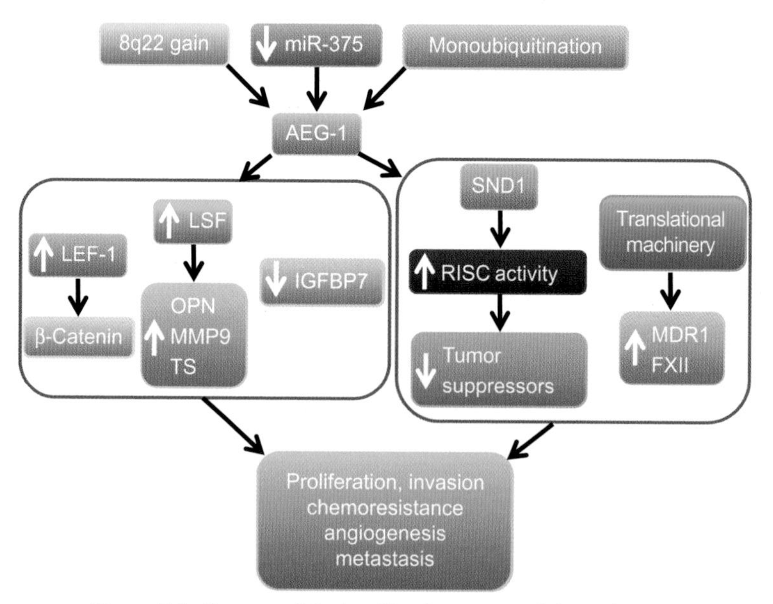

Figure 7.1, Devanand Sarkar (See Page 204 of this volume.)